Sample Size Calculations
Practical Methods for Engineers and Scientists

Sample Size Calculations
Practical Methods for Engineers and Scientists

Paul Mathews

Sample Size Calculations: Practical Methods for Engineers and Scientists
Paul Mathews
paul@mmbstatistical.com

Copyright ©2010 Paul Mathews

All rights reserved. No part of this publication may be reproduced or stored in any form or by any means without the prior written permission of the publisher.

Published by:
Mathews Malnar and Bailey, Inc.
217 Third Street, Fairport Harbor, OH 44077
Phone: 440-350-0911
Fax: 440-350-7210
Web: www.mmbstatistical.com

ISBN 978-0-615-32461-6

Original release: 24 February 2010

Contents

1 Fundamentals   1
- 1.1 Motivation for Sample Size Calculations 1
- 1.2 Rationale for Sample Size and Power Calculations 2
  - 1.2.1 Rationale for Confidence Intervals 2
  - 1.2.2 Rationale for Hypothesis Tests 6
- 1.3 Practical Considerations 11
  - 1.3.1 Roles of the Investigator and the Statistician 11
  - 1.3.2 Preference for Approximate Methods 12
  - 1.3.3 Factors that Affect Sample Size 12
  - 1.3.4 Graphical Presentations 18
- 1.4 Problems and Solutions 20
  - 1.4.1 When These Methods Fail 20
  - 1.4.2 When the Calculated Sample Size Is Too Large 21
  - 1.4.3 Bad Practices 22
  - 1.4.4 Good Practices 23
- 1.5 Software 25

2 Means   27
- 2.1 Assumptions 27
- 2.2 One Mean 28
  - 2.2.1 Confidence Interval for the Mean 28
  - 2.2.2 Test for the Mean 30
  - 2.2.3 Paired Observations 33
- 2.3 Two Independent Means 35
  - 2.3.1 Confidence Interval for the Difference Between Two Means 36
  - 2.3.2 Test for the Difference Between Two Means 39
- 2.4 Equivalence Tests 44
  - 2.4.1 Equivalence Test for One Mean 45
  - 2.4.2 Equivalence Test for Two Means 47
- 2.5 Contrasts 48
- 2.6 Multiple Comparisons Tests 50
  - 2.6.1 Bonferroni's Method 50
  - 2.6.2 All Possible Comparisons 51

	2.6.3	Comparisons with a Control	54
3	**Standard Deviations**		**57**
3.1	One Standard Deviation		57
	3.1.1	Confidence Interval for the Standard Deviation	57
	3.1.2	Tests for One Standard Deviation	60
3.2	Two Standard Deviations		64
	3.2.1	Confidence Intervals for Two Standard Deviations	64
	3.2.2	Tests for Two Standard Deviations	66
3.3	Coefficient of Variation		69
	3.3.1	Confidence Interval for the Coefficient of Variation	70
	3.3.2	Tests for the Coefficient of Variation	70
4	**Proportions**		**73**
4.1	One Proportion (Large Population)		73
	4.1.1	Confidence Interval for One Proportion	74
	4.1.2	Tests for One Proportion	78
4.2	One Proportion (Small Population)		85
	4.2.1	Approximations to the Hypergeometric Distribution	85
	4.2.2	Confidence Intervals for One Proportion	86
	4.2.3	Tests for One Proportion	90
4.3	Two Proportions		92
	4.3.1	Confidence Intervals for Two Proportions	92
	4.3.2	Tests for Two Proportions	97
4.4	Equivalence Tests		106
	4.4.1	Equivalence Test for One Proportion	106
	4.4.2	Equivalence Test for Two Independent Proportions	108
4.5	Chi-square Tests		110
	4.5.1	General Case	110
	4.5.2	Test for Independence in a Two-way Contingency Table	111
	4.5.3	Chi-square Goodness of Fit Test	113
5	**Poisson Counts**		**115**
5.1	One Poisson Count		115
	5.1.1	Confidence Intervals for the Poisson Mean	115
	5.1.2	Tests for the Poisson Mean	118
5.2	Two Poisson Counts		120
	5.2.1	Confidence Intervals for Two Poisson Means	120
	5.2.2	Tests for Two Poisson Means	123
5.3	Tests for Many Poisson Counts		128
	5.3.1	Test for Differences Between Many Poisson Means (Square-Root Transform)	128
	5.3.2	Test for Differences Between Many Poisson Means (Chi-square Method)	129
5.4	Correcting for Background Counts		130

6 Regression — 133
- 6.1 Linear Regression — 133
  - 6.1.1 Confidence Interval for the Slope — 134
  - 6.1.2 Test for the Slope — 137
- 6.2 Logistic Regression — 139
  - 6.2.1 Dichotomous Independent Variable — 139
  - 6.2.2 Normally Distributed Independent Variable — 140

7 Correlation and Agreement — 141
- 7.1 Pearson's Correlation — 141
  - 7.1.1 Confidence Interval for Pearson's Correlation — 142
  - 7.1.2 One-Sample Test for Pearson's Correlation — 143
  - 7.1.3 Two-Sample Test for Pearson's Correlation — 144
  - 7.1.4 Multiple Correlation — 145
- 7.2 Intraclass Correlation — 147
  - 7.2.1 Confidence Interval for the Intraclass Correlation — 149
  - 7.2.2 Test for the Intraclass Correlation — 151
- 7.3 Cohen's Kappa — 153
  - 7.3.1 Confidence Interval for Cohen's Kappa — 154
  - 7.3.2 Test for Cohen's Kappa — 155
- 7.4 Receiver Operating Characteristic (ROC) Curves — 158
  - 7.4.1 Confidence Interval for the ROC Curve's AUC — 159
  - 7.4.2 Test for the ROC Curve's AUC — 160
- 7.5 Bland-Altman Plots — 162

8 Designed Experiments — 163
- 8.1 One-Way Fixed Effects ANOVA — 163
  - 8.1.1 Balanced One-Way Design — 163
  - 8.1.2 Unbalanced One-Way Design — 166
- 8.2 Randomized Block Design — 167
- 8.3 Balanced Full Factorial Design with Fixed Effects — 169
- 8.4 Random and Mixed Models — 172
  - 8.4.1 Fixed Effects in Mixed Models — 172
  - 8.4.2 Random Effects in Mixed and Random Models — 174
  - 8.4.3 Confidence Intervals for Variance Components — 175
- 8.5 Nested Designs — 175
- 8.6 Two-Level Factorial Designs — 176
  - 8.6.1 Test for a Main Effect — 177
  - 8.6.2 Confidence Interval for a Regression Coefficient — 180
  - 8.6.3 Two-Level Fractional Factorial Designs — 181
  - 8.6.4 Plackett-Burman Designs — 183
- 8.7 Two-Level Factorial Designs with Centers — 184
  - 8.7.1 Main Effects — 184
  - 8.7.2 Lack of Fit Test — 185
- 8.8 Response Surface Designs — 187

9 Reliability and Survival — 191

- 9.1 Reliability Parameter Estimation 191
  - 9.1.1 Exponential Reliability 191
  - 9.1.2 Weibull Reliability 194
  - 9.1.3 Normal Reliability 197
- 9.2 Reliability Demonstration Tests 199
  - 9.2.1 Tests for Location Parameters 200
  - 9.2.2 Tests for Specified Reliability 202
  - 9.2.3 Tests for Percentiles 205
- 9.3 Two-Sample Reliability Tests 206
  - 9.3.1 Two-Sample Test for Mean Exponential Life 207
  - 9.3.2 Two-Sample Log-Rank Test 208
- 9.4 Interference 211
  - 9.4.1 Normal-Normal Interference 212
  - 9.4.2 Exponential–Exponential Interference 216
  - 9.4.3 Weibull-Weibull Interference 217

10 Statistical Quality Control — 221

- 10.1 Statistical Process Control 221
  - 10.1.1 Control Chart Run Rules 221
  - 10.1.2 Power and Sample Size for Control Charts 224
- 10.2 Process Capability 229
  - 10.2.1 Confidence Intervals for c_p and c_{pk} 229
  - 10.2.2 Tests for c_p and c_{pk} 231
- 10.3 Tolerance Intervals 233
  - 10.3.1 Nonparametric Tolerance Intervals 233
  - 10.3.2 Normal Tolerance Intervals 235
- 10.4 Acceptance Sampling 236
  - 10.4.1 Single Sampling Plans for Attributes 237
  - 10.4.2 Rectifying Inspection for Attributes 246
  - 10.4.3 Variables Sampling Plans for Defectives 251
- 10.5 Gage R&R Studies 256

11 Resampling Methods — 261

- 11.1 Software Requirements 262
- 11.2 Monte Carlo 262
  - 11.2.1 Sample Size for Confidence Intervals 262
  - 11.2.2 Power and Sample Size for Hypothesis Tests 265
- 11.3 Bootstrap 268
  - 11.3.1 Bootstrap Confidence Intervals 268
  - 11.3.2 Sample Size for Bootstrap Confidence Intervals 268
  - 11.3.3 Power and Sample Size for Bootstrap Hypothesis Tests . . 269

A Notation — 273

CONTENTS ix

B	**Glossary**	**277**
C	**Greek Alphabet**	**281**
D	**Probability Distributions**	**283**
	D.1 Noncentral Distributions	283
	D.2 Hypergeometric Distribution	284
	D.3 Binomial Distribution	285
	D.4 Poisson Distribution	286
	D.5 Normal Distribution	287
	D.6 Student's t Distribution	287
	D.7 Chi-square (χ^2) Distribution	288
	D.8 F Distribution	289
E	**Probability Tables**	**291**
	E.1 Larson's Nomogram for the Cumulative Binomial Distribution	292
	E.2 Cumulative Poisson Probability	293
	E.3 Standard Normal Probabilities	294
	E.4 Quantiles of Student's t Distribution	295
	E.5 Quantiles of the Chi-square (χ^2) Distribution	296
	E.6 Quantiles of the F Distribution	297
	E.7 One- and Two-sided Tolerance Factors for Normal Distributions	298
	E.8 Software	299
	E.8.1 Microsoft Excel	299
	E.8.2 MINITAB	299
	E.8.3 PASS	299
	E.8.4 R	299
	E.8.5 Piface	300
F	**Identities and Approximations**	**301**
	F.1 Identities	301
	F.2 Approximations	302
G	**The Delta Method**	**305**
	G.1 One Unknown Parameter	305
	G.2 Two or More Unknown Parameters	306
	G.3 Applications of the Delta Method	306
	G.3.1 Arcsine Transform for the Binomial Proportion	306
	G.3.2 Log Transform for the Binomial Proportion	307
	G.3.3 Log Odds Transform for the Binomial Proportion	307
	G.3.4 Log Transform for the Risk Ratio	308
	G.3.5 Log Transform for the Odds Ratio	308
	G.3.6 Square-Root Transform for Poisson Counts	309
	G.3.7 Difference Between Two Independent Poisson Counts	309
	G.3.8 Ratio of Two Independent Poisson Counts	310

G.3.9 Standard Normal $\hat{z}$ Statistic 311
G.3.10 Normal Probability at Specified x Value 311
G.3.11 Sample Standard Deviation 312
G.3.12 Logarithmic Transform for the Standard Deviation 312
G.3.13 Logarithmic Transform for the Ratio of Two Standard Deviations 313
G.3.14 Coefficient of Variation 314
G.3.15 Process Capability Statistic c_{pk} 314

Preface

The purpose of this book is to present methodologies for calculating sample size and power for: means, standard deviations, proportions, counts, regression, correlation and agreement, ANOVA for fixed and random effects, reliability, acceptance sampling, process capability, and gage error studies. The book was written for engineers, scientists, statisticians, technical managers, and quality engineering professionals who are responsible for recommending sample sizes for activities that involve data collection and analysis in their organizations. Readers are expected to have a basic understanding of the theory and practice of inferential statistical methods. Knowledge of advanced statistical methods is not required.

My education is in physics, where I was taught that factors of two are only important in matters of salary. That lack of precision is excessive with respect to sample size calculations, but it sets the approach that I've taken in the book. Because sample size and power calculations are usually performed when there is significant uncertainty about important inputs to the calculations, the results of these calculations should always be treated as approximations. Consequently, the book emphasizes approximate methods using large-sample approximations and variable transformations. Exact methods are presented where space allows or when the approximate methods fail.

The notation required for this book presented special problems. Readers should review the notation conventions that are included in the appendices, but be prepared to see deviations from those conventions in difficult cases. For example, it is easy to distinguish between a population parameter and the statistic that estimates it, like μ and $\bar{x}$, respectively, but it is difficult to unambiguously indicate a parameter estimate used in a sample size and power calculation. In most cases I have tried to use the caret notation, like $\hat{\mu}$, but that was not always possible.

To keep the book small and easy to maintain, it does not contain instructions for using sample size calculation software. Software solutions to the example problems in the book using PASS, MINITAB, Piface, and R are posted at *www.mmbstatistical.com/SampleSize.html*. Homework problems, such as for self study or a college course, and errata are also posted at the web site. If you find any new errors, please report them to me at *paul@mmbstatistical.com*.

My customers and students deserve credit for forcing me to master this material, so my thanks go to them for their hard questions and the support they provided. Thanks to MINITAB Inc. for providing me with a copy of their software, to Russ Lenth for making Piface available, and to George Pearson at MacKichan Software Inc. for his help with Scientific Workplace and LaTeX. I thank my assistant, Rebecca Malnar, for helping prepare much of this material; Mary Keane, who did the copy editing; and Skip Malm at Activities Press, Inc., for production help.

This book would not have been possible without the support, encouragement, patience, and love of my wife, Kathy. This book is dedicated to her.

Chapter 1

Fundamentals

1.1 Motivation for Sample Size Calculations

Whether you work in science, engineering, manufacturing, business, or the health or service industries, the decisions that you make are based on data. In most cases, 100% inspection is impractical or impossible, so we must use sampling methods and tolerate or manage the risks associated with making decisions from limited data.

Figure 1.1 presents a cost model for making decisions using sample data. The figure plots resources consumed versus sample size, where resources consumed are due to sampling costs and costs associated with incorrect decisions. Sampling costs increase with sample size. Decisions made from large samples tend to be correct, so costs associated with incorrect decisions decrease with sample size. The combined costs are given by the *Total Costs* curve in the figure. The optimal sample size is the one that minimizes the total costs.

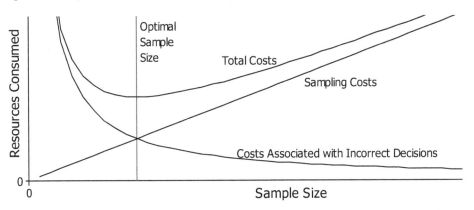

Figure 1.1: Quality cost model for resources consumed versus sample size.

1

1.2 Rationale for Sample Size and Power Calculations

We use statistics calculated from sample data to estimate the unknown parameters of the populations that we are sampling. For example, we hope that the sample mean $\bar{x}$ determined from a sample of size n is close enough to the unknown population mean μ that we make an accurate interpretation of the situation. If $\bar{x}$ falls close to μ, then we will probably take the correct action on the process, but if $\bar{x}$ falls too far from μ, then we might take an inappropriate action. How close or far $\bar{x}$ falls from μ is a matter of estimation precision, which depends upon - among other things - the sample size. Larger sample sizes provide better precision and reduce the chances of making mistakes.

A statistic like $\bar{x}$ is called a *point estimate* because it provides a single-number estimate for its associated parameter μ. Too much emphasis is placed on point estimates; their weakness is that they do not account for the precision of the estimate. There are two statistical methods that integrate point estimates and their estimation precision: confidence intervals and hypothesis tests. Unfortunately, many people do not understand or are unwilling to acknowledge the weakness of point estimates and insist on making decisions based on them rather than on confidence intervals and hypothesis tests.

The purpose of this section is to present the rationale for data analysis and interpretation using confidence intervals and hypothesis tests and to identify the role played by sample size in these methods. The problems of estimating and testing a population mean are used to demonstrate these methods. More complicated problems involving population means and problems involving proportions, counts, and other process metrics are considered in detail in later chapters.

1.2.1 Rationale for Confidence Intervals

A statistic provides a point estimate for an unknown population parameter, but a confidence interval provides a range of values within which the unknown population parameter is statistically likely to fall. For example, the confidence interval for the unknown population mean μ has the form

$$P(LCL < \mu < UCL) = 1 - \alpha \tag{1.1}$$

where $P()$ is the probability function, UCL and LCL are the upper and lower confidence limits, respectively, and $1 - \alpha$ is the confidence level. We read Equation 1.1 as, "We can be $(1 - \alpha)\,100\%$ confident that the true but unknown value of the population mean μ falls between LCL and UCL." The confidence level is usually taken to be be close to 1; values of $1 - \alpha = 0.95, 0.99$, and 0.90 are common. The confidence limits are calculated from the sample data and have the form

$$UCL/LCL = \bar{x} \pm \delta \tag{1.2}$$

where $\bar{x}$ is the sample mean and the confidence interval half-width δ indicates the precision of the estimate. Narrow confidence intervals are preferred because they are less ambiguous than wide ones.

1.2. Rationale for Sample Size and Power Calculations

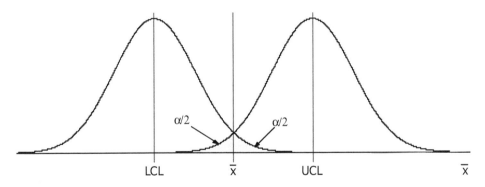

Figure 1.2: Relationship between $\bar{x}$ and the confidence limits for μ.

Example 1.1 Express the confidence interval $P(3.1 < \mu < 3.7) = 0.95$ in words.

Solution: The confidence interval indicates that we can be 95% confident that the true but unknown value of the population mean μ falls between $\mu = 3.1$ and $\mu = 3.7$. Apparently, the mean of the sample used to construct the confidence interval is $\bar{x} = 3.4$ and the confidence interval half-width is $\delta = 0.3$.

The upper and lower limits of a confidence interval indicate the values of a parameter that would be statistically likely to produce the sample data. Figure 1.2 shows the confidence limits for the population mean, UCL and LCL, calculated from a sample mean $\bar{x}$ as determined from a sample of size n. The figure also shows the corresponding $\bar{x}$ distributions if the true population mean is $\mu = LCL$ or $\mu = UCL$. These distributions indicate that if μ were any less than LCL or any greater than UCL, then the sample mean $\bar{x}$ would be an unexpected result. By design, UCL and LCL are calculated to limit the probability that the true population mean falls outside the confidence limits to α, thus forcing the probability to be $1 - \alpha$ that the true population mean is between LCL and UCL.

A confidence interval can be interpreted by considering the number of management actions indicated over the range of the interval. If a confidence interval is sufficiently narrow that a single management action is indicated over its entire range, then the interval is useful. If a confidence interval is so wide that two or more management actions are indicated over its range, then the interval is ambiguous.

The specific formula used to calculate confidence interval half-width δ depends on the parameter being estimated and the situation. When the distribution of x values is normal with known standard deviation σ_x,[1] then the half-width of

---

[1] That is, known from experience and/or historical data.

the confidence interval for the population mean μ is given by

$$\delta = z_{\alpha/2} \frac{\sigma_x}{\sqrt{n}} \tag{1.3}$$

where $z_{\alpha/2}$ is the standard normal distribution z-score with tail area $\alpha/2$. This equation shows that the confidence interval half-width

- increases as the confidence level increases.
- increases as the standard deviation increases.
- decreases as the sample size increases.

These dependencies are general; the half-widths of confidence intervals for other parameters behave the same way.

Equation 1.3 allows us to calculate confidence interval half-width as an experimental result,[2] but from a planning perspective the value of δ should always be chosen in advance, before any data are collected, to guarantee that the experiment will deliver a sufficiently narrow confidence interval. Because the confidence level should also be chosen in advance and σ_x is assumed to be known, then the required sample size for the experiment is

$$n = \left(\frac{z_{\alpha/2}\sigma_x}{\delta}\right)^2. \tag{1.4}$$

Equation 1.4 shows that the sample size for an experiment to estimate the population mean is determined by three factors, which may be chosen before any data are collected: the confidence level, the population standard deviation, and the confidence interval half-width.

Example 1.2 Data are to be collected for the purpose of estimating the mean of a mechanical measurement. Data from a similar process suggest that the standard deviation will be $\sigma_x = 0.003mm$. Determine the sample size required to estimate the value of the population mean with a 95% confidence interval of half-width $\delta = 0.002mm$.

Solution: With $z_{\alpha/2} = z_{0.025} = 1.96$ in Equation 1.4, the required sample size is

$$n = \left(\frac{1.96 \times 0.003}{0.002}\right)^2 = 8.64.$$

The sample size must be an integer; therefore, we round the calculated value of n up to $n = 9$.

---

[2] The word *experiment* is used here, as it will be throughout, to indicate *any* data collection activity for the purpose of estimating or testing one or more population parameters.

1.2. Rationale for Sample Size and Power Calculations

Example 1.3 What is the new sample size in Example 1.2 if the process owner prefers a 99% confidence interval?
Solution: With $z_{\alpha/2} = z_{0.005} = 2.575$ in Equation 1.4, the required sample size is

$$n = \left(\frac{2.575 \times 0.003}{0.002}\right)^2 = 15.$$

Example 1.4 What is the new sample size in Example 1.2 if the process owner prefers a 95% confidence level with $\delta = 0.001mm$ half-width?
Solution: With $z_{0.025} = 1.96$ and $\delta = 0.001mm$ in Equation 1.4, the required sample size is

$$n = \left(\frac{1.96 \times 0.003}{0.001}\right)^2 = 35.$$

The confidence interval given by Equation 1.1 is called a *two-sided interval* because the interval provides both upper and lower limits for the population mean. Sometimes a two-sided interval is inappropriate and a one-sided interval, with either an upper or a lower limit, is required. A one-sided lower confidence interval has the form

$$P\left(\bar{x} - \delta < \mu < \infty\right) = 1 - \alpha \tag{1.5}$$

where δ is given by

$$\delta = z_\alpha \frac{\sigma_x}{\sqrt{n}}. \tag{1.6}$$

Similarly, a one-sided upper confidence interval has the form

$$P\left(-\infty < \mu < \bar{x} + \delta\right) = 1 - \alpha. \tag{1.7}$$

In one-sided intervals, it is common to drop the reference to the limit that goes to plus or minus infinity. For example, the intervals in Equations 1.5 and 1.7 may also be written as $P\left(\bar{x} - \delta < \mu\right) = 1 - \alpha$ and $P\left(\mu < \bar{x} + \delta\right) = 1 - \alpha$, respectively. The definition of δ as the confidence interval half-width breaks down in these cases, but δ is still a useful measure of confidence interval width.

For both the one-sided upper and one-sided lower confidence interval cases, the corresponding sample size equation is

$$n = \left(\frac{z_\alpha \sigma_x}{\delta}\right)^2. \tag{1.8}$$

Example 1.5 Determine the sample size required to estimate the mean of a population when $\sigma_x = 30$ is known and the population mean must not exceed the sample mean by more than $\delta = 10$ with 95% confidence.

Solution: A one-sided upper 95% confidence interval is required of the form

$$P(\mu < \bar{x} + \delta) = 0.95.$$

With $z_{0.05} = 1.645$ in Equation 1.8, the necessary sample size is

$$n = \left(\frac{1.645 \times 30}{10}\right)^2 = 25.$$

1.2.2 Rationale for Hypothesis Tests

A hypothesis test is performed by forming two complementary statements about the value of a population parameter and then testing to see which of the two statements is supported by the sample data. The two statements or hypotheses being tested are called the *null hypothesis* (H_0) and the *alternate hypothesis* (H_A).

Hypothesis testing is more about the alternate hypothesis than the null hypothesis. It is good practice to make H_A the hypothesis that represents the opportunity of interest or value and to make H_0 the complement of H_A. If the data support H_A, we say that we reject H_0, that we accept H_A, or that the experimental result is statistically significant. It is bad practice to accept H_0, so, if the data do not support H_A, we make one of the following statements: we cannot reject H_0, we cannot accept H_A, the test is inconclusive, or the experimental result is not statistically significant. The philosophy of crafting and interpreting hypothesis tests is concisely expressed in Carl Sagan's statement, "Extraordinary claims require extraordinary evidence," where the extraordinary claim is H_A and there is no opportunity to accept H_0. Sagan's statement also makes it clear that there must be substantial evidence in favor of the alternate hypothesis if we are to accept it.[3]

The hypotheses for a test of the population mean are

$$H_0 : \mu = \mu_0 \text{ versus } H_A : \mu \neq \mu_0 \tag{1.9}$$

where μ_0 is a specified value. An obvious choice of statistic to test these hypotheses is the sample mean $\bar{x}$. If $\bar{x}$ falls sufficiently close to μ_0, then the data do not support H_A. If $\bar{x}$ falls sufficiently far from μ_0, then the data do support H_A. Whether $\bar{x}$ is statistically close to or far from μ_0 is a matter of estimation precision, which the hypothesis testing method is designed to manage.

In the test for the population mean using these hypotheses, if the population is normally distributed with known standard deviation σ_x, then under

---

[3]The language of accepting and rejecting hypotheses follows strict guidelines that sometimes cause confusion. The practice adopted here is to follow these guidelines as closely as possible except where I've chosen to violate them for the sake of convenience. For those who can't abide by my choices, please substitute "Can't reject H_0" for every incidence of "Accept H_0".

1.2. Rationale for Sample Size and Power Calculations

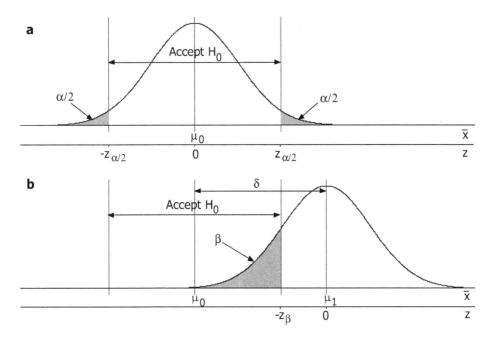

Figure 1.3: Distributions of $\bar{x}$ under $H_0 : \mu = \mu_0$ and $H_A : \mu \neq \mu_0$.

$H_0 : \mu = \mu_0$ the expected distribution of sample means ($\bar{x}$) determined for samples of size n is normal in shape by the central limit theorem with mean $\mu_{\bar{x}} = \mu_0$ and standard deviation $\sigma_{\bar{x}} = \sigma_x/\sqrt{n}$ as shown in Figure 1.3a, where the limits $\bar{x}_{A/R-}$ and $\bar{x}_{A/R+}$ define the acceptance and rejection regions for H_0. These limits are chosen so that the probability of obtaining a sample mean that falls inside the acceptance region for H_0 is $1 - \alpha$. The complementary probability of obtaining a sample $\bar{x}$ that falls outside the acceptance region for H_0 is α, also known as the *type I error rate*, the *false alarm rate*, or the *false positive rate*. A type I error occurs when we reject a null hypothesis that is true. The value of α is chosen to be small, often $\alpha = 0.05$ or 0.01, depending on the consequences of committing a type I error.

Figure 1.3b shows the distribution of sample means under $H_A : \mu \neq \mu_0$ when the population mean has shifted from μ_0 to $\mu_1 = \mu_0 + \delta$ where $\delta = \mu_1 - \mu_0$ is the size of the shift in the population mean, better known as the *effect size*. Because H_0 is false, the correct decision here is to reject it, which, according to the figure, happens with probability $\pi = 1 - \beta$, where π is the statistical power and β, the complement of the power, is the type II error rate. A type II error occurs when we do not reject a null hypothesis that is false. High values of power are desirable; the power is often taken to be $\pi = 0.90, 0.95,$ or 0.80. A power value $\pi = 1 - \beta$ is *always* paired with an effect size δ.

From Figure 1.3, $\bar{x}_{A/R+}$ is related to μ_0 and μ_1 by

$$\bar{x}_{A/R+} = \mu_0 + z_{\alpha/2}\sigma_{\bar{x}} = \mu_1 - z_\beta \sigma_{\bar{x}} \tag{1.10}$$

so the effect size is given by

$$\begin{aligned} \delta &= \mu_1 - \mu_0 \\ &= \left(z_{\alpha/2} + z_\beta\right) \sigma_{\bar{x}} \\ &= \frac{\left(z_{\alpha/2} + z_\beta\right) \sigma_x}{\sqrt{n}}. \end{aligned} \tag{1.11}$$

This equation shows that the effect size δ that an experiment can detect with corresponding power π or type II error rate $\beta = 1 - \pi$ is determined by the population standard deviation σ_x and the sample size n.

Equation 1.11 allows us to calculate the effect size that an experiment can detect with specified power, but, from an experiment planning perspective, the effect size δ determines whether rejecting H_0 in favor of H_A has any practical value. From that standpoint, if δ is very small then there is little value in accepting H_A, but if δ is very large then there is value in accepting H_A. Somewhere in between very small and very large δ values there is a critical value of δ, called the *smallest practically significant effect size*, which an experiment should be able to detect with high probability or power $\pi = 1 - \beta$. The value of δ must be chosen by a process expert qualified to interpret its size. Then, from Equation 1.11, the sample size required for an experiment to detect a practically significant effect δ with specified power $\pi = 1 - \beta$ is

$$n = \left(\frac{\left(z_{\alpha/2} + z_\beta\right) \sigma_x}{\delta}\right)^2. \tag{1.12}$$

The benefit of selecting an experiment's sample size in this way is that it guarantees that a statistically significant experimental result is also practically significant. All sample size calculations for hypothesis tests equate statistical significance with practical significance.

Example 1.6 An experiment is planned to test the hypotheses $H_0 : \mu = 3200$ versus $H_A : \mu \neq 3200$. The process is known to be normally distributed with standard deviation $\sigma_x = 400$. What sample size is required to detect a practically significant shift in the process mean of $\delta = 300$ with power $\pi = 0.90$?
Solution: With $\beta = 1 - \pi = 0.10$ and assuming $\alpha = 0.05$ in Equation 1.12, the sample size required to detect a shift from $\mu = 3200$ to $\mu = 2900$ or $\mu = 3500$

1.2. Rationale for Sample Size and Power Calculations

with 90% power is

$$n = \left(\frac{(z_{0.025} + z_{0.10})\sigma_x}{\delta}\right)^2$$

$$= \left(\frac{(1.96 + 1.282)400}{300}\right)^2$$

$$= 19$$

where the calculated value of n was rounded up to the nearest integer value.

The alternative hypothesis in Equation 1.9 describes a two-tailed test because we can reject $H_0 : \mu = \mu_0$ if there is evidence that the true population mean is either too low or too high with respect to μ_0. Sometimes a two-tailed test is not appropriate and a one-tailed test is required. There are two one-tailed alternative hypotheses available: $H_A : \mu > \mu_0$ and $H_A : \mu < \mu_0$. If it bothers you that these alternative hypotheses are not truly complementary with the null hypothesis $H_0 : \mu = \mu_0$, then you could rewrite the null hypothesis as $H_0 : \mu \leq \mu_0$ or $H_0 : \mu \geq \mu_0$. However, because the goal in hypothesis testing is always to accept H_A, the null hypotheses may be written either way.

Power and sample size may be calculated for one-tailed tests using the formulas for two-tailed tests by replacing $\alpha/2$ with α.

Example 1.7 An experiment will be performed to test $H_0 : \mu = 8.0$ versus $H_A : \mu > 8.0$. What sample size is required to reject H_0 with 90% power when $\mu = 8.2$? The process is known to be normally distributed with $\sigma_x = 0.2$.

Solution: For the one-tailed hypothesis test with $\alpha = 0.05$, $\beta = 1 - \pi = 0.10$, and $\delta = 0.2$ the required sample size is

$$n = \left(\frac{(z_\alpha + z_\beta)\sigma_x}{\delta}\right)^2 \tag{1.13}$$

$$= \left(\frac{(z_{0.05} + z_{10})\sigma_x}{\delta}\right)^2$$

$$= \left(\frac{(1.645 + 1.282)0.2}{0.2}\right)^2$$

$$= 9.$$

After a random sample of the predetermined sample size has been collected and measured, the test statistic can be calculated and used to decide if the null hypothesis can be rejected. The test statistic can be compared directly to appropriate critical values, such as by comparing the sample mean $\bar{x}$ to $\bar{x}_{A/R-}$ and $\bar{x}_{A/R+}$ in Figure 1.3a, but the preferred method for making the decision is to calculate the significance level or p value associated with the test statistic. The p value is the probability of obtaining an experimental result equal to or more extreme than the observed result under H_0. The p value gets contributions from both tails of the H_0 distribution in a two-tailed test and from one tail in a one-tailed test. If the experimental p value is less than the predetermined α value, then the experimental result is inconsistent with H_0, so H_0 can be rejected. If the p value is larger than α, then H_0 cannot be rejected.

The p value provides another way to interpret the meaning of the power of a hypothesis test. The power of a test is the probability of obtaining an experimental p value such that $p < \alpha$ for a specified value of effect size δ under H_A.

Example 1.8 Calculate the p value for the test performed under the conditions of Example 1.6 if the sample mean was $\bar{x} = 3080$.
Solution: Figure 1.4 shows the contributions to the p value from the two tails of the $\bar{x}$ distribution under H_0. The z test statistic that corresponds to $\bar{x}$ is

$$\begin{aligned} z &= \frac{\bar{x} - \mu_0}{\sigma_{\bar{x}}} \\ &= \frac{\bar{x} - \mu_0}{\sigma_x/\sqrt{n}} \\ &= \frac{3080 - 3200}{400/\sqrt{19}} \\ &= -1.31, \end{aligned}$$

so the p value is

$$\begin{aligned} p &= 1 - \Phi\left(-1.31 < z < 1.31\right) \\ &= 0.19. \end{aligned}$$

Because $(p = 0.19) > (\alpha = 0.05)$, the observed sample mean is statistically consistent with $H_0 : \mu = 3200$, so we can not reject H_0.

Example 1.9 Calculate the p value for the test performed under the conditions of Example 1.7 if the sample mean was $\bar{x} = 8.39$.
Solution: Figure 1.5 shows the single contribution to the p value from the right tail of the $\bar{x}$ distribution under H_0. The z test statistic that corresponds to $\bar{x}$ is

$$\begin{aligned} z &= \frac{8.39 - 8.2}{0.2/\sqrt{9}} \\ &= 2.85, \end{aligned}$$

so the p value is

$$\begin{aligned} p &= \Phi\left(2.85 < z < \infty\right) \\ &= 0.0022. \end{aligned}$$

Because $(p = 0.0022) < (\alpha = 0.05)$, the observed sample mean is an improbable result under $H_0 : \mu = 8.2$, so we must reject H_0.

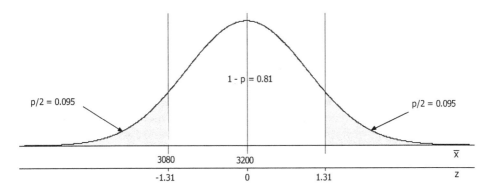

Figure 1.4: Normal (z) distribution showing two-tailed p value.

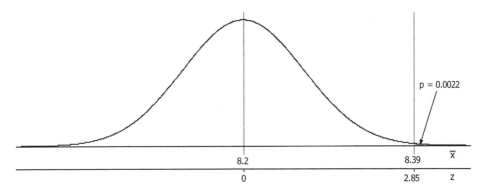

Figure 1.5: Normal (z) distribution showing one-tailed p value.

1.3 Practical Considerations

The purpose of this section is to discuss the practical issues involved in determining sample size and power for experiments.

1.3.1 Roles of the Investigator and the Statistician

Collecting the information for a sample size or power calculation for an experiment requires a dialog between two subject matter experts - the process expert who knows the process intimately and has a vested interest in the experiment, and the statistical methods expert who ensures that the statistical requirements are satisfied. Even if one person plays both roles, they both must be fairly and diligently represented.

The process expert is responsible for identifying the response to be studied and the goals or acceptance criteria for the response. The process expert is also responsible for all other aspects of the process, including identification of secondary responses and the independent process variables, covariates, gage error and process capability conditions, physical operation of the process, execution of the experiment, and data collection. The process expert must provide the statistical methods expert the information about the process required to complete the sample size or power calculation while being open and honest with the facts, so that issues that could compromise the experiment or the analysis are not overlooked.

The statistical methods expert is responsible for confirming that the experiment design chosen by the process expert is consistent with the goals of the experiment and, if necessary, assists in redesigning the experiment to meet those goals. The statistical methods expert must combine information about the process and the goals of the experiment with the anticipated method of analysis to determine the sample size or power. The statistical methods expert is often involved only as a consultant, with little or no direct knowledge, experience, or personal contact with the process, the response, or the goals or acceptance criteria that apply to the response, so must ask lots of questions of the process expert to make sure that issues that could compromise the experiment or the analysis are not overlooked.

1.3.2 Preference for Approximate Methods

Many of the exact sample size and power calculation methods are computationally intensive, so the only practical way to obtain exact results is from published tables or with specialized software. This book emphasizes approximate methods over exact ones. Approximate calculation methods can be nearly as accurate as exact ones; the errors they introduce are usually smaller than the errors introduced by uncertainties in the inputs to the calculations. Exact methods are presented when they are simple or when the approximate methods fail. Approximate methods also provide a good starting point for exact methods when the exact method requires iterations to determine the solution.

1.3.3 Factors that Affect Sample Size

Section 1.2 presented the rationale for inferential methods. This section goes into more detail about the factors that affect those methods, to prepare the investigator to collect the information required to calculate sample size and power for experiments. The example of inference for the population mean is used to support this discussion because it is simple, well known, and includes all of the factors that need to be considered. Inference for other parameters involves the same or similar factors.

1.3. Practical Considerations

1.3.3.1 Factors that Affect Confidence Intervals

The $(1 - \alpha)\,100\%$ two-sided confidence interval for the population mean has the form
$$P\left(\bar{x} - \delta < \mu < \bar{x} + \delta\right) = 1 - \alpha \tag{1.14}$$
where the confidence interval half-width is
$$\delta = \frac{z_{\alpha/2}\sigma_x}{\sqrt{n}}. \tag{1.15}$$

The sample size required to obtain a specified value of the confidence interval half-width is
$$n = \left(\frac{z_{\alpha/2}\sigma_x}{\delta}\right)^2. \tag{1.16}$$

The factors that affect the sample size are

- the confidence interval half-width (δ).
- the population standard deviation (σ_x).
- the confidence level (α).
- whether the interval is one- or two-sided.

Practical constraints on the confidence level (it must be high) and the population standard deviation (it is what it is) make these factors less flexible than others, so the primary trade-off between factors occurs between the confidence interval half-width and the sample size.

1.3.3.1.1 Sample Size
Sample size is an input to the confidence interval half-width calculation during experimental data analysis, but it is an output of the experiment planning process. Relative to the other factors like population standard deviation and confidence level, the sample size has fewer constraints, so it is the primary determinant of confidence interval half-width.

Sometimes practical limitations, such as cost, time, or the number of experimental subjects available, prohibit the use of a precalculated sample size for an experiment. In such situations it is necessary to compromise by stating the confidence interval half-width that the experiment can resolve for the fixed sample size.

1.3.3.1.2 Confidence Interval Half-Width
Confidence interval half-width is calculated as an output or result from the analysis of experimental data, but it is an input for the sample size calculation in the experiment planning process.[4] The

---

[4] The exception to the statement is when practical limitations, such as cost, time, or the number of experimental subjects available, prohibit the use of a precalculated sample size and it becomes necessary to compromise by stating the confidence interval half-width that an experiment can resolve. See Parker and Berman [52] for a discussion of these issues.

confidence interval half-width indicates the precision of the estimate for the population parameter. If the confidence interval half-width is too large, the precision of the estimate is poor and the confidence interval will be too wide to be useful. If the confidence interval half-width is very small, the precision of the estimate is excellent but usually at the expense of a very large sample size. The process expert will need to choose a value for the precision that is small enough to be useful but not smaller than necessary to keep the sample size to a minimum.

Sometimes the process expert struggles to select a value for the confidence interval half-width and needs coaching from the statistical methods expert to determine an appropriate value. To begin the process, it helps to consider outrageously wide and outrageously narrow confidence intervals and then refine the value of the half-width. Suppose that the process expert wants to study the mean resting heart rate of a group of human test subjects. If he cannot decide what value of confidence interval half-width to use, the statistical methods expert might suggest $\delta = 200$ beats per minute, that is, $UCL/LCL = \bar{x} \pm 200$, which is probably fatal at both extremes, and then $\delta = 2$ beats per minute, that is, $UCL/LCL = \bar{x} \pm 2$, which is probably more precise than necessary. With further discussion, both people should be able to negotiate an appropriate value for δ that lies between $\delta = 200$ and $\delta = 2$ beats per minute.

Equation 1.14 expresses the confidence interval half-width in the same physical units as the measurements. For example, if x and $\bar{x}$ are in inches, then δ is also in inches. The confidence interval half-width is sometimes expressed in relative terms with respect to the sample mean:

$$P\left(\bar{x}\left(1-\delta\right) < \mu < \bar{x}\left(1+\delta\right)\right) = 1-\alpha \qquad (1.17)$$

or the sample standard deviation:

$$P\left(\bar{x} - \delta s < \mu < \bar{x} + \delta s\right) = 1-\alpha, \qquad (1.18)$$

where in both cases δ is a unitless number. The former method, Equation 1.17, is generally accepted and in widespread use. The latter method, Equation 1.18, should be avoided unless the expected value of the standard deviation is explicitly stated, which is the same as specifying the $\delta \times s$ product, and takes us back to the preferred confidence interval half-width formulation from Equation 1.14.

1.3.3.1.3 Population Standard Deviation

Equation 1.15 shows that the standard deviation, which is a characteristic of the process being studied, must be known from prior experience with the process to calculate the confidence interval half-width or the sample size. When the standard deviation is not known, it will need to be estimated. Some possible sources for the estimate are historical data, data from a similar process, information from a subject-matter expert, or published values from similar experiments.[5] When none of these sources is

---

[5]Published standard deviations tend to be smaller than what should be expected because experiments that have larger standard deviations often do not yield statistically significant results, so they do not get published.

1.3. Practical Considerations

available, it will be necessary to perform a pilot study to estimate the standard deviation. The obvious question is: What sample size is required for the pilot study to obtain a sufficiently accurate estimate of the standard deviation to use in the sample size calculation for the primary experiment? If δ is the maximum allowable relative error in the sample size of the primary experiment with associated confidence level $1 - \alpha$, that is,

$$P\left(\hat{n}\left(1-\delta\right)<n<\hat{n}\left(1-\delta\right)\right)=1-\alpha, \tag{1.19}$$

then, by the methods of Section 3.1.1.1, the sample size of the preliminary experiment must be approximately

$$n \simeq 2 \left(\frac{z_{\alpha/2}}{\delta}\right)^2. \tag{1.20}$$

The result of this calculation can be quite depressing, which only emphasizes the importance of entering into a sample size calculation with a good estimate for the standard deviation.

Example 1.10 What sample size is required for a pilot study to estimate the standard deviation to be used in the sample size calculation for a primary experiment if the sample size for the primary experiment should be within 20% of the correct value with 90% confidence?

Solution: With $\delta = 0.20$ and $\alpha = 0.10$ in Equation 1.20, the required sample size for the preliminary experiment to estimate the standard deviation is

$$n \simeq 2 \left(\frac{1.645}{0.20}\right)^2$$
$$\simeq 136.$$

1.3.3.1.4 Confidence Level The confidence level associated with a confidence interval determines the probability that the confidence interval will contain the population parameter that it is estimating, so a $(1 - \alpha)\,100\%$ confidence interval has $(1 - \alpha)\,100\%$ probability of containing the parameter and $100\alpha\%$ probability of not containing the parameter. Relative to other factors that affect confidence intervals, like the confidence interval half-width, sample size, and standard deviation, the confidence level is usually easy to choose. Confidence levels of 95, 90, and 99% are most common, but other values are possible.

1.3.3.1.5 One-sided or Two-sided Interval Confidence intervals can be *one-sided* or *two-sided*, where one-sided intervals can be either *one-sided upper* or *one-sided lower* intervals. The context of the problem being considered determines which type of sided-ness is appropriate for the situation. If there is no clear directionality about the value of the population parameter, then a two-sided interval is indicated. For example, research on the effect of a drug on resting heart rate might be interested in both increases and decreases in heart rate, so a two-sided interval is appropriate, but management is probably more concerned about a project's cost being underestimated than overestimated, so a one-sided budget estimate is more appropriate.

1.3.3.2 Factors that Affect Hypothesis Tests

The purpose of this section is to present some of the factors that affect the sample size in hypothesis tests. Many of these factors are the same as or are related to the factors that affect confidence intervals; see Section 1.3.3.1 for another point of view.

Equation 1.12 gives the sample size required to reject $H_0 : \mu = \mu_0$ in favor of $H_A : \mu \neq \mu_0$ with power π when $\mu = \mu_0 + \delta$:

$$n = \left(\frac{(z_{\alpha/2} + z_\beta) \sigma_x}{\delta} \right)^2. \tag{1.21}$$

The factors that affect the sample size are

- the effect size ($\delta = \mu - \mu_0$).
- the power ($\pi = 1 - \beta$ where β is the type II error rate).
- the population standard deviation (σ_x).
- the type I error rate (α).
- whether the test is one- or two-sided.

Practical constraints on the type I error rate (it must be low), the power (it must be high), and the population standard deviation (it is what it is) make these three factors less flexible than others, so the primary trade-off between factors occurs between the effect size and the sample size.

1.3.3.2.1 Effect Size

Effect size is an output of the analysis of experimental data, but it is an input to the sample size calculation in the experiment planning process. Effect size quantifies the difference in the value of the parameter being tested under the null and alternative hypotheses:

$$\delta = \mu - \mu_0. \tag{1.22}$$

Relatively small effect sizes are not of practical value, but δ should not be chosen to be too large, otherwise a small but practically significant δ might go undetected. The appropriate value of δ to be used in the sample size calculation is the smallest value that is still considered to be practically significant. This value must be chosen by the process expert, sometimes in agreement with other people

1.3. Practical Considerations

who have a vested interest in the experiment, and perhaps with some coaching from the statistical methods expert.

The effect size as defined in Equation 1.22 is the absolute effect size, expressed in the same measurement units as the measurements, the population mean, and the population standard deviation. Effect size can also be expressed in relative terms, either with respect to the mean,

$$\delta = \frac{\mu - \mu_0}{\mu_0}, \qquad (1.23)$$

or with respect to the standard deviation,

$$\delta = \frac{\mu - \mu_0}{\sigma_x} \qquad (1.24)$$

where these relative δs are unitless numbers. The former method, Equation 1.23, is in widespread use and is an accepted method of specifying the effect size when the population mean under H_0 is stated. The latter method, Equation 1.24, was proposed by Cohen [13] and offers the apparent advantage of not requiring that the standard deviation be specified. This method is also in widespread use in some disciplines *even though its use is strongly discouraged*. The problem with Cohen's method is that it does not acknowledge that standard deviations can vary from experiment to experiment, so absolute effect sizes as given by Equation 1.22, which are the correct ones to be concerned about, can be uncorrelated to the relative effect sizes of Equation 1.24. *Do not* use Cohen's method to specify the effect size.[6]

1.3.3.2.2 Power
Statistical power is the probability of correctly rejecting the null hypothesis when it is false, that is, when the effect size is not equal to 0. Power is a function of effect size, so a target value of power must be chosen for the smallest practically significant effect size δ that the experiment is intended to detect. The most common choices for power are $\pi = 0.90, 0.80$, and 0.95, but other values are used in some situations.

Some software packages can calculate sample sizes for hypothesis tests and not for confidence intervals, but they can usually be tricked into doing the confidence interval calculations. For example, Equations 1.16 and 1.21 differ only by the z_β factor in the latter equation; however, Equation 1.21 reduces to Equation 1.16 when $\pi = \beta = 0.5$ because $z_{0.5} = 0$. Calculate the sample size required for a confidence interval using software for calculating sample size for a hypothesis test by setting the test power to 50% and setting the effect size equal to the desired confidence interval half-width. The confidence level $(1 - \alpha)\,100\%$ is determined by the hypothesis test's type I error rate α.

---

[6]Even Cohen recognized the problems with this method and warned against its use, but unfortunately those problems have been selectively ignored. See Russ Lenth's discussion [36].

1.3.3.2.3 Significance Level The hypothesis test's significance level or type I error rate α corresponds to the probability of rejecting the null hypothesis when it is true. It can also be interpreted as the upper limit for experimental p values for which the null hypothesis will be rejected. Relative to other factors that affect the sample size, like the effect size and power, the significance level is relatively easy to choose, with values of $\alpha = 0.05$ and 0.01 being most common.

1.3.3.2.4 Population Standard Deviation The population standard deviation is a characteristic of the process and response being studied, so the only issue that pertains to it is determining its value. See Section 1.3.3.1 for details about methods for estimating the population standard deviation.

1.3.3.2.5 One- or Two-sided Test Hypothesis tests can be one-sided or two-sided, as determined by the context of the problem. If there is no clear directionality about the value of the population parameter, so that the null hypothesis should be rejected when $\mu < \mu_0$ or when $\mu > \mu_0$, then a two-sided test is indicated. See Section 1.3.3.1 for a discussion of the analogous issues involved in one-sided and two-sided confidence intervals.

1.3.4 Graphical Presentations

Most of the methods presented in this book describe how to calculate a single value of effect size, sample size, or power as a function of the other factors; however, in most experiment planning, some of these factors tend to be fixed while others are more flexible. In such cases, it is helpful to create a plot or plots of the flexible factors while holding other factors fixed. Examples of such plots are

- power (π) versus effect size (δ) for a fixed sample size (n).
- power (π) versus sample size (n) for a fixed effect size (δ).
- sample size (n) versus effect size (δ) for a fixed power (π).
- sample size (n) versus power (π) for a fixed effect size (δ).
- effect size (δ) versus sample size (n) for a fixed power (π).
- effect size (δ) versus power (π) for a fixed sample size (n).

In some specialized hypothesis testing situations, it is common to plot the complement of the power, also known as the type II error probability $\beta = 1 - \pi$, instead of the power. Plots of β versus a quality characteristic are called *operating characteristic* (OC) curves. For example, in acceptance sampling using attribute inspection for defectives, the preferred plot to characterize a sampling plan is the OC curve of β versus the defective rate p. Sometimes the β scale is identified as the probability of accepting the null hypothesis, P_A or $P_A(H_0)$.

1.3. Practical Considerations

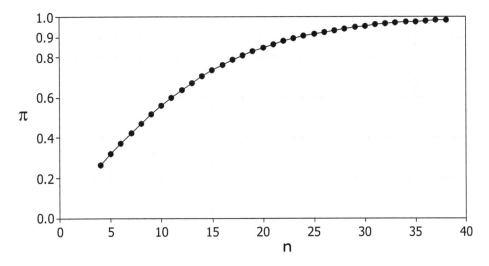

Figure 1.6: Power versus sample size for $\delta = 400$.

Example 1.11 An engineer must obtain approval from his manager to test a certain number of units to determine the mean response for a validation study. The standard deviation of the response is $\sigma_x = 600$ and the smallest practically significant shift in the mean that the experiment should detect is understood to be $\delta = 400$. What graph should the engineer use to present his case?

Solution: The value of the effect size of interest is firm at $\delta = 400$. The sample size is going to affect the power of the test, so an appropriate graph is power versus sample size. The sample size required to obtain a specified value of power for the test of $H_0 : \delta = 0$ versus $H_A : \delta \ne 0$ is given by Equation 1.12. Figure 1.6 shows the resulting power curve. The sample size required to obtain 80% power is $n = 18$ and the sample size required for 90% power is $n = 24$.

Example 1.12 Suppose that the manager in Example 1.11 approves the use of $n = 24$ units in the validation study. What power does the study have to reject H_0 when the effect size is $\delta = 200$, 400, and 600?[7]

Solution: The power is given by

$$\pi = \Phi\left(-z_\beta < z < \infty\right)$$

where z_β is determined from Equation 1.12:

$$z_\beta = \sqrt{n}\frac{\delta}{\sigma_x} - z_{\alpha/2}.$$

Figure 1.7 shows the power as a function of effect size. The power to reject H_0 when $\delta = 200$ is $\pi \simeq 0.37$, when $\delta = 400$ is $\pi \simeq 0.90$, and when $\delta = 600$ is $\pi \simeq 1$.

---

[7] This problem is premature. Its method of solution is presented in Section 2.2.2.1, but it is included here to demonstrate the use of power curves.

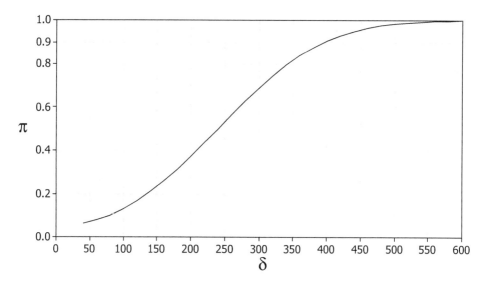

Figure 1.7: Power versus effect size for $n = 24$.

1.4 Problems and Solutions

The purpose of this section is to describe some of the problems that can compromise the accuracy of the sample size and power calculations presented in this book and methods to solve them.

1.4.1 When These Methods Fail

The results from the sample size and power calculation methods presented in this book are only as accurate as the inputs to the calculations. The following list of some common mistakes that can lead to incorrect sample size or power is provided as a warning and checklist to help avoid making those mistakes:

- Inappropriate or inaccurate standard error (σ_ϵ) estimate
- Inappropriate or inaccurate confidence interval half-width (δ)
- Inappropriate or inaccurate effect size (δ) and/or power (π)
- Analysis method or model used to calculate the sample size or power different from the method or model that is actually used to analyze the data.
- Violations in distribution assumption
    - Actual distribution differs from the assumed distribution.

1.4. Problems and Solutions

- Two or more populations are assumed to be homoscedastic (equal variances) when in fact they are heteroscedastic (unequal variances).
- Two or more populations are assumed to have the same distribution shape when in fact they do not.
- Variable transformation does not completely solve a distribution problem.
- Large-sample approximation is inaccurate.

- Failure to plan for missing observations
- Deviation from the experimental plan, e.g., introducing a randomization restriction
- Failure to anticipate and account for the effect of a lurking variable or covariate

1.4.2 When the Calculated Sample Size Is Too Large

One of the most frustrating aspects of sample size calculation occurs when the sample size required for a beautifully designed experiment turns out to be impractically large. In such cases, it might be necessary to compromise the original experiment by revising the acceptance requirements or to identify one or more methods of improving the precision of the experiment. Some methods that might be used to reduce sample sizes include the following:

- Improve the measurement repeatability and reproducibility.
- Introduce a blocking variable to reduce the precision error.
- Use a different response variable, one that has better precision.
- Replace an attribute response with an ordinal or measurement response.
- Identify and measure or control fluctuating process variables (i.e., covariates) that inflate the precision error of the response.
- For confidence intervals, reduce the confidence level or increase the confidence interval half-width.
- For hypothesis tests, reduce the target power or increase the smallest practically significant effect size.
- Use a repeated measures design to provide a homogeneous environment within subjects to improve sensitivity to treatment effects while controlling for differences between subjects.
- Use paired-sample methods instead of two-independent-sample methods.

- In stress testing (e.g., burst pressure testing) where specimen cost is low but the cost of performing a single test is very high, test many units simultaneously until the first unit fails and treat the surviving units as censored observations.

1.4.3 Bad Practices

Many organizations have developed guidelines for determining sample size that are not founded on sound statistical principles, so those guidelines may give incorrect results. The following list of bad practices and misconceptions is intended to help you recognize and avoid common mistakes. Explanations are presented in the less-than-obviously silly cases.

- **Bad Practice #1:** There is one magic sample size for all situations.

- **Bad Practice #2:** Use $n = 30$ because Student's t distribution becomes approximately normal for that sample size.

- **Bad Practice #3:** Use a sample of size $n = \sqrt{N} + 1$ when single sampling for defectives from a large lot of size N.

- **Bad Practice #4:** ANSI/ASQ Z1.4 and Z1.9 provide a comprehensive approach to sample size selection.

- **Bad Practice #5:** Use Cohen's small, medium, or large relative effect sizes in power and/or sample size calculations.

    Explanation: Cohen's [13] relative effect sizes are expressed as a fraction of the experimental precision, which makes the often-false assumptions that 1) similar experiments have similar standard deviations; and 2) a relative effect size that is practically significant in one experiment is practically significant in all experiments.

- **Bad Practice #6:** Sample size and power calculations are exact.

    Explanation: Sample size and power calculations are almost always done in the presence of significant uncertainty, so the results of these calculations should be interpreted only as approximations. Do not quibble over small differences in sample size.

- **Bad Practice #7:** There are some experiments which do not require an *a priori* sample size calculation.

    Explanation: If an experiment is worth doing, it should use the right sample size.

- **Bad Practice #8:** We do not know the process standard deviation well enough to perform the sample size calculation.

Explanation: In most cases, an appropriate estimate for the standard deviation may be obtained from historical process data, data from a similar process, the recommendation of a subject matter expert, or published data taken under similar circumstances. If none of those sources is available, perform a pilot study.

- **Bad Practice #9:** Zero acceptance number attribute sampling plans are superior to plans that allow defectives in samples.

    Explanation: Zero acceptance number sampling plans can hide problems by using small sample sizes and are often misunderstood by the managers supporting their use. Sometimes zero acceptance number sampling plans are appropriate, but sometimes they are not.

- **Bad Practice #10:** Postexperiment power is a useful indicator of the value of an experiment.

    Explanation: Postexperiment power calculations erroneously assume that the parameter estimates obtained in an experiment are the true parameter values and that they are practically significant. A high value of postexperiment power *might* support the goals of the experiment, but high postexperiment power by itself is not a sufficient condition.

- **Bad Practice #11:** Special software is required to perform sample size and power calculations.

    Explanation: Many exact sample size and power calculations involve special probability tables or complex calculations that are best done with software, but most problems can be solved approximately with pencil, paper, a pocket calculator, and tables of common statistical distributions. There are also many free sample size calculators available on the Internet.

- **Bad Practice #12:** We do not need to know the method of analysis or the decision criteria before calculating the experimental sample size.

    Explanation: Every sample size and power calculation is matched to an analysis method and decision criteria.

- **Bad Practice #13:** An experiment does not need to produce a statistically significant result for the effect to be practically significant.

    Explanation: For an experimental result to be useful, it must be both practically and statistically significant. By design, an experiment that uses the appropriate sample size to detect the predetermined practically-significant effect size with high power is highly likely to produce a statistically significant result when that result is also practically significant.

1.4.4 Good Practices

The following practices are recommended for calculating sample size or power for an experiment:

- Calculate the sample size, power, and/or the effect size before collecting data to make sure that there will be sufficient data to meet the goals of the experiment.

- If necessary, perform a preliminary experiment or pilot study to estimate the standard deviation of the process.

- If the calculated sample size is too large or the calculated power for a specified effect size is too small, take appropriate action to

    - improve the measurement repeatability and/or reproducibility.
    - improve the experiment's resolution by changing the experiment design.
    - identify, record, and incorporate covariates into the analysis to reduce the noise.
    - replace the original response variable with another one that has better precision.

- Increase the sample size to compensate for anticipated experimental runs lost at random.

- Before collecting your sample data, make sure that your experiment design and analysis can meet the goals of the experiment. Enter the experiment design into your software, create a fictional (e.g., random normal) response, and confirm that you can run the intended analysis. Note the important outputs and review the acceptance criteria that apply to them.

- Hypothesis testing tells you whether or not you can reject H_0 with a corresponding p value, but p values are often too abstract and of limited practical value. Plan to report and interpret the results in terms of effect size.

- Use a power calculation to plan an experiment, but use a confidence interval when reporting results.

- Use at least two methods to determine the sample size for an experiment, such as by manual calculation and with software, to reduce the chances of making a mistake. There are lots of ways to mismanage experiments, but using the wrong sample size should not be one of them.

- Write up a summary statement explaining your sample size solution including the assumptions you made, the sources of the information you used as inputs to the calculations, and the planned statistical analysis methods and decision criteria, etc. You might need to review that statement later if the experiment doesn't deliver the expected results.

1.5 Software

There are some excellent sample size and power calculation software packages available. The ones that I run are Power and Sample Size (PASS), MINITAB, Piface, and R, although there are certainly other packages that you might consider. I use four packages because not any one of them can do all of the problems that I encounter and some do a better job than others on certain types of problems. The following factors determine which package I use for a particular problem: whether the package supports the method that I need, flexibility in configuring the particular method, ease of use, content of the output, and accuracy.

PASS (*www.ncss.com*) is commercial software and is one of the most comprehensive specialty sample size and power calculation packages available. PASS is more for the statistics or quality engineering professional rather than the occasional user. It is not the easiest of the packages to use, but it supports a wide range of methods and is fantastically well documented with theory, detailed user instructions, and solutions to problems taken from the technical literature. The value of the PASS documentation alone is worth the cost of the package.

MINITAB (*www.minitab.com*) is general-use commercial statistical software that has sample size, power, and effect size calculators for many common problems. Many of these calculators are in the **Stat> Power and Sample Size** menu, however, MINITAB also provides calculations for reliability experiments from the **Stat> Reliability/Survival> Test Plans** menu and for acceptance sampling from the **Stat> Quality Tools> Acceptance Sampling by Attributes** and **Stat> Quality Tools> Acceptance Sampling by Variables** menus. There are also macros available at the MINITAB web site and on the Internet to do other calculations that are not available in the standard MINITAB package. MINITAB's sample size, effect size, and power calculators are among the easiest to configure.

Piface (*www.stat.uiowa.edu/~rlenth/Power/*) is Russ Lenth's free java applet that runs from his website or can be downloaded and run locally on your computer [37]. Piface is a wonderful little program with good scope, flexibility, and ease of use. Its documentation is sparse though, so it might take you some time to figure out how to configure a calculation. Piface has an exceptionally long list of preprogrammed ANOVA models and you can also specify your own models for balanced designs with fixed, random, mixed, and nested variables. The quality, scope, and cost of Piface make it a must-have software package for anyone doing frequent sample size and power calculations.

R (*www.r-project.org*) is my fourth software package for sample size and power calculations. R is free open-source software that may be downloaded from the Internet. R is a command line environment; it does not have a graphical user interface like NCSS, MINITAB, and Piface, but it is very popular and many people have contributed to it, so it is probably the single most comprehensive sample size and power calculation package available today. Unfortunately, many of the calculators were written by different people so there are inconsistencies in usage styles and the quality of the documentation. My own R skills are marginal, so I use R only when none of the other software packages will solve my problem.

Details about the use of and outputs from software packages are not presented in this book to keep it as small as possible and to avoid having to update it frequently to keep up with software revisions; however, solutions to selected example problems using PASS, MINITAB, Piface, and R are posted at *www.mmbstatistical.com/SampleSize.html*.

Two final words of warning with respect to software:

1. There are many sample size and power calculators available on Internet web sites. Most of them have limited capabilities and their calculation assumptions may or may not be disclosed, so be very careful to validate any online calculator that you intend to use by comparing its output to the solutions of known problems.

2. Even with easy-to-configure software, it is still possible to misconfigure a calculation, which could have substantial consequences. My own personal practice, which I highly recommend, is to calculate an exact or approximate sample size using manual calculation methods and then confirm the answer using one or more software solutions.

Chapter 2

Means

This chapter presents sample size calculations for confidence intervals and hypothesis tests for means for quantitative responses for one, two, or more populations. Multiple comparisons test methods are presented here for differences between the means of three or more populations, but ANOVA methods are covered in Section 8.1.

2.1 Assumptions

The sample size calculations presented in this chapter are based on the assumption that the test statistics follow the normal or Student's t distributions; consequently, the populations being sampled are assumed to be normal, at least approximately normal, or to have been mathematically transformed to at least approximate normality. When two or more populations are considered, both the homoscedastic (equal variances) and heteroscedastic (unequal variances) cases are presented.

The population size is assumed to be very large compared to the sample size. When this assumption is not satisfied, the standard deviation of the sampling distribution of $\bar{x}$ given by $\sigma_{\bar{x}} = \sigma_x/\sqrt{n}$ should be corrected using the finite population correction factor

$$\sigma_{\bar{x}} = \frac{\sigma_x}{\sqrt{n}}\sqrt{1 - \frac{n}{N}}. \tag{2.1}$$

Inspection of the correction factor shows that it can be ignored when the sample size n is small compared to the population size N, as when $n < N/20$, which is a reasonable assumption in most cases.

2.2 One Mean

This section addresses sample size and power calculations for problems involving one sample mean including the σ-known and σ-unknown cases. The paired-sample t test is included here as a special case of the one-sample t test for the mean.

2.2.1 Confidence Interval for the Mean

Sample size calculations are presented for confidence intervals for the mean in the σ-known and σ-unknown cases. The effect of measurement precision error on sample size is also considered.

2.2.1.1 Confidence Interval for the Mean (σ Known)

The two-sided $(1-\alpha)\,100\%$ confidence interval for the unknown population mean μ based on a sample of size n when σ is known has the form

$$\Phi\left(\bar{x} - \delta < \mu < \bar{x} + \delta\right) = 1 - \alpha \tag{2.2}$$

where

$$\delta = \frac{z_{\alpha/2}\sigma}{\sqrt{n}}. \tag{2.3}$$

Then, to obtain a confidence interval of specified half-width δ, the required sample size is

$$n \geq \left(\frac{z_{\alpha/2}\sigma}{\delta}\right)^2. \tag{2.4}$$

Example 2.1 Find the sample size required to estimate the unknown mean of a population to within ± 3 with 95% confidence if the population standard deviation is known to be $\sigma = 5$.
Solution: With $\alpha = 0.05$, $z_{0.025} = 1.96$, and $\delta = 3$ in Equation 2.4, the required sample size is

$$\begin{aligned} n &\geq \left(\frac{1.96 \times 5}{3}\right)^2 \\ &\geq 11. \end{aligned}$$

2.2.1.2 Confidence Interval for the Mean (σ Unknown)

The two-sided $(1-\alpha)\,100\%$ confidence interval for the unknown population mean μ based on a sample of size n when σ is unknown has the form

$$P\left(\bar{x} - \delta < \mu < \bar{x} + \delta\right) = 1 - \alpha \tag{2.5}$$

2.2. One Mean

where

$$\delta = \frac{t_{\alpha/2}\widehat{\sigma}}{\sqrt{n}} \quad (2.6)$$

and the t distribution has $df_\epsilon = n - 1$ degrees of freedom. The sample size required to obtain a confidence interval of specified half-width δ is the smallest value of n that meets the condition

$$n \geq \left(\frac{t_{\alpha/2}\widehat{\sigma}}{\delta}\right)^2. \quad (2.7)$$

This equation is transcendental because $t_{\alpha/2}$ depends on the sample size, so it must be solved iteratively to find the correct value of n. Use $t_{\alpha/2} \simeq z_{\alpha/2}$ in the first iteration.

Example 2.2 Find the sample size required to estimate the unknown mean of a population to within $\delta = 3$ measurement units with 95% confidence if the estimated population standard deviation is $\widehat{\sigma} = 5$.
Solution: From Equation 2.7 with $t_{0.025} \simeq (z_{0.025} = 1.96)$ in the first iteration,

$$n \geq \left(\frac{1.96 \times 5}{3}\right)^2 = 11.$$

In the second iteration with $t_{0.025,10} = 2.228$,

$$n \geq \left(\frac{2.228 \times 5}{3}\right)^2 = 14.$$

Another iteration indicates that $n = 13$ is the smallest sample size that satisfies the sample size condition.

2.2.1.3 Confidence Interval for the Mean with Measurement Precision Error

When there is precision error in individual measurements, the confidence interval for the population mean is given by

$$P(\bar{x} - \delta < \mu < \bar{x} + \delta) = 1 - \alpha \quad (2.8)$$

where

$$\delta = t_{\alpha/2}\sqrt{\frac{\sigma_x^2 + \sigma_\epsilon^2}{n}} \quad (2.9)$$

where σ_x is the standard deviation of the population, σ_ϵ is the measurement precision error, n is the sample size, and the t distribution has $df_\epsilon = n - 1$ degrees of freedom.

When σ_ϵ^2 is not negligible compared to σ_x^2, the effect of the measurement precision error may be reduced by averaging over several repeated measurements taken on each unit. When k such repeated measurements are taken on each of n units, the confidence interval half-width becomes

$$\delta = t_{\alpha/2} \sqrt{\frac{\sigma_x^2 + \frac{\sigma_\epsilon^2}{k}}{n}} \qquad (2.10)$$

where the t distribution still has $df_\epsilon = n - 1$ degrees of freedom.

2.2.2 Test for the Mean

Sample size and power calculations are presented for hypothesis tests for the population mean for the σ-known and σ-unknown cases.

2.2.2.1 Test for the Mean (σ Known)

The hypotheses to be tested are $H_0 : \mu = \mu_0$ versus $H_A : \mu \neq \mu_0$ where σ is known and the sample mean $\bar{x}$ is determined from a sample of size n. The distributions of $\bar{x}$ under H_0, when the true mean is μ_0, and H_A, when the true mean is $\mu_1 = \mu_0 + \delta$, are shown in Figure 1.3a and b, respectively. The decision to reject H_0 or not is based on the statistic

$$z = \frac{\bar{x} - \mu_0}{\sigma/\sqrt{n}} \qquad (2.11)$$

and the acceptance interval for H_0 is

$$\Phi\left(-z_{\alpha/2} < z < z_{\alpha/2}\right) = 1 - \alpha. \qquad (2.12)$$

The effect size that can be detected with sample size n and power $\pi = 1 - \beta$ is given by

$$\begin{aligned} \delta &= \mu_1 - \mu_0 \\ &= \left(z_{\alpha/2} + z_\beta\right) \sigma_{\bar{x}} \\ &= \frac{\left(z_{\alpha/2} + z_\beta\right) \sigma}{\sqrt{n}}. \end{aligned} \qquad (2.13)$$

The power to reject H_0 when the true mean is $\mu_1 = \mu_0 + \delta$ is given by

$$\pi = \Phi\left(-\infty < z < z_\beta\right) \qquad (2.14)$$

where

$$z_\beta = \frac{|\delta|}{\sigma/\sqrt{n}} - z_{\alpha/2}. \qquad (2.15)$$

The sample size required to obtain power $\pi = 1 - \beta$ for a specified value of the effect size δ is given by

$$n = \left(z_{\alpha/2} + z_\beta\right)^2 \left(\frac{\sigma}{\delta}\right)^2. \qquad (2.16)$$

2.2. One Mean

Example 2.3 For the one-sample test of $H_0 : \mu = 30$ versus $H_A : \mu \neq 30$ when the population is known to be normal with $\sigma = 3$, what sample size is required to detect a shift to $\mu = 32$ with 90% power?
Solution: By Equation 2.16 with $\delta = 2$, $z_{0.025} = 1.96$, and $z_{0.10} = 1.28$, the necessary sample size is

$$n \geq (1.96 + 1.28)^2 \left(\frac{3}{2}\right)^2 = 24.$$

2.2.2.2 Test for the Mean (σ Unknown)

The hypotheses to be tested are $H_0 : \mu = \mu_0$ versus $H_A : \mu \neq \mu_0$ where σ is unknown and the sample mean $\bar{x}$ and standard deviation s are determined from a sample of size n. The decision to reject H_0 or not is based on the statistic

$$t = \frac{\bar{x} - \mu_0}{s/\sqrt{n}}. \tag{2.17}$$

The acceptance interval for H_0 is

$$P\left(-t_{\alpha/2} < t < t_{\alpha/2}\right) = 1 - \alpha \tag{2.18}$$

where the t distribution has $df_e = n - 1$ degrees of freedom.

The approximate power to reject H_0 when the true mean is shifted to $\mu = \mu_0 + \delta$ is given by

$$\pi = P\left(-\infty < t < t_\beta\right) \tag{2.19}$$

where

$$t_\beta = \frac{|\delta|}{\hat{\sigma}/\sqrt{n}} - t_{\alpha/2}. \tag{2.20}$$

Then the sample size required to obtain power $\pi = 1 - \beta$ is given by the smallest value of n that meets the condition

$$n \geq (t_{\alpha/2} + t_\beta)^2 \left(\frac{\hat{\sigma}}{\delta}\right)^2. \tag{2.21}$$

Example 2.4 For the one-sample test of $H_0 : \mu = 30$ versus $H_A : \mu \neq 30$, what sample size is required to detect a shift to $\mu = 32$ with 90% power? The population standard deviation is unknown but expected to be $\sigma \simeq 1.5$.
Solution: The sample size condition given by Equation 2.21 is transcendental, so the correct value of n must be determined iteratively. With $t \simeq z$ as a first guess, $z_{0.025} = 1.96$, $z_{0.10} = 1.282$, and

$$n = (1.96 + 1.282)^2 \left(\frac{1.5}{2}\right)^2 = 6.$$

Then with $df_\epsilon = 5$, $t_{0.025,5} = 2.571$, and $t_{0.10,5} = 1.476$ the new sample size estimate is

$$n \geq (2.571 + 1.476)^2 \left(\frac{1.5}{2}\right)^2 = 9.21.$$

Further iterations are required because $(n = 6) \not\geq 9.21$. Another iteration indicates that $n = 9$ delivers the desired power.

Equations 2.19 and Equation 2.21 are approximations to the exact solutions for the power and sample size. The approximations are accurate for large samples; they are so convenient, however, that they are often used without regard to sample size. The exact relationship that determines the power and sample size for the two-sided one-sample t test is

$$t_{\alpha/2} = t_{\beta,\phi} \tag{2.22}$$

where $t_{\beta,\phi}$ is the noncentral t distribution with noncentrality parameter

$$\phi = \frac{|\delta|}{\hat{\sigma}/\sqrt{n}} \tag{2.23}$$

and both t values have $df_\epsilon = n - 1$ degrees of freedom. When the familiar central t distribution is symmetric about its mean at $t = 0$, the noncentral t distribution is shifted away from $t = 0$, is asymmetric, and has larger variance. As the degrees of freedom get very large, the noncentral t distribution approaches normality.

In the past, the barrier to using Equation 2.22 for power calculations was the lack of easy access to tables for the noncentral t distribution; today, however, most statistical software packages have this capability.

Example 2.5 Find the approximate and exact power for the solution obtained for Example 2.4.

Solution: With $n = 9$ and $t_{0.025,8} = 2.306$ the approximate power by Equation 2.19 is

$$\begin{aligned}
\pi &= P\left(-\infty < t < \frac{\delta}{\hat{\sigma}/\sqrt{n}} - t_{\alpha/2}\right) \\
&= P\left(-\infty < t < \frac{2}{1.5/\sqrt{9}} - 2.306\right) \\
&= P(-\infty < t < 1.694) \\
&= 0.9356.
\end{aligned}$$

From Equation 2.23 the t distribution noncentrality parameter is

$$\phi = \frac{2}{1.5/\sqrt{9}} = 4.00,$$

2.2. One Mean

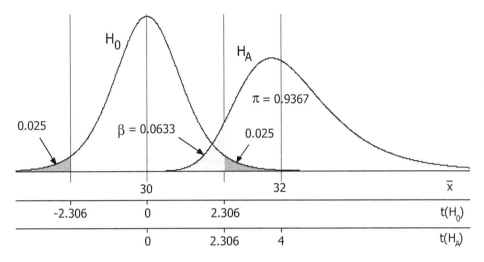

Figure 2.1: Central and noncentral t distributions for t test power calculation.

so, from Equation 2.22,
$$t_{0.025} = 2.306 = t_{\beta, 4.0},$$
which is satisfied by $\beta = 0.0633$ and power $\pi = 1 - \beta = 0.9367$. This value is in excellent agreement with the value obtained by the approximate method even though the sample size is relatively small. Figure 2.1 shows the corresponding central t distribution under $H_0 : \mu = 30$ and the noncentral t distribution when $\mu = 32$. The asymmetry of the noncentral t distribution is substantial because the sample size is so small.

2.2.3 Paired Observations

The paired-sample t test method is used when the same experimental units are exposed to two different treatments for the purpose of determining if there is a bias between the treatments. This method is more efficient at quantifying that bias than the two-independent-samples method.[1]

The quantity of interest with paired observations is the difference between the paired data values
$$\Delta x_i = x_{1i} - x_{2i} \tag{2.24}$$
where x_i indicates a measurement on the ith unit and the subscripts 1 and 2 distinguish the two treatments. The estimated mean difference between the treat-

---
[1] The two-independent-samples method is presented in Section 2.3, but the paired observations case is presented here because it is a special case of the one-sample method.

ments determined from a sample of n units is then

$$\overline{\Delta x} = \frac{1}{n}\sum_{i=1}^{n}\Delta x_i \tag{2.25}$$

and the standard deviation of the Δx_i is

$$s_{\Delta x} = \sqrt{\frac{1}{n-1}\sum_{i=1}^{n}\left(\Delta x_i - \overline{\Delta x}\right)^2}. \tag{2.26}$$

These statistics are analogous to $\bar{x}$ and s in the one-sample situation, so the one-sample t distribution-based confidence interval and hypothesis test methods can be applied to the paired-observations case without modification.

The higher efficiency of the paired-samples design versus the two-independent-samples design comes from the difference in their characteristic standard deviations. In the independent samples case, the standard deviation of the observations within a treatment gets contributions from the actual variability between units combined with the measurement precision error. If σ_x represents the variation in units and σ_ϵ represents the measurement precision, then their combined variation is

$$\sigma_{independent} = \sqrt{\sigma_x^2 + \sigma_\epsilon^2}. \tag{2.27}$$

In the paired samples case, the biases of each unit from the treatment means cancel when we calculate Δx_i, so the variation in Δx_i comes only from the measurement precision error. In effect, when a unit is measured twice, once under each treatment, it serves as its own control. Because two measurements are required to determine each Δx_i, the standard deviation of the Δx_i is

$$\sigma_{\Delta x} = \sqrt{\sigma_\epsilon^2 + \sigma_\epsilon^2} = \sqrt{2}\sigma_\epsilon. \tag{2.28}$$

If the sample sizes for the two-sample and paired-sample designs are sufficiently large to ignore the difference in the t distributions' degrees of freedom, the ratio of the two-sample to paired-sample designs' sample sizes is

$$\begin{aligned}\frac{n_{two-sample}}{n_{paired-sample}} &\simeq \frac{2\sigma_{independent}}{\sigma_{\Delta x}} \\ &\simeq \frac{2\sqrt{\sigma_x^2+\sigma_\epsilon^2}}{\sqrt{2}\sigma_\epsilon} \\ &\simeq \sqrt{2}\sqrt{\left(\frac{\sigma_x}{\sigma_\epsilon}\right)^2 + 1}.\end{aligned} \tag{2.29}$$

Equation 2.29 shows that if the measurement precision error σ_ϵ is relatively small compared to the unit variation σ_x, then the two-sample design will require a much larger sample size than the paired-samples design.

2.3. Two Independent Means

Example 2.6 Compare the sample sizes for the two-independent-samples experiment and the paired-sample experiment if they must detect a bias between two treatments of $\Delta\mu = 2$ with 90% power when the standard deviation of individual units is $\hat{\sigma}_x = 2$ and the measurement precision error is $\hat{\sigma}_\epsilon = 0.5$.

Solution: For the two-independent-sample t test the characteristic standard deviation for each treatment is (from Equation 2.27)

$$\hat{\sigma}_{independent} = \sqrt{2^2 + 0.5^2} = 2.062.$$

Then, from Equation 2.62, the required sample size for each treatment is

$$
\begin{aligned}
n &\geq 2\left(t_{0.025} + t_{0.10}\right)^2 \left(\frac{\hat{\sigma}_{independent}}{\Delta\mu}\right)^2 \\
&\geq 2\left(t_{0.025} + t_{0.10}\right)^2 \left(\frac{2.062}{2}\right)^2 \\
&\geq 24.
\end{aligned}
$$

For the paired-sample t test, the characteristic standard deviation for the Δx_i can be estimated from Equation 2.28:

$$\hat{\sigma}_{\Delta x} = \sqrt{2}\hat{\sigma}_\epsilon = \sqrt{2} \times 0.5 = 0.707.$$

Then, from Equation 2.21, the required sample size is approximately

$$
\begin{aligned}
n &\geq \left(t_{0.025} + t_{0.10}\right)^2 \left(\frac{\hat{\sigma}_{\Delta x}}{\Delta\mu}\right)^2 \\
&\geq \left(t_{0.025} + t_{0.10}\right)^2 \left(\frac{0.707}{2}\right)^2 \\
&\geq 4
\end{aligned}
$$

and further iterations confirm that $n = 4$. When the independent-samples design requires two samples of size $n = 24$ units each, for a total of 48 measurements, the paired-sample design requires only $n = 4$ units for a total of 8 measurements!

2.3 Two Independent Means

Sample size and power calculations for three types of two-sample mean problems are presented in this section: σ_1 and σ_2 known, σ_1 and σ_2 unknown but assumed equal, and σ_1 and σ_2 unknown and unequal. In all cases the populations being studied are assumed to be normal, approximately normal, or to have been transformed to approximate normality.

2.3.1 Confidence Interval for the Difference Between Two Means

Sample size calculations are presented for confidence intervals for three cases: σ_1 and σ_2 known, σ_1 and σ_2 unknown but assumed to be equal, and σ_1 and σ_2 unknown and unequal.

2.3.1.1 Confidence Interval for the Difference Between Two Means (σ_1 and σ_2 Known)

The $(1 - \alpha)\,100\%$ confidence interval for the difference between two population means $\Delta\mu = \mu_1 - \mu_2$ when σ_1 and σ_2 are known is given by

$$\Phi\left(\Delta\bar{x} - \delta < \Delta\mu < \Delta\bar{x} + \delta\right) = 1 - \alpha \tag{2.30}$$

where $\Delta\bar{x} = \bar{x}_1 - \bar{x}_2$ and

$$\delta = z_{\alpha/2}\sqrt{\frac{\sigma_1^2}{n_1} + \frac{\sigma_2^2}{n_2}}. \tag{2.31}$$

To obtain a confidence interval of specified half-width δ with $n_1 = n_2 = n$, the required sample size is

$$n = z_{\alpha/2}^2 \frac{\sigma_1^2 + \sigma_2^2}{\delta^2}. \tag{2.32}$$

When $\sigma_1 \neq \sigma_2$, Equation 2.32 does not optimally allocate observations between the two samples resulting in a larger-than-necessary total sample size. The optimal allocation of observations that minimizes the total sample size is

$$n_1 = z_{\alpha/2}^2 \frac{\sigma_1(\sigma_1 + \sigma_2)}{\delta^2} \tag{2.33a}$$

$$n_2 = n_1 \left(\frac{\sigma_2}{\sigma_1}\right). \tag{2.33b}$$

Equation 2.33b shows that the sample sizes should be determined in proportion to their population standard deviations:

$$(n_1 : n_2) = (\sigma_1 : \sigma_2). \tag{2.34}$$

Example 2.7 Find the sample sizes required for the a) equal-allocation and b) optimal-allocation conditions if the 95% two-sided confidence interval for $\Delta\mu$ must have half-width $\delta = 0.003$ when $\sigma_1 = 0.003$ and $\sigma_2 = 0.006$. Compare the total sample sizes required by the two methods.
Solution:
a) By Equation 2.32 the sample size required for equal allocation is

$$n = (1.96)^2 \frac{(0.003)^2 + (0.006)^2}{(0.003)^2} = 20.$$

2.3. Two Independent Means

b) By Equations 2.33a and b, the sample sizes required for optimal allocation are

$$n_1 = (1.96)^2 \frac{(0.003)(0.003+0.006)}{(0.003)^2} = 12$$

$$n_2 = 11.5 \left(\frac{0.006}{0.003}\right) = 24.$$

For the equal-allocation method, the total sample size is $2n = 40$, and for the optimal-allocation method, the total sample size is $n_1 + n_2 = 36$ - a 10% savings in sample size.

2.3.1.2 Confidence Interval for the Difference Between Two Means (σ_1 and σ_2 Unknown but Equal)

The $(1 - \alpha)100\%$ confidence interval for the difference between two population means $\Delta\mu = \mu_1 - \mu_2$ when σ_1 and σ_2 are unknown but assumed to be equal is given by

$$P(\Delta\bar{x} - \delta < \Delta\mu < \Delta\bar{x} + \delta) = 1 - \alpha \tag{2.35}$$

with

$$\delta = t_{\alpha/2}\sqrt{\frac{(n_1-1)s_1^2 + (n_2-1)s_2^2}{n_1+n_2-2}\left(\frac{1}{n_1}+\frac{1}{n_2}\right)} \tag{2.36}$$

where $\Delta\bar{x} = \bar{x}_1 - \bar{x}_2$ and the t distribution has $df_\epsilon = n_1 + n_2 - 2$ degrees of freedom.

Under the assumption that $\sigma_1 = \sigma_2 = \sigma_\epsilon$, the optimal allocation of observations is to use equal sample sizes $n_1 = n_2 = n$, which gives

$$\delta = t_{\alpha/2}\hat{\sigma}_\epsilon\sqrt{\frac{2}{n}}. \tag{2.37}$$

Then the value of n required to obtain a confidence interval of specified half-width δ is

$$n = 2\left(\frac{t_{\alpha/2}\hat{\sigma}_\epsilon}{\delta}\right)^2. \tag{2.38}$$

This expression is transcendental in n; however, $t_{\alpha/2} \simeq z_{\alpha/2}$ provides a convenient starting point for the first iteration.

Example 2.8 Determine the sample size required to obtain a confidence interval half-width $\delta = 50$ when $\hat{\sigma}_1 = \hat{\sigma}_2 = 80$.
Solution: With $t_{0.025} \simeq z_{0.025}$ for the first iteration, the sample size is

$$n = 2\left(\frac{1.96 \times 80}{50}\right)^2 = 20. \tag{2.39}$$

Another iteration with $t_{0.025,38} = 2.024$ gives

$$n = 2\left(\frac{2.024 \times 80}{50}\right)^2 = 21. \tag{2.40}$$

A third iteration (not shown) confirms that $n = 21$ is the necessary sample size.

2.3.1.3 Confidence Interval for the Difference Between Two Means (σ_1 and σ_2 Unknown and Unequal)

When σ_1 and σ_2 are unknown and expected to be unequal, the contributions from the two samples to the t distribution degrees of freedom must be σ_i-weighted. This method, called the *Satterthwaite* or *Welch* method, gives effective t distribution degrees of freedom equal to

$$df_\epsilon = \frac{\left(\frac{s_1^2}{n_1} + \frac{s_2^2}{n_2}\right)^2}{\frac{s_1^4}{n_1^2(n_1+1)} + \frac{s_2^4}{n_2^2(n_2+1)}} - 2, \tag{2.41}$$

which may be used exactly or rounded up to the nearest integer value. The required confidence interval and sample size are obtained using the methods of Section 2.3.1.1 by substituting the σ_i estimates for the known σ_i values and replacing z values with t values with df_ϵ degrees of freedom. As in that section, the optimal allocation of units to the two samples is to set sample sizes in proportion to their standard deviations.

Example 2.9 What optimal sample sizes are required to determine a confidence interval for the difference between two population means with confidence interval half-width $\delta = 15$ when $\hat{\sigma}_1 = 24$ and $\hat{\sigma}_2 = 8$?

Solution: From Equations 2.33a and b, initial guesses for the sample sizes are

$$n_1 \simeq (1.96)^2 \frac{24(24+8)}{15^2} \simeq 14$$

and

$$n_2 \simeq 14\left(\frac{8}{24}\right) \simeq 5.$$

To obtain optimal sample size allocation n_1 and n_2 must be in the ratio

$$(n_1 : n_2) = (\hat{\sigma}_1 : \hat{\sigma}_2) = (24 : 8) = (3 : 1),$$

so reasonable choices for the sample sizes are $n_1 = 15$ and $n_2 = 5$. By Equation 2.41, the t distribution degrees of freedom will be

$$df_\epsilon = \frac{\left(\frac{24^2}{15} + \frac{8^2}{5}\right)^2}{\frac{24^4}{15^2(15+1)} + \frac{8^4}{5^2(5+1)}} - 2 = 20.$$

2.3. Two Independent Means

With $t_{0.025,20} = 2.086$ the next iteration on the sample sizes gives

$$n_1 \simeq (2.086)^2 \frac{24(24+8)}{15^2} \simeq 15$$

and

$$n_2 \simeq 15 \left(\frac{8}{24}\right) = 5$$

which must be the correct values.

2.3.2 Test for the Difference Between Two Means

Power and sample size calculations are presented for hypothesis tests for the difference between two means for three cases: σ_1 and σ_2 known, σ_1 and σ_2 unknown but assumed to be equal, and σ_1 and σ_2 unknown and unequal. The methods presented are limited to the hypotheses $H_0 : \mu_1 = \mu_2$ versus $H_A : \mu_1 \neq \mu_2$ or equivalently $H_0 : \mu_1 - \mu_2 = 0$ versus $H_A : \mu_1 - \mu_2 \neq 0$. However, these methods can be easily adapted to test for a specified non-zero difference between two means by appropriate modification of the numerator of the z or t test statistic. For example, when the test of $H_0 : \mu_1 - \mu_2 = 0$ versus $H_A : \mu_1 - \mu_2 \neq 0$ has a z statistic of the form

$$z = \frac{\bar{x}_1 - \bar{x}_2}{\sigma_{\bar{x}_1 - \bar{x}_2}}, \qquad (2.42)$$

the test of $H_0 : \mu_1 - \mu_2 = \Delta\mu$ versus $H_A : \mu_1 - \mu_2 \neq \Delta\mu$ where $\Delta\mu \neq 0$ has a z statistic of the form

$$z = \frac{\bar{x}_1 - \bar{x}_2 - \Delta\mu}{\sigma_{\bar{x}_1 - \bar{x}_2}} \qquad (2.43)$$

where the sign of $\Delta\mu$ must be chosen so that $\bar{x}_1 - \bar{x}_2 - \Delta\mu = 0$ under H_0.

2.3.2.1 Test for the Difference Between Two Means (σ_1 and σ_2 Known)

The hypotheses to be tested are $H_0 : \mu_1 = \mu_2$ versus $H_A : \mu_1 \neq \mu_2$, where σ_1 and σ_2 are known and not necessarily equal. The experimental data consist of two independent random samples of size n_1 and n_2 from which the sample means $\bar{x}_1$ and $\bar{x}_2$ are determined. The decision to reject H_0 or not is based on the statistic

$$z = \frac{\bar{x}_1 - \bar{x}_2}{\sqrt{\frac{\sigma_1^2}{n_1} + \frac{\sigma_2^2}{n_2}}} \qquad (2.44)$$

where the acceptance interval for H_0 is given by

$$P\left(-z_{\alpha/2} < z < z_{\alpha/2}\right) = 1 - \alpha. \qquad (2.45)$$

For a specified difference between the population means $\Delta\mu = |\mu_1 - \mu_2|$, the power of the test is

$$\pi = P\left(-\infty < z < z_\beta\right) \qquad (2.46)$$

where
$$z_\beta = \frac{\Delta\mu}{\sqrt{\frac{\sigma_1^2}{n_1} + \frac{\sigma_2^2}{n_2}}} - z_{\alpha/2}. \tag{2.47}$$

The sample size n_1 required to obtain power $\pi = 1 - \beta$ for a difference between the means $\Delta\mu$ is

$$n_1 = \left(\frac{z_{\alpha/2} + z_\beta}{\Delta\mu}\right)^2 \left(\sigma_1^2 + \frac{n_1}{n_2}\sigma_2^2\right) \tag{2.48}$$

where n_1/n_2 is the ratio of the two sample sizes. The optimal sample size ratio is given by Equation 2.34.

Equations 2.47 and 2.48 express the general power and sample size solutions, but they simplify in the special cases of $\sigma_1 = \sigma_2$ and/or $n_1 = n_2$:

- When $\sigma_1 = \sigma_2 = \sigma_\epsilon$ and $n_1 = n_2$, the power is given by Equation 2.46 with

$$z_\beta = \sqrt{\frac{n}{2}}\frac{\Delta\mu}{\sigma_\epsilon} - z_{\alpha/2} \tag{2.49}$$

and the sample size required to obtain a specified power value is

$$n_1 = n_2 = 2\left(z_{\alpha/2} + z_\beta\right)^2 \left(\frac{\sigma_\epsilon}{\Delta\mu}\right)^2. \tag{2.50}$$

- When $\sigma_1 = \sigma_2 = \sigma_\epsilon$ and $n_1 \neq n_2$, the power is given by Equation 2.46 with

$$z_\beta = \frac{\Delta\mu}{\sigma_\epsilon\sqrt{\frac{1}{n_1} + \frac{1}{n_2}}} - z_{\alpha/2} \tag{2.51}$$

and the sample size n_1 required to obtain a specified power value is

$$n_1 = \left(1 + \frac{n_1}{n_2}\right)\left(z_{\alpha/2} + z_\beta\right)^2 \left(\frac{\sigma_\epsilon}{\Delta\mu}\right)^2 \tag{2.52}$$

where n_1/n_2 is the specified sample size ratio. The optimal sample size allocation is $n_1 = n_2$.

- When $\sigma_1 \neq \sigma_2$ and $n_1 = n_2$, the power is given by Equation 2.46 with

$$z_\beta = \frac{\sqrt{n}\Delta\mu}{\sqrt{\sigma_1^2 + \sigma_2^2}} - z_{\alpha/2} \tag{2.53}$$

and the sample size required to obtain a specified power value is

$$n_1 = n_2 = \left(\frac{z_{\alpha/2} + z_\beta}{\Delta\mu}\right)^2 \left(\sigma_1^2 + \sigma_2^2\right). \tag{2.54}$$

2.3.2.2 Test for the Difference Between Two Means (σ_1 and σ_2 Unknown but Equal)

The hypotheses to be tested are $H_0 : \mu_1 = \mu_2$ versus $H_A : \mu_1 \neq \mu_2$ where σ_1 and σ_2 are unknown but assumed to be equal. The experimental data consist of two independent random samples of size n_1 and n_2 from which the sample means $\bar{x}_1$ and $\bar{x}_2$ and sample standard deviations s_1 and s_2 are determined. The decision to reject H_0 or not is based on the statistic

$$t = \frac{\bar{x}_1 - \bar{x}_2}{\sqrt{\frac{(n_1-1)s_1^2+(n_2-1)s_2^2}{n_1+n_2-2}\left(\frac{1}{n_1}+\frac{1}{n_2}\right)}} \tag{2.55}$$

where the t distribution has $df_\epsilon = n_1+n_2-2$ degrees of freedom. The acceptance interval for H_0 is

$$P\left(-t_{\alpha/2} \leq t \leq t_{\alpha/2}\right) = 1 - \alpha. \tag{2.56}$$

For the purpose of sample size calculations, because σ_1 and σ_2 are unknown but assumed to be equal, Equation 2.55 simplifies to

$$t = \frac{\bar{x}_1 - \bar{x}_2}{\hat{\sigma}_\epsilon \sqrt{\frac{1}{n_1}+\frac{1}{n_2}}} \tag{2.57}$$

where $\hat{\sigma}_\epsilon = \hat{\sigma}_1 = \hat{\sigma}_2$. The power to reject H_0 when $|\mu_1 - \mu_2| = \Delta\mu$ is given by

$$\pi = P\left(-\infty < t < t_\beta\right) \tag{2.58}$$

where, for different sample sizes

$$t_\beta = \frac{1}{\sqrt{\frac{1}{n_1}+\frac{1}{n_2}}} \left(\frac{\Delta\mu}{\hat{\sigma}_\epsilon}\right) - t_{\alpha/2} \tag{2.59}$$

or when the sample sizes are equal

$$t_\beta = \sqrt{\frac{n}{2}} \left(\frac{\Delta\mu}{\hat{\sigma}_\epsilon}\right) - t_{\alpha/2}. \tag{2.60}$$

From Equation 2.59 the sample size n_1 required to reject H_0 with power π when $|\mu_1 - \mu_2| = \Delta\mu$ and sample size ratio n_1/n_2 is given by

$$n_1 = \left(1 + \frac{n_1}{n_2}\right) (t_{\alpha/2} + t_\beta)^2 \left(\frac{\hat{\sigma}_\epsilon}{\Delta\mu}\right)^2. \tag{2.61}$$

Because the two standard deviations are expected to be equal, the optimal allocation of samples is $n_1 = n_2$, so

$$n_1 = n_2 = 2 \left(t_{\alpha/2} + t_\beta\right)^2 \left(\frac{\hat{\sigma}_\epsilon}{\Delta\mu}\right)^2. \tag{2.62}$$

These sample size equations are transcendental in n; however, $t \simeq z$ provides a convenient starting point for the first iteration.

The approximate sample size and power calculations already presented are reasonably accurate for large samples but do less well for small ones. The exact relationship that determines the power and sample size is

$$t_{\alpha/2} = t_{\beta,\phi} \qquad (2.63)$$

where the t distribution noncentrality parameter is

$$\phi = \frac{\Delta\mu}{\hat{\sigma}_\epsilon \sqrt{\frac{1}{n_1} + \frac{1}{n_2}}} \qquad (2.64)$$

and both t values have $df_\epsilon = n_1 + n_2 - 2$ degrees of freedom. When the sample sizes are equal ($n_1 = n_2 = n$), the noncentrality parameter simplifies to

$$\phi = \sqrt{\frac{n}{2}} \frac{\Delta\mu}{\hat{\sigma}_\epsilon}. \qquad (2.65)$$

The errors inherent in the approximate methods tend to be relatively small compared to the uncertainties in the values of σ_ϵ and the practically significant effect size, so the approximations are still widely used.

Example 2.10 Calculate the sample size for the two-sample t test to reject H_0 with 90% power when $|\mu_1 - \mu_2| = 5$. Assume that the sample sizes will be equal and that the two populations have equal standard deviations estimated to be $\hat{\sigma}_\epsilon = 3$. Compare the approximate and exact powers.

Solution: With $\Delta\mu = 5$ and $\hat{\sigma}_\epsilon = 3$ in Equation 2.62, the sample size predicted in the first iteration with $t \simeq z$ is

$$n = 2\left(\frac{(1.96 + 1.282)3}{5}\right)^2 = 8.$$

A second and third iteration indicate that the required sample size is $n = 9$.

With $n = 9$ for both samples, $df_\epsilon = 18 - 2 = 16$ and the approximate power is given by Equations 2.58 and 2.60:

$$\begin{aligned}
\pi &= P\left(-\infty < t < \sqrt{\frac{n}{2}}\frac{\Delta\mu}{\hat{\sigma}_\epsilon} - t_{0.025,16}\right) \\
&= P\left(-\infty < t < \sqrt{\frac{9}{2}}\frac{5}{3} - 2.12\right) \\
&= P(-\infty < t < 1.416) \\
&= 0.912.
\end{aligned}$$

2.3. Two Independent Means

The t distribution noncentrality parameter is given by Equation 2.64:

$$\phi = \sqrt{\frac{9}{2}\frac{5}{3}} = 3.536.$$

The exact power is determined by Equation 2.63 with $\alpha = 0.05$:

$$t_{0.025} = 2.120 = t_{\beta,3.536},$$

which is satisfied by $\beta = 0.087$, so the exact power is $\pi = 0.913$. The exact power is in excellent agreement with the approximate power despite the somewhat small sample size.

2.3.2.3 Test for the Difference Between Two Means (σ_1 and σ_2 Unknown and Unequal)

The hypotheses to be tested are $H_0 : \mu_1 = \mu_2$ versus $H_A : \mu_1 \neq \mu_2$, where σ_1 and σ_2 are unknown and assumed to be different. The t test statistic is given by

$$t = \frac{\bar{x}_1 - \bar{x}_2}{\sqrt{\frac{s_1^2}{n_1} + \frac{s_2^2}{n_2}}} \quad (2.66)$$

where the t distribution's degrees of freedom are given by Satterthwaite's method in Equation 2.41. Power and sample size calculations are performed using the methods of Section 2.3.2.1 using the standard deviation estimates instead of the known values and replacing z values with t values with df_ϵ degrees of freedom. As in that section, the optimal allocation of units to the two samples is to set the sample sizes in proportion to their standard deviations.

2.3.2.4 Test for the Difference Between Two Means (One Sample Size Fixed)

In the preceding sections, power and sample size calculations were presented for two-sample tests for the unequal-n and equal-n cases. If one of the sample sizes is fixed, then the other sample size must be chosen to obtain the desired power value. The second sample's size may be determined by iterating until the target power is obtained, but if the optimal sample sizes n_1' and n_2' are calculated first, where $n_1'/n_2' = \sigma_1/\sigma_2$, then for a given n_1 value the second sample size is

$$n_2 = \frac{(\sigma_2/\sigma_1)^2}{\frac{1+\sigma_2/\sigma_1}{n_1'} - \frac{1}{n_1}}. \quad (2.67)$$

When the two standard deviations are equal, the optimal sample size allocation is $n_1' = n_2' = n'$ and the second sample's size simplifies to

$$n_2 = \frac{1}{\frac{2}{n'} - \frac{1}{n_1}} = \frac{n_1 n'}{2n_1 - n'}. \quad (2.68)$$

If the calculated value of n_2 is negative, then n_1 is too small for the experiment and a larger value will be required.

Equations 2.67 and 2.68 are exact for the two-sample z test where σ_1 and σ_2 are known, but they are only approximate for the two-sample t test because the t distribution's degrees of freedom depend on the sample sizes.

Example 2.11 Determine the size of the second sample under the conditions described in Example 2.10 if the first sample size must be $n = 6$.

Solution: From Example 2.10, the optimal equal sample sizes are $n' = 9$. If $n_1 = 6$ is fixed, then, from Equation 2.68, the approximate value of the second sample size must be

$$n_2 = \frac{6 \times 9}{(2 \times 6) - 9} = 18.$$

In the equal-n solution, we had $n_1 + n_2 = 18$ and $df_\epsilon = 16$ with 91% power; therefore, we know that $n_1 + n_2 = 6 + 18 = 24$ and $df_\epsilon = 22$ will give a slightly larger power, so the next guess for n_2 can be a value less than $n_2 = 18$. By appropriate guesses and iterations, the required value of n_2 is determined to be $n_2 = 15$ with approximate power

$$\pi = P\left(-\infty < t < \frac{1}{\sqrt{\frac{1}{n_1} + \frac{1}{n_2}}} \left(\frac{\Delta\mu}{\hat{\sigma}}\right) - t_{0.025, 19}\right)$$

$$= P\left(-\infty < t < \frac{1}{\sqrt{\frac{1}{6} + \frac{1}{15}}} \left(\frac{5}{3}\right) - 2.093\right)$$

$$= P(-\infty < t < 1.357)$$

$$= 0.905.$$

2.4 Equivalence Tests

The hypothesis tests for means considered in the preceding sections were *significance* tests in which the alternative hypothesis described a significant difference between a population mean and its specified value in the one-sample case (e.g., $H_0 : \mu = \mu_0$ versus $H_A : \mu \neq \mu_0$) or between a pair of population means in the two-sample case (e.g., $H_0 : \mu_1 = \mu_2$ versus $H_A : \mu_1 \neq \mu_2$). Tests with the goal of demonstrating that the population mean is *equal to* a specified value or that two population means are *equal to* each other are called *equivalence* tests; therefore, the null and alternate hypotheses in equivalence tests are reversed relative to the hypotheses in significance tests. That is, the hypotheses in an equivalence test are $H_0 : \mu \neq \mu_0$ versus $H_A : \mu = \mu_0$ for the one-sample case and $H_0 : \mu_1 \neq \mu_2$ versus $H_A : \mu_1 = \mu_2$ in the two-sample case.

2.4. Equivalence Tests

It is impossible to demonstrate the exact equality of the two means involved in an equivalence test, so we are forced to compromise by specifying an upper limit on the difference between the means such that, if the true difference is smaller in magnitude than this upper limit, then we conclude that the two means are *practically equivalent* to each other. This upper limit on the magnitude of the difference between the means, called the *limit of practical equivalence* (δ), must be chosen by the process expert before the equivalence test sample size calculation is performed. In terms of the limit of practical equivalence, the modified hypotheses for the one-sample test for means are $H_0 : \mu < \mu_0 - \delta$ or $\mu > \mu_0 + \delta$ versus $H_A : \mu_0 - \delta < \mu < \mu_0 + \delta$ where $\delta > 0$.

The purpose of this section is to present sample size and power calculations for the one-sample and two-independent-samples equivalence tests for means. The σ-known methods are presented here, but when σ is unknown and must be estimated with the sample standard deviation s, the approximate central t distribution-based methods or the exact noncentral t distribution-based methods of Sections 2.2.2.2 and 2.3.2.2 should be used instead.

Superiority and noninferiority tests are closely related to equivalence tests, but they are not presented in this book. See Chow, Shao, and Wang [12] for consideration of these types of tests.

2.4.1 Equivalence Test for One Mean

The hypotheses for the one-sample equivalence test for the population mean may be written

$$H_0 : |\Delta\mu| \geq \delta \text{ versus } H_A : |\Delta\mu| < \delta$$

where $\Delta\mu = \mu - \mu_0$ and δ, the limit of practical equivalence, is sufficiently small that, if the data support H_A, then μ is practically equivalent to μ_0.

Equivalence tests are performed using the *two one-sided tests* (TOST) method. The TOST method employs two one-sided significance tests constructed by breaking up the absolute values in the original hypotheses:

$$H_{01} : \Delta\mu \leq -\delta \text{ versus } H_{A1} : \Delta\mu > -\delta$$
$$H_{02} : \Delta\mu \geq \delta \text{ versus } H_{A2} : \Delta\mu < \delta.$$

If both H_{01} and H_{02} can be rejected, then we can accept the claim of the equivalence of μ and μ_0 within the practical limits $-\delta < \Delta\mu < \delta$.

The TOST null and alternate hypotheses are shown graphically in Figure 2.2. The test statistics for H_{01} and H_{02} are, respectively

$$z_1 = \frac{\Delta\bar{x} + \delta}{\sigma/\sqrt{n}} \tag{2.69}$$

and

$$z_2 = \frac{\Delta\bar{x} - \delta}{\sigma/\sqrt{n}}, \tag{2.70}$$

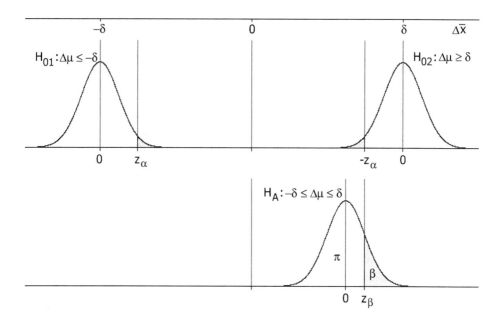

Figure 2.2: Two one-sided test approach to equivalence testing.

where σ is the known population standard deviation. The rejection conditions for H_{01} and H_{02} are, respectively

$$z_1 > z_\alpha \tag{2.71}$$

and

$$z_2 < -z_\alpha. \tag{2.72}$$

The approximate power $\pi = 1 - \beta$ for a specified value of $\Delta\mu$ corresponds to the probability of $\Delta\bar{x}$ falling to the right of z_α under H_{01} and to the left of $-z_\alpha$ under H_{02}:

$$\pi = \Phi\left(\frac{-\delta - \Delta\mu}{\sigma/\sqrt{n}} + z_\alpha < z < \frac{\delta - \Delta\mu}{\sigma/\sqrt{n}} - z_\alpha\right). \tag{2.73}$$

When $\Delta\mu \neq 0$, one of the bounds on z is relatively large in magnitude, so the other bound will essentially determine the power. By equating z_β with that bound (with an appropriate + or − sign) and solving for the sample size we obtain

$$n = \left(\frac{(z_\alpha + z_\beta)\sigma}{\delta - \Delta\mu}\right)^2. \tag{2.74}$$

2.4. Equivalence Tests

Example 2.12 Determine the sample size required for a one-sample equivalence test of the hypotheses $H_0 : \mu < 490$ or $\mu > 510$ versus $H_A : 490 < \mu < 510$ if the experiment must have 90% power to reject H_0 when $\mu = 504$ and $\sigma = 4$.
Solution: With $\mu_0 = 500$, $\mu = 505$, and $\delta = 10$, the sample size given by Equation 2.74 is

$$n = \left(\frac{(z_{0.05} + z_{0.10})\sigma}{\delta - \Delta\mu}\right)^2$$

$$= \left(\frac{(1.645 + 1.282)4}{10 - 5}\right)^2$$

$$= 6.$$

2.4.2 Equivalence Test for Two Means

For the two-sample z test for equivalence with equal sample size ($n_1 = n_2 = n$) and unknown but equal standard deviations ($\sigma_1 = \sigma_2 = \sigma_\epsilon$), the test statistics for H_{01} and H_{02} are, respectively

$$z_1 = \frac{\Delta\bar{x} + \delta}{\sqrt{\frac{2}{n}}\sigma_\epsilon} \tag{2.75}$$

and

$$z_2 = \frac{\Delta\bar{x} - \delta}{\sqrt{\frac{2}{n}}\sigma_\epsilon}. \tag{2.76}$$

The rejection conditions for H_{01} and H_{02} are, respectively

$$z_1 > z_\alpha \tag{2.77}$$

and

$$z_2 < -z_\alpha. \tag{2.78}$$

The power of the equivalence test for a specified value of $\Delta\mu$ corresponds to the probability of $\Delta\bar{x}$ falling to the right of z_α under H_{01} and to the left of $-z_\alpha$ under H_{02}:

$$\pi = \Phi\left(\frac{-\delta - \Delta\mu}{\sqrt{\frac{2}{n}}\sigma_\epsilon} + z_\alpha < z < \frac{\delta - \Delta\mu}{\sqrt{\frac{2}{n}}\sigma_\epsilon} - z_\alpha\right). \tag{2.79}$$

When $\Delta\mu \neq 0$, one of the bounds on z will be larger in magnitude than the other bound, so the smaller one will substantially determine the power. By equating z_β with that bound and solving for the sample size, we obtain

$$n = 2\left(\frac{(z_\alpha + z_\beta)\sigma_\epsilon}{\delta - \Delta\mu}\right)^2. \tag{2.80}$$

Example 2.13 Determine the power of the two independent-sample equivalence test where μ_1 and μ_2 are considered to be practically equivalent if $|\Delta\mu| < 2$ when $\Delta\mu = 0.2$, $\sigma_1 = \sigma_2 = 2$, and $n_1 = n_2 = 20$.
Solution: With $\delta = 2$ as the limit of practical equivalence, the hypotheses to be tested are

$$H_{01} : \Delta\mu \leq -2 \text{ versus } H_{A1} : \Delta\mu > -2$$
$$H_{02} : \Delta\mu \geq 2 \text{ versus } H_{A2} : \Delta\mu < 2.$$

From Equation 2.79 with $\Delta\mu = 0.2$, the power of the equivalence test is

$$\begin{aligned}\pi &= \Phi\left(\frac{-2-0.2}{\sqrt{\frac{2}{20}2}} + 1.645 < z < \frac{2-0.2}{\sqrt{\frac{2}{20}2}} - 1.645\right) \\ &= \Phi(-1.83 < z < 1.20) \\ &= 0.85.\end{aligned}$$

Example 2.14 What sample size is required in Example 2.13 to obtain 90% power?
Solution: From Equation 2.80 with $\beta = 0.10$, the sample size is

$$\begin{aligned}n &= 2\left(\frac{(1.645 + 1.282)\,2}{2 - 0.2}\right)^2 \\ &= 22.\end{aligned}$$

2.5 Contrasts

A contrast μ_c is a linear combination of k population means of the form

$$\mu_c = \sum_{i=1}^{k} c_i \mu_i \qquad (2.81)$$

where the c_i are constants. A point estimate for a contrast can be obtained from the sample means

$$\widehat{\mu}_c = \sum_{i=1}^{k} c_i \bar{x}_i. \qquad (2.82)$$

If the individual treatments are normal and homoscedastic with common standard deviation σ_ϵ, then the distribution of the sample contrast is also normal and

2.5. Contrasts

its standard error is given by

$$s_c = \sqrt{\sum_{i=1}^{k} c_i^2 s_{\bar{x}}^2}$$

$$= \sqrt{\sum_{i=1}^{k} c_i^2 \frac{s_\epsilon^2}{n_i}}$$

$$= s_\epsilon \sqrt{\sum_{i=1}^{k} \frac{c_i^2}{n_i}} \qquad (2.83)$$

where s_ϵ is the pooled standard error, usually estimated from the standard error of the one-way ANOVA. If the sample sizes for the k treatment groups are all equal, $n_i = n$, then

$$s_c = \frac{s_\epsilon}{\sqrt{n}} \sqrt{\sum_{i=1}^{k} c_i^2}. \qquad (2.84)$$

The two-sided confidence interval for the contrast has the form

$$P\left(\hat{\mu}_c - \delta < \mu_c < \hat{\mu}_c + \delta\right) = 1 - \alpha \qquad (2.85)$$

where the confidence interval half-width is

$$\delta = t_{\alpha/2} s_c$$

$$= \frac{t_{\alpha/2} s_\epsilon}{\sqrt{n}} \sqrt{\sum_{i=1}^{k} c_i^2} \qquad (2.86)$$

and the t distribution has df_ϵ degrees of freedom from the ANOVA. Thus, the sample size required to obtain a specified confidence interval half-width δ is

$$n = \left(\frac{t_{\alpha/2} s_\epsilon}{\delta}\right)^2 \sum_{i=1}^{k} c_i^2. \qquad (2.87)$$

Example 2.15 How many observations per treatment group are required to estimate the contrast

$$\mu_c = \left(\frac{\mu_1 + \mu_2 + \mu_3}{3}\right) - \mu_4$$

to within $\delta = 80$ measurement units with 95% confidence if the one-way ANOVA standard error is $s_\epsilon = 200$?

Solution: The goal is to obtain a 95% confidence interval for the contrast of the form given in Equation 2.85 with a confidence interval half-width of $\delta = 80$.

The contrast coefficients are $c_i = \{\frac{1}{3}, \frac{1}{3}, \frac{1}{3}, -1\}$. If there are sufficient error degrees of freedom so that $t \simeq z$, then, from Equation 2.87, the approximate sample size is

$$n \simeq \left(\frac{1.96 \times 200}{80}\right)^2 \left(\left(\frac{1}{3}\right)^2 + \left(\frac{1}{3}\right)^2 + \left(\frac{1}{3}\right)^2 + (-1)^2\right)$$
$$\simeq 33.$$

With $df_\epsilon = k(n-1) = 4(32) = 128$ error degrees of freedom the $t \simeq z$ approximation is satisfied, so the sample size is accurate.

2.6 Multiple Comparisons Tests

Multiple comparisons tests (MCT) for differences between treatment means are used when a study contains three or more treatments and tests must be performed for differences between some or all possible pairs of treatment means. The treatments are assumed to be homoscedastic and the sample sizes are preferred to be equal, although the latter condition may be relaxed. At first glance we might consider applying two-sample t tests to the relevant pairs of treatment means; however, the use of multiple t tests increases the likelihood of committing at least one type I error. The risk of committing at least one type1 error in the many tests is approximately equal to the sum of the type I errors for the individual tests. Multiple comparisons tests are designed specifically to limit the overall risk of a type I error for the complete set, or *family*, of tests.

There are many different multiple comparisons tests available, each with its unique sample size and power calculations. Instead of getting into the details of all of these tests, a single general method, called *Bonferroni's method*, is presented here. Bonferroni's method is very important because it can be used to control the family error rate in any multiple testing situation; however, Bonferroni's method becomes too conservative when the number of tests is large, so in such cases other methods, such as Sidak's or Dunn's method, should be used instead.

2.6.1 Bonferroni's Method

In a multiple comparisons testing situation that requires K tests, Bonferroni's method advises that to restrict the family or *experiment-wise* type I error rate to α_{family}, the individual tests must be performed using a reduced value of α, called the *Bonferroni-corrected* or *comparison-wise* α value:

$$\alpha = \frac{\alpha_{family}}{K}. \qquad (2.88)$$

After the Bonferroni-corrected α value has been determined, the power and sample size analyses for multiple comparisons tests can be performed using the

2.6. Multiple Comparisons Tests

methods developed for two-sample t tests (Section 2.3.2.2) with some minor modifications. For the hypothesis test of $H_0 : \mu_i = \mu_j$ versus $H_A : \mu_i \neq \mu_j$, the decision to reject H_0 or not is based on the t test statistic

$$t_{ij} = \frac{\bar{x}_i - \bar{x}_j}{\hat{\sigma}_\epsilon \sqrt{\frac{1}{n_i} + \frac{1}{n_j}}} \tag{2.89}$$

with H_0 acceptance interval

$$P\left(-t_{\alpha/2, df_\epsilon} < t < t_{\alpha/2, df_\epsilon}\right) = 1 - \alpha \tag{2.90}$$

where α is the Bonferroni-corrected α for individual tests from Equation 2.88, and $\hat{\sigma}_\epsilon$ and df_ϵ are determined from the pooled error information from all of the treatments. The latter information is usually taken from the standard error s_ϵ and df_ϵ from a one-way ANOVA table.

The four most common multiple comparisons testing situations are

- specific predetermined comparisons between treatments.
- all possible comparisons between treatments.
- multiple comparisons with a control treatment.
- multiple comparisons to the best (largest or smallest) treatment.

Short explanations and examples for the second and third situations follow. Be aware that Bonferroni's correction is conservative, so the sample sizes determined by this method are slightly larger or the power is slightly lower compared to the preferred multiple comparisons methods that are more likely to be used for data analysis. The most popular among these alternatives are Tukey's method for all possible comparisons, Dunnett's method for comparisons between treatment groups and a control group, and Hsu's method for comparisons to the best group.

Some statisticians advise that ANOVA should be used to determine if there are any significant differences between treatments before any multiple comparisons tests are performed. Although many consider this to be good practice, it is not strictly necessary for at least some of the multiple comparisons methods. Advantages of performing the ANOVA first, however, are that it provides the pooled standard deviation estimate and error degrees of freedom required for all of the multiple comparisons tests and the usual ANOVA diagnostics can also be used to confirm the validity of the normality and homoscedasticity assumptions that are required of both methods.

2.6.2 All Possible Comparisons

If an experiment has k treatments and all possible two-treatment comparisons are of interest, then there will be $K = \binom{k}{2} = k(k-1)/2$ multiple comparisons tests and the Bonferroni-corrected α for individual tests will be

$$\alpha = \frac{\alpha_{family}}{K} = \frac{2\alpha_{family}}{k(k-1)}. \quad (2.91)$$

The sample size and power for such tests are given by the usual two-sample t test methods (Section 2.3.2.2) using this α value.

A popular alternative to Bonferroni-corrected two-sample t tests for all possible comparisons is Tukey's honest significant difference (HSD) test. Tukey's test is preferred because it is more sensitive to small differences between treatment means while still protecting the experiment-wise type I error rate. Although Tukey's HSD test is performed using the usual two-sample t test statistic, it uses special critical values instead of the usual $t_{\alpha/2}$ values for hypothesis test decisions and power and sample size calculations. This close relationship between the two methods makes Tukey's test easy to use and understand because it takes advantage of all of the methods developed for two-sample t tests. For example, the power for the two-sample t test is calculated using Equations 2.58 with 2.60, but the power for Tukey's test uses

$$t_\beta = \sqrt{\frac{n}{2}} \frac{\Delta\mu}{\widehat{\sigma}_\epsilon} - \frac{q_{\alpha,k,df_\epsilon}}{\sqrt{2}} \quad (2.92)$$

where the q_{α,k,df_ϵ} values are obtained from a table of Tukey HSD test critical values.

Example 2.16 Determine the sample size required per treatment to detect a difference $\Delta\mu = 200$ between two treatment means using Bonferroni-corrected two-sample t tests for all possible pairs of five treatments with 90% power. Assume that the five populations are normal and homoscedastic with $\widehat{\sigma}_\epsilon = 100$.
Solution: With $k = 5$ treatments there will be $K = \binom{5}{2} = 10$ two-sample t tests to perform. To restrict the family error rate to $\alpha_{family} = 0.05$, the Bonferroni-corrected error rate for individual tests is

$$\alpha = \frac{0.05}{10} = 0.005.$$

By Equation 2.62 with $t \simeq z$, the sample size is

$$\begin{aligned} n &= 2\left(\frac{(z_{0.0025} + z_{0.10})\widehat{\sigma}_\epsilon}{\delta}\right)^2 \\ &= 2\left(\frac{(2.81 + 1.282)\,100}{200}\right)^2 = 9. \end{aligned}$$

There will be $df_\epsilon = df_{total} - df_{model} = (5 \times 9 - 1) - (4) = 40$ degrees of freedom to estimate σ_ϵ from the pooled treatment standard deviations, so the approximation $t \simeq z$ is justified.

2.6. Multiple Comparisons Tests

Example 2.17 Determine the approximate power for the sample size calculated in Example 2.16.

Solution: The approximate power for the test is given by Equations 2.58 and 2.60 with $\alpha = 0.005$:

$$\begin{aligned}
\pi &= P(-\infty < t < t_\beta) \\
&= P\left(-\infty < t < \left(\sqrt{\frac{n}{2}}\frac{\Delta\mu}{\widehat{\sigma}} - t_{\alpha/2}\right)\right) \\
&= P\left(-\infty < t < \left(\sqrt{\frac{9}{2}\frac{200}{100}} - t_{0.0025,40}\right)\right) \\
&= P(-\infty < t < 1.273) \\
&= 0.895.
\end{aligned}$$

Example 2.18 Bonferroni's method becomes very conservative when the number of tests gets very large. A less conservative method for determining α for individual tests is given by Sidak's method:

$$\alpha = 1 - (1 - \alpha_{family})^{1/K}. \tag{2.93}$$

Compare the sample sizes determined using Bonferroni's and Sidak's methods for multiple comparisons between all possible pairs of fifteen treatments when the tests must detect a difference of $\Delta\mu = 8$ with 90% power when $\widehat{\sigma}_\epsilon = 6$.

Solution: The number of multiple comparisons tests required is

$$\binom{15}{2} = \frac{15 \times 14}{2} = 105.$$

By Bonferroni's method with $\alpha_{family} = 0.05$, the α for individual tests is

$$\alpha = \frac{0.05}{105} = 0.000476,$$

so with $t \simeq z$ in Equation 2.62 the sample size is

$$\begin{aligned}
n &= 2\left(\frac{(z_{0.000476/2} + z_{0.10})\widehat{\sigma}_\epsilon}{\delta}\right)^2 \\
&= 2\left(\frac{(3.494 + 1.282)6}{8}\right)^2 = 26.
\end{aligned}$$

By Sidak's method (Equation 2.93), the α for individual tests is

$$\alpha = 1 - (1 - 0.05)^{1/105} = 0.000488,$$

so the sample size is

$$n = 2\left(\frac{(z_{0.000488/2} + z_{0.10})\hat{\sigma}_\epsilon}{\delta}\right)^2$$
$$= 2\left(\frac{(3.487 + 1.282)6}{8}\right)^2 = 26.$$

Even with over 100 multiple comparisons, the sample sizes by the two calculation methods are still equal.

2.6.3 Comparisons with a Control

For multiple comparisons between K treatments and a control group, the hypotheses to be tested are $H_{0i} : \mu_i = \mu_0$ versus $H_{Ai} : \mu_i \neq \mu_0$, where $i = 1$ to K and μ_0 is the control group's mean. The K tests may be performed using Bonferroni-corrected two-sample t tests with test statistics

$$t_{i0} = \frac{\bar{x}_i - \bar{x}_0}{\hat{\sigma}_\epsilon \sqrt{\frac{1}{n_i} + \frac{1}{n_0}}} \tag{2.94}$$

where the groups are assumed to be homoscedastic with standard deviation $\hat{\sigma}_\epsilon$ and the t distribution's degrees of freedom are equal to the pooled error degrees of freedom:

$$df_\epsilon = (n_0 - 1) + \sum_{i=1}^{K}(n_i - 1).$$

Then, by Bonferroni's method, the α value for individual tests must be

$$\alpha = \frac{\alpha_{family}}{K} \tag{2.95}$$

with H_0 acceptance interval given by Equation 2.90. The sample size and power for such tests are given by the usual two-sample t test methods (Section 2.3.2.2) using this α value.

When several treatment groups are to be compared to a single control group, the relative importance of the control group is greater and therefore it deserves to have a larger sample size than the treatment groups. It can be shown that if there are K treatments to be compared to the control,[2] then the power for multiple comparisons tests is maximized when

$$n_0 = n_i \sqrt{K} \tag{2.96}$$

---

[2]Maximize the value of t_{i0} (Equation 2.94) by minimizing the value of $(1/n_0 + 1/n_i)$ with respect to n_0 subject to the constraint that the total sample size $N = n_0 + Kn_i$ is constant.

2.6. Multiple Comparisons Tests

where n_0 is the sample size for the control and n_i is the common sample size for the treatments.[3] Substitution of this result into Equation 2.94 gives

$$t_{i0} = \frac{\bar{x}_i - \bar{x}_0}{\hat{\sigma}_\epsilon \sqrt{\frac{1}{n_i}\left(1 + \frac{1}{\sqrt{K}}\right)}}, \qquad (2.97)$$

which leads to a sample size condition analogous to Equation 2.62:

$$n_i = \left(1 + \frac{1}{\sqrt{K}}\right)\left(\frac{(t_{\alpha/2} + t_\beta)\hat{\sigma}_\epsilon}{\Delta\mu}\right)^2. \qquad (2.98)$$

Note that for $K = 1$, this equation reduces to the usual two-sample t test sample size.

When n_i and n_0 are specified, the power for Bonferroni-corrected two-sample t tests between treatments and the control is given by

$$\pi = P\left(-\infty < t < t_\beta\right) \qquad (2.99)$$

where

$$t_\beta = \frac{\delta}{\hat{\sigma}_\epsilon \sqrt{\frac{1}{n_i} + \frac{1}{n_0}}} - t_{\frac{\alpha}{2K}}. \qquad (2.100)$$

Bonferroni-corrected two-sample t tests for multiple comparisons with a control are less sensitive than Dunnett's test, which is the preferred method of analysis for this situation; however, the sample size and power values obtained using the Bonferroni method are usually accurate enough for planning an experiment, even when the data will be analyzed by Dunnett's method.

The relationship between Dunnett's test for multiple comparisons with a control and two-sample t tests is analogous to the relationship between Tukey's test for all possible comparisons and two-sample t tests – Dunnett's test simply replaces $t_{\alpha/2}$ in the two-sample t test with special critical values that provide better sensitivity to small differences between means while preserving the experiment-wise type I error rate.

Example 2.19 An experiment will be performed to compare four treatment groups to a control group. Determine the sample size required to detect a difference $\delta = 200$ between the treatments and the control using Bonferroni-corrected two-sample t tests with 90% power. Use a balanced design with the same number of observations in each of the five groups and assume that the five populations are normal and homoscedastic with $\hat{\sigma}_\epsilon = 100$.

---

[3] If the multiple comparisons test is Dunnett's test, Equation 2.96 slightly overestimates the value of n_0 that maximizes the power.

Solution: To restrict the family error rate to $\alpha_{family} = 0.05$ with $K = 4$ tests, the Bonferroni-corrected error rate for individual tests is

$$\alpha = \frac{0.05}{4} = 0.0125.$$

By Equation 2.62 with $t \simeq z$, the sample size is

$$n = 2\left(\frac{(z_{0.0125/2} + z_{0.10})\hat{\sigma}_\epsilon}{\delta}\right)^2$$

$$= 2\left(\frac{(2.50 + 1.282)\,100}{200}\right)^2 = 8.$$

Despite the small treatment-group sample size, the approximation $t \simeq z$ is justified because there will be $df_\epsilon = df_{total} - df_{model} = (5 \times 8 - 1) - (4) = 35$ degrees of freedom to estimate $\hat{\sigma}_\epsilon$ from the five pooled treatment standard deviations.

Example 2.20 Repeat Example 2.19 using the optimal allocation of units to treatments and controls.

Solution: From Equation 2.98 with $t \simeq z$ and $K = 4$,

$$n_i = \left(1 + \frac{1}{\sqrt{4}}\right)\left(\frac{(2.50 + 1.282)\,100}{200}\right)^2 = 6$$

and

$$n_0 = n_i\sqrt{K} = 6\sqrt{4} = 12.$$

The approximation $t \simeq z$ is still justified because the error degrees of freedom will be $df_\epsilon = (4 \times 6 + 12) - 4 = 32$. The original experiment required $5 \times 8 = 40$ units, but the optimal experiment requires only $4 \times 6 + 12 = 36$ units to obtain the same power.

Chapter 3

Standard Deviations

The purpose of this chapter is to present sample size and power calculations for confidence intervals and hypothesis tests for the population standard deviation or variance and the coefficient of variation for quantitative responses. The chi-square (χ^2) distribution, on which most of these methods are based, is very sensitive to the normality assumption that is assumed to be satisfied throughout the chapter. Although exact solutions are available for most types of problems, large-sample approximations are easier to use and are sufficient for most practical situations, so they are the methods emphasized here.

3.1 One Standard Deviation

Exact and approximate sample size calculations are given for the one-sample standard deviation problem for confidence intervals and hypothesis tests. The exact methods, based on the χ^2 distribution, are easiest to perform using a table, a spreadsheet, or appropriate software; however, the large-sample approximations are easy to use, reasonably accurate, and provide a good starting point for further iterations if a more exact result is required.

3.1.1 Confidence Interval for the Standard Deviation

Two methods, one exact and one approximate, are presented for determining the sample size for a confidence interval for the population standard deviation.

3.1.1.1 Confidence Interval for the Standard Deviation (Exact Method)

When the distribution of x is normal with population variance σ^2, then the distribution of

$$\chi^2 = \frac{(n-1)s^2}{\sigma^2} \qquad (3.1)$$

is χ^2 with $n-1$ degrees of freedom where n is the sample size and s^2 is the sample variance. Therefore, the distribution of sample variances must meet the condition

$$P\left(\chi^2_{\alpha/2} < \frac{(n-1)s^2}{\sigma^2} < \chi^2_{1-\alpha/2}\right) = 1-\alpha \tag{3.2}$$

where the χ^2 distribution is indexed by its left tail area. This expression may be manipulated to give the following exact $(1-\alpha)\,100\%$ confidence interval for the population variance:

$$P\left(\frac{(n-1)s^2}{\chi^2_{1-\alpha/2}} < \sigma^2 < \frac{(n-1)s^2}{\chi^2_{\alpha/2}}\right) = 1-\alpha \tag{3.3}$$

or, for the population standard deviation:

$$P\left(s\sqrt{\frac{n-1}{\chi^2_{1-\alpha/2}}} < \sigma < s\sqrt{\frac{n-1}{\chi^2_{\alpha/2}}}\right) = 1-\alpha. \tag{3.4}$$

Factors for 95% confidence limits for σ as a function of the degrees of freedom $df = n-1$ as determined from Equation 3.4 are shown in Table 3.1.[1] Inspection of the tabulated values shows that confidence intervals for σ are asymmetric, especially when n is small, which compromises the concept of confidence interval half-width.

Example 3.1 Determine the sample size required to construct the 95% confidence interval for σ based on a random sample of size n drawn from a normal population if the confidence interval half-width must be about 10% of the sample standard deviation.
Solution: From Table 3.1 the sample size must be about $n = 200$. The lower and upper confidence limits will fall at about -9% and +11% relative to the sample standard deviation, so the asymmetry for this relatively large sample size is not too severe.

3.1.1.2 Confidence Interval for the Standard Deviation (Approximate Method)

The mean of the χ^2 distribution with $\nu = n-1$ degrees of freedom is $\mu_{\chi^2} = \nu$ and the standard deviation is $\sigma_{\chi^2} = \sqrt{2\nu}$. As ν becomes large, the χ^2 distribution approaches normality, so Equation 3.2 may be approximated with

$$\begin{aligned}P\left(\mu_{\chi^2} - z_{\alpha/2}\sigma_{\chi^2} < \chi^2 < \mu_{\chi^2} + z_{\alpha/2}\sigma_{\chi^2}\right) &= 1-\alpha \\ P\left(\nu - z_{\alpha/2}\sqrt{2\nu} < \chi^2 < \nu + z_{\alpha/2}\sqrt{2\nu}\right) &= 1-\alpha.\end{aligned} \tag{3.5}$$

---
[1] Factors for one-sided upper 95% confidence limits are given in Table 10.6.

3.1. One Standard Deviation

df	$\sqrt{\frac{df}{\chi^2_{0.975}}}$	$\sqrt{\frac{df}{\chi^2_{0.025}}}$	df	$\sqrt{\frac{df}{\chi^2_{0.975}}}$	$\sqrt{\frac{df}{\chi^2_{0.025}}}$
1	0.446	31.91	25	0.784	1.380
2	0.521	6.285	30	0.799	1.337
3	0.566	3.729	35	0.811	1.304
4	0.599	2.874	40	0.821	1.280
5	0.624	2.453	45	0.829	1.260
6	0.644	2.202	50	0.837	1.243
7	0.661	2.035	60	0.849	1.217
8	0.675	1.916	70	0.858	1.198
9	0.688	1.826	80	0.866	1.183
10	0.699	1.755	90	0.873	1.171
11	0.708	1.698	100	0.879	1.161
12	0.717	1.651	120	0.888	1.145
13	0.725	1.611	140	0.895	1.133
14	0.732	1.577	160	0.901	1.123
15	0.739	1.548	180	0.907	1.115
16	0.745	1.522	200	0.911	1.109
17	0.751	1.499	250	0.920	1.096
18	0.756	1.479	300	0.926	1.087
19	0.760	1.461	400	0.935	1.074
20	0.765	1.444	500	0.942	1.066

Table 3.1: 95% confidence limit factors for the population standard deviation.

This equation may be manipulated to give the following approximate $(1 - \alpha) 100\%$ confidence interval for σ:

$$P(s(1 - \delta) < \sigma < s(1 + \delta)) = 1 - \alpha \qquad (3.6)$$

where the confidence interval's relative half-width is

$$\delta = \frac{z_{\alpha/2}}{\sqrt{2n}}. \qquad (3.7)$$

Then the sample size required to obtain a specified relative confidence interval half-width is

$$n = \frac{1}{2}\left(\frac{z_{\alpha/2}}{\delta}\right)^2. \qquad (3.8)$$

Example 3.2 Use the large sample approximation method to determine the sample size for the situation in Example 3.1.
Solution: The required confidence interval has the form

$$P(s(1 - 0.10) < \sigma < s(1 + 0.10)) = 0.95.$$

With $\alpha = 0.05$ and $\delta = 0.10$ in Equation 3.8, the sample size required to obtain a confidence interval of the desired half-width is

$$n = \frac{1}{2}\left(\frac{1.96}{0.10}\right)^2 = 193,$$

which is in excellent agreement with the original solution.

3.1.1.3 Confidence Interval for the Standard Deviation (Log-transform Method)

An approximate large-sample confidence interval for the log-transformed standard deviation is given by

$$P\left(\ln(s) - \delta < \ln(\sigma) < \ln(s) + \delta\right) = 1 - \alpha \tag{3.9}$$

where the confidence interval half-width is given by the delta method (see Appendix G.3.12):

$$\delta = \frac{z_{\alpha/2}}{\sqrt{2n}}. \tag{3.10}$$

This interval is easy to calculate for sample data, but it is not practical for sample size calculations because of the difficulty in specifying the confidence interval half-width in terms of logarithmic units.

3.1.2 Tests for One Standard Deviation

3.1.2.1 Test for One Standard Deviation (Exact Method)

The hypotheses to be tested are $H_0 : \sigma^2 = \sigma_0^2$ versus $H_A : \sigma^2 > \sigma_0^2$. The test is performed by determining the sample variance s^2 from a random sample of size n drawn from a normal population and calculating the χ^2 statistic:

$$\chi^2 = \frac{(n-1)s^2}{\sigma_0^2}, \tag{3.11}$$

which follows the χ^2 distribution with $n-1$ degrees of freedom. We reject H_0 when $\chi^2 > \chi^2_{1-\alpha}$.

Figure 3.1 shows the sampling distributions of s^2 under H_0, where $\sigma = \sigma_0$, and H_A, where $\sigma = \sigma_1$. At the critical value of the sample variance $s^2_{A/R}$ that defines the accept/reject boundary for H_0, we have

$$s^2_{A/R} = \frac{\chi^2_{1-\alpha}\sigma_0^2}{n-1} = \frac{\chi^2_\beta \sigma_1^2}{n-1} \tag{3.12}$$

from which

$$\chi^2_\beta = \chi^2_{1-\alpha}\frac{\sigma_0^2}{\sigma_1^2}. \tag{3.13}$$

3.1. One Standard Deviation

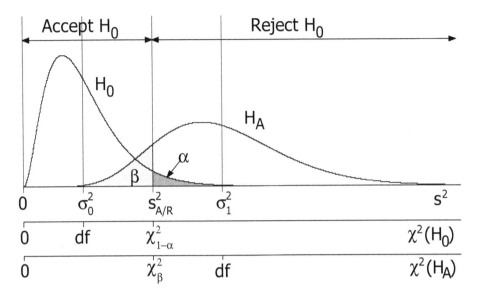

Figure 3.1: Distributions of sample variances under H_0 and H_A.

The corresponding power of the test is given by

$$\pi = P\left(\chi_\beta^2 < \chi^2 < \infty\right)$$
$$= P\left(\chi_{1-\alpha}^2 \frac{\sigma_0^2}{\sigma_1^2} < \chi^2 < \infty\right). \quad (3.14)$$

The following sample size condition may be derived from Equation 3.12:

$$\frac{\chi_{1-\alpha}^2}{\chi_\beta^2} \leq \frac{\sigma_1^2}{\sigma_0^2} \quad (3.15)$$

where the smallest value of n that satisfies this condition provides power $\pi = 1 - \beta$ to reject H_0 in favor of H_A. While this condition appears to be tedious to solve, it is actually quite easy to do with an appropriate χ^2 table or spreadsheet.

When the hypotheses to be tested are $H_0 : \sigma^2 = \sigma_0^2$ versus $H_A : \sigma^2 < \sigma_0^2$, the power is given by

$$\pi = P\left(0 < \chi^2 < \chi_\alpha^2 \frac{\sigma_0^2}{\sigma_1^2}\right) \quad (3.16)$$

and the sample size that provides power $\pi = 1 - \beta$ for a specified value of σ is the smallest value of n that satisfies

$$\frac{\chi_{1-\beta}^2}{\chi_\alpha^2} \geq \frac{\sigma_0^2}{\sigma_1^2} \quad (3.17)$$

df	$\chi^2_{0.95}/\chi^2_{0.10}$	$\chi^2_{0.90}/\chi^2_{0.05}$	df	$\chi^2_{0.95}/\chi^2_{0.10}$	$\chi^2_{0.90}/\chi^2_{0.05}$
2	28.43	44.89	20	2.524	2.618
3	13.37	17.77	25	2.286	2.353
4	8.920	10.95	30	2.125	2.177
5	6.875	8.063	40	1.919	1.954
6	5.713	6.509	50	1.791	1.817
7	4.965	5.545	70	1.636	1.653
8	4.444	4.890	100	1.510	1.521
9	4.059	4.416	200	1.400	1.407
10	3.763	4.057	300	1.269	1.272
15	2.925	3.072	500	1.203	1.204

Table 3.2: Factors for the one-sample test for the variance.

where the χ^2 distributions both have $n-1$ degrees of freedom.

Table 3.2 gives values of the χ^2 ratios for Equations 3.15 and 3.17.

Example 3.3 For the test of $H_0 : \sigma^2 = 10$ versus $H_A : \sigma^2 > 10$, find the power associated with $\sigma^2 = 20$ when the sample size is $n = 20$ using $\alpha = 0.05$.

Solution: From Equation 3.14 the power is given by

$$\begin{aligned} \pi &= P\left(\chi^2_{0.95}\left(\frac{10}{20}\right) < \chi^2 < \infty\right) \\ &= P\left(15.1 < \chi^2 < \infty\right) \\ &= 0.72. \end{aligned}$$

Example 3.4 Find the sample size required to reject $H_0 : \sigma^2 = 40$ with 90% power when $\sigma^2 = 100$ using $H_A : \sigma^2 > 40$ with $\alpha = 0.05$.

Solution: From Equation 3.15 with $\sigma^2_0 = 40$ and $\sigma^2_1 = 100$, the necessary sample size is the smallest value of n that meets the requirement

$$\frac{\chi^2_{0.95}}{\chi^2_{0.10}} \leq \frac{100}{40}$$
$$\leq 2.5.$$

By inspecting Table 3.2 and a table of χ^2 values, the required sample size is $n = 22$ for which

$$\left(\frac{\chi^2_{0.95}}{\chi^2_{0.10}} = \frac{32.67}{13.24} = 2.469\right) \leq 2.5.$$

3.1. One Standard Deviation

Example 3.5 Find the sample size required to reject $H_0 : \sigma = 0.003$ in favor of $H_A : \sigma < 0.003$ with 90% power when in fact $\sigma = 0.001$.
Solution: With $\alpha = 0.05$ and $\beta = 1 - \pi = 0.10$, the sample size condition given by Equation 3.17 is

$$\frac{\chi^2_{0.90}}{\chi^2_{0.05}} \geq \left(\frac{0.003}{0.001}\right)^2$$
$$\geq 9.0,$$

which, from Table 3.2, is satisfied by $n = 5$.

3.1.2.2 Test for One Standard Deviation (Approximate Method)

When n is large, by the delta method (see Appendix G.3.12) the distribution of $\ln(s)$ is approximately normal with mean $\mu_{\ln(s)} = \ln(\sigma)$ and approximate standard deviation $\hat{\sigma}_{\ln(s)} = 1/\sqrt{2n}$. Then in terms of $\ln(s)$, the critical accept/reject value for the test of $H_0 : \ln(\sigma) = \ln(\sigma_0)$ versus $H_A : \ln(\sigma) > \ln(\sigma_0)$ is

$$(\ln(s))_{A/R} = \ln(\sigma_0) + \frac{z_\alpha}{\sqrt{2n}} = \ln(\sigma_1) - \frac{z_\beta}{\sqrt{2n}}, \qquad (3.18)$$

which may be solved for the power

$$\pi = \Phi(-z_\beta < z < \infty) \qquad (3.19)$$

where

$$z_\beta = \sqrt{2n} \ln\left(\frac{\sigma_1}{\sigma_0}\right) - z_\alpha. \qquad (3.20)$$

The sample size required to obtain a specified value of the power $\pi = 1 - \beta$ is determined from Equation 3.20 as

$$n = \frac{1}{2}\left(\frac{z_\alpha + z_\beta}{\ln\left(\frac{\sigma_1}{\sigma_0}\right)}\right)^2. \qquad (3.21)$$

Example 3.6 Compare the power determined by the large-sample approximation method to the exact power determined in Example 3.3.
Solution: The null hypothesis may be written as $H_0 : \ln(\sigma) = \ln(\sqrt{10})$ and we wish to find the power to reject H_0 when $\ln(\sigma) = \ln(\sqrt{20})$ with $n = 20$. From Equation 3.20 we have

$$z_\beta = \sqrt{2 \times 20} \ln\left(\sqrt{\frac{20}{10}}\right) - z_{0.05} = 0.547$$

and by Equation 3.19 the approximate power is

$$\pi = \Phi(-0.547 < z < \infty)$$
$$= 0.71.$$

This result is still in good agreement with the exact power of 72% despite the rather small sample size.

Example 3.7 Compare the sample size determined by the large-sample approximation method to the exact sample size determined in Example 3.4.
Solution: The problem is to find the sample size to reject $H_0 : \ln(\sigma) = \ln(\sqrt{40})$ with 90% power when $\ln(\sigma) = \ln(\sqrt{100})$. With $\alpha = 0.05$ and $\beta = 0.10$ in Equation 3.21 the approximate sample size required is

$$n = \frac{1}{2}\left(\frac{1.645 + 1.282}{\ln\left(\sqrt{\frac{100}{40}}\right)}\right)^2 = 21,$$

which is in good agreement with the exact sample size of $n = 22$.

3.2 Two Standard Deviations

Exact and approximate methods are presented for calculating sample sizes for experiments to determine confidence intervals and to perform hypothesis tests for the ratio of two independent population standard deviations.

3.2.1 Confidence Intervals for Two Standard Deviations

3.2.1.1 Confidence Interval for the Ratio of Two Standard Deviations (Exact Method)

For two independent normal populations with standard deviations σ_1 and σ_2, where σ_1 and σ_2 may be different, the distribution of

$$\left(\frac{s_1/\sigma_1}{s_2/\sigma_2}\right)^2 \tag{3.22}$$

follows the F distribution with $n_1 - 1$ numerator and $n_2 - 1$ denominator degrees of freedom, where s_1 and s_2 are sample standard deviations determined from samples of size n_1 and n_2, respectively. Then the exact $(1 - \alpha)\,100\%$ confidence interval for the ratio of the population standard deviations σ_1/σ_2 is given by

3.2. Two Standard Deviations

$$P\left(\frac{s_1}{s_2}\sqrt{F_{\alpha,n_1-1,n_2-1}} < \frac{\sigma_1}{\sigma_2} < \frac{s_1}{s_2}\sqrt{F_{1-\alpha,n_1-1,n_2-1}}\right) = 1-\alpha. \quad (3.23)$$

This interval is not very helpful for calculating the sample sizes to obtain a confidence interval of desired width. The approximation that follows is usually sufficient to determine the sample sizes and is a good starting point if the accuracy of Equation 3.23 is required.

3.2.1.2 Confidence Interval for the Ratio of Two Standard Deviations (Large-Sample Approximation)

The distribution of $F = (s_1/s_2)^2$ is skewed, but when n_1 and n_2 are both large, by the delta method (see Appendix G.3.13) the distribution of $\ln\left(\sqrt{F}\right) = \ln(s_1/s_2)$ is approximately normal with mean

$$\mu_{\ln(s_1/s_2)} = \ln\left(\frac{\sigma_1}{\sigma_2}\right) \quad (3.24)$$

and standard deviation

$$\sigma_{\ln(s_1/s_2)} = \sqrt{\frac{1}{2}\left(\frac{1}{n_1} + \frac{1}{n_2}\right)}. \quad (3.25)$$

Then the two-sided $(1-\alpha)\,100\%$ confidence interval for $\ln(\sigma_1/\sigma_2)$ is approximately

$$P\left(\ln\left(\frac{s_1}{s_2}\right) - \delta < \ln\left(\frac{\sigma_1}{\sigma_2}\right) < \ln\left(\frac{s_1}{s_2}\right) + \delta\right) = 1-\alpha \quad (3.26)$$

where the confidence interval half-width is

$$\begin{aligned}\delta &= z_{\alpha/2}\sigma_{\ln(s_1/s_2)} \\ &= z_{\alpha/2}\sqrt{\frac{1}{2}\left(\frac{1}{n_1} + \frac{1}{n_2}\right)}.\end{aligned} \quad (3.27)$$

The corresponding confidence interval for σ_1/σ_2 is

$$P\left(\frac{s_1}{s_2}e^{-\delta} < \frac{\sigma_1}{\sigma_2} < \frac{s_1}{s_2}e^{\delta}\right) = 1-\alpha. \quad (3.28)$$

Because the sample sizes are assumed to be large, δ will be near zero and the confidence interval for σ_1/σ_2 may be approximated by[2]

$$P\left(\frac{s_1}{s_2}(1-\delta) < \frac{\sigma_1}{\sigma_2} < \frac{s_1}{s_2}(1+\delta)\right) = 1-\alpha. \quad (3.29)$$

---

[2] see Appendix F, Equation F.18.

Then in an experiment to determine a confidence interval for σ_1/σ_2 with specified relative half-width δ and sample size ratio n_1/n_2, the sample size must be

$$n_1 = \frac{1}{2}\left(1 + \frac{n_1}{n_2}\right)\left(\frac{z_{\alpha/2}}{\delta}\right)^2.$$

For $n_1 = n_2$, the sample size reduces to

$$n_1 = n_2 = \left(\frac{z_{\alpha/2}}{\delta}\right)^2. \tag{3.30}$$

Example 3.8 What equal-n sample size is required by an experiment to deliver a confidence interval for the ratio of two independent population standard deviations if the true ratio should fall within 20% of the experimental ratio with 95% confidence?

Solution: The goal of the experiment is to determine an interval of the form

$$P\left(\frac{s_1}{s_2}(1 - 0.2) < \frac{\sigma_1}{\sigma_2} < \frac{s_1}{s_2}(1 + 0.2)\right) = 0.95.$$

Then, from Equation 3.30 with $\delta = 0.2$, the required sample sizes are

$$n_1 = n_2 = \left(\frac{z_{0.025}}{\delta}\right)^2 = \left(\frac{1.96}{0.20}\right)^2 = 97.$$

3.2.2 Tests for Two Standard Deviations

3.2.2.1 Test for the Variance Ratio (Exact Method)

The hypotheses to be tested are $H_0 : \sigma_1/\sigma_2 = 1$ versus $H_A : \sigma_1/\sigma_2 > 1$ or equivalently $H_0 : \sigma_1^2 = \sigma_2^2$ versus $H_A : \sigma_1^2 > \sigma_2^2$. The test is performed by determining the sample variances s_1^2 and s_2^2 from random samples of size n_1 and n_2, respectively, drawn from independent normal populations, and calculating the F statistic:

$$F = \left(\frac{s_1}{s_2}\right)^2. \tag{3.31}$$

Figure 3.2 shows the sampling distributions of F under H_0 and H_A. Under H_0, the F statistic follows the F distribution with $\nu_1 = n_1 - 1$ numerator and $\nu_2 = n_2 - 1$ denominator degrees of freedom. We reject H_0 when $F > F_{1-\alpha}$. Under H_A, the distribution of

$$F = \left(\frac{\sigma_2}{\sigma_1}\right)^2 \left(\frac{s_1}{s_2}\right)^2 \tag{3.32}$$

3.2. Two Standard Deviations

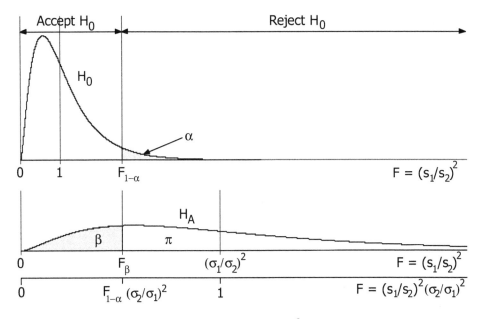

Figure 3.2: Distributions of $F = (s_1/s_2)^2$ under H_0 and H_A.

also follows the F distribution, so we have

$$F_\beta = \left(\frac{\sigma_2}{\sigma_1}\right)^2 F_{1-\alpha} \tag{3.33}$$

and the power to reject H_0 is

$$\begin{aligned}\pi &= P(F_\beta < F < \infty) \\ &= P\left(\left(\frac{\sigma_2}{\sigma_1}\right)^2 F_{1-\alpha} < F < \infty\right).\end{aligned} \tag{3.34}$$

Example 3.9 Find the power to reject $H_0 : \sigma_1^2 = \sigma_2^2$ in favor of $H_A : \sigma_1^2 > \sigma_2^2$ if $n_1 = n_2 = 26$, $\sigma_1^2 = 15$, and $\sigma_2^2 = 5$ using $\alpha = 0.05$.
Solution: From Equation 3.34 the power is

$$\begin{aligned}\pi &= P\left(\left(\frac{\sigma_2}{\sigma_1}\right)^2 F_{1-\alpha} < F < \infty\right) \\ &= P\left(\left(\frac{5}{15}\right) F_{0.95, 25, 25} < F < \infty\right) \\ &= P(0.652 < F < \infty) \\ &= 0.854.\end{aligned}$$

Example 3.10 What equal sample size is required to detect a factor of two difference between two population standard deviations with 90% power and $\alpha = 0.05$?

Solution: A factor of two difference in population standard deviation corresponds to a factor of four difference in variance, so we need to determine the sample size such that

$$P\left(\frac{1}{4}F_{1-\alpha} < F < \infty\right) = 0.90.$$

By iterating through several values of sample size, we find that when $n_1 = n_2 = 20$, $F_{0.95,19,19} = 2.168$ and $F_{0.096} = 2.168/4 = 0.542$, which satisfies the problem statement.

3.2.2.2 Test for the Ratio of Two Standard Deviations (Approximate Method)

The hypotheses to be tested are $H_0 : \sigma_1/\sigma_2 = 1$ versus $H_A : \sigma_1/\sigma_2 > 1$. When n_1 and n_2 are large, by the delta method (Section G.3.13) the distribution of $\ln(s_1/s_2)$ is approximately normal with mean $\mu_{\ln(s_1/s_2)} = \ln(\sigma_1/\sigma_2)$ and standard deviation $\sigma_{\ln(s_1/s_2)} = \sqrt{\frac{1}{2}\left(\frac{1}{n_1} + \frac{1}{n_2}\right)}$. Then for the specified value of $\ln(\sigma_1/\sigma_2)$ taken under H_A, the critical accept/reject value for the test is

$$(\ln(s_1/s_2))_{A/R} = z_\alpha\sqrt{\frac{1}{2}\left(\frac{1}{n_1} + \frac{1}{n_2}\right)} = \ln(\sigma_1/\sigma_2) - z_\beta\sqrt{\frac{1}{2}\left(\frac{1}{n_1} + \frac{1}{n_2}\right)}, \quad (3.35)$$

which leads to the approximate power

$$\pi = \Phi\left(-z_\beta < z < \infty\right) \quad (3.36)$$

where

$$z_\beta = \frac{\ln\left(\frac{\sigma_1}{\sigma_2}\right)}{\sqrt{\frac{1}{2}\left(\frac{1}{n_1} + \frac{1}{n_2}\right)}} - z_\alpha. \quad (3.37)$$

Then the sample size n_1 required to obtain power $\pi = 1 - \beta$ to reject H_0 for specified σ_1/σ_2 with sample size ratio n_1/n_2 is

$$n_1 = \frac{1}{2}\left(1 + \frac{n_1}{n_2}\right)\left(\frac{z_\alpha + z_\beta}{\ln\left(\frac{\sigma_1}{\sigma_2}\right)}\right)^2. \quad (3.38)$$

For $n_1 = n_2$, the approximate required sample size is

$$n_1 = n_2 = \left(\frac{z_\alpha + z_\beta}{\ln\left(\frac{\sigma_1}{\sigma_2}\right)}\right)^2. \quad (3.39)$$

3.3. Coefficient of Variation

Example 3.11 Repeat Example 3.9 using the large-sample approximation method.
Solution: From the information given in the example problem statement

$$z_\beta = \frac{\ln\left(\sqrt{\frac{15}{5}}\right)}{\sqrt{\frac{1}{2}\left(\frac{1}{26}+\frac{1}{26}\right)}} - 1.645 = 1.16$$

so the power is

$$\pi = \Phi(-1.16 < z < \infty) = 0.877,$$

which is in good agreement with the exact solution of $\pi = 0.854$.

Example 3.12 Repeat Example 3.10 using the large-sample approximation method.
Solution: From the information given in the example problem statement

$$n_1 = n_2 = \left(\frac{1.645 + 1.282}{\ln(2)}\right)^2 = 18,$$

which slightly underestimates the exact solution $n = 20$.

3.3 Coefficient of Variation

Sometimes the variability of a process is reported in terms of the coefficient of variation given by

$$\widehat{CV} = \frac{s}{\bar{x}} \qquad (3.40)$$

where the population being sampled is normal with mean μ and standard deviation σ. The purpose of this section is to present sample size calculations for constructing a confidence interval and for performing hypothesis tests for the coefficient of variation parameter $CV = \sigma/\mu$.

More accurate methods for CV may be obtained by considering the reciprocal of CV

$$\frac{1}{\widehat{CV}} = \frac{\bar{x}}{s} \qquad (3.41)$$

which follows a noncentral Student's t distribution. However, the approximate methods shown here are usually sufficient for sample size and power calculations.

3.3.1 Confidence Interval for the Coefficient of Variation

For large samples drawn from a normal population, the distribution of $\widehat{CV}$ is approximately normal with mean $CV = \sigma/\mu$ and approximate standard deviation given by the delta method (see Appendix G.3.14):

$$\hat{\sigma}_{\widehat{CV}} \simeq \widehat{CV}\sqrt{\frac{1}{n}\left(\widehat{CV}^2 + \frac{1}{2}\right)}. \tag{3.42}$$

The two-sided $(1-\alpha)\,100\%$ confidence interval for CV has the form

$$P\left(\widehat{CV}\,(1-\delta) < CV < \widehat{CV}\,(1+\delta)\right) = 1 - \alpha \tag{3.43}$$

where the confidence interval's relative half-width δ is

$$\begin{aligned}\delta &= z_{\alpha/2}\hat{\sigma}_{\widehat{CV}} \\ &= z_{\alpha/2}\sqrt{\frac{1}{n}\left(\widehat{CV}^2 + \frac{1}{2}\right)}.\end{aligned} \tag{3.44}$$

Then the sample size required to obtain a confidence interval for CV with specified relative half-width δ is given by

$$n = \left(\frac{z_{\alpha/2}}{\delta}\right)^2 \left(\widehat{CV}^2 + \frac{1}{2}\right). \tag{3.45}$$

Example 3.13 Determine the sample size required to estimate the population coefficient of variation to within $\pm 25\%$ with 95% confidence if the coefficient of variation is expected to be about 30%.

Solution: With $\alpha = 0.05$, $\delta = 0.25$, and $\widehat{CV} = 0.3$ in Equation 3.45, the sample size must be

$$n = \left(\frac{1.96}{0.25}\right)^2 \left((0.3)^2 + \frac{1}{2}\right) = 37.$$

3.3.2 Tests for the Coefficient of Variation

3.3.2.1 Test for One Coefficient of Variation

The hypotheses to be tested are $H_0 : CV = CV_0$ versus $H_A : CV \neq CV_0$. The test is performed by drawing a random sample of size n and calculating $\widehat{CV}$. If the sample size is sufficiently large and the population being sampled is normal, then by the delta method (see Appendix G.3.14) the test statistic

$$\begin{aligned}z &= \frac{\widehat{CV} - CV_0}{\hat{\sigma}_{CV_0}} \\ &= \frac{\widehat{CV} - CV_0}{CV_0\sqrt{\frac{1}{n}\left(CV_0^2 + \frac{1}{2}\right)}}\end{aligned} \tag{3.46}$$

3.3. Coefficient of Variation

is approximately normal and the acceptance interval for H_0 is $-z_{\alpha/2} < z < z_{\alpha/2}$. Under H_A when $CV = CV_1$, the probability of rejecting H_0 is given by

$$\pi = P(-\infty < z < -z_\beta) \tag{3.47}$$

where

$$z_\beta = \frac{|CV_1 - CV_0| - z_{\alpha/2}\sigma_{CV_0}}{\sigma_{CV_1}}$$

$$= \frac{|CV_1 - CV_0| - z_{\alpha/2}CV_0\sqrt{\frac{1}{n}\left(CV_0^2 + \frac{1}{2}\right)}}{CV_1\sqrt{\frac{1}{n}\left(CV_1^2 + \frac{1}{2}\right)}}. \tag{3.48}$$

Then the sample size required to reject H_0 with specified power $\pi = 1 - \beta$ is given by

$$n = \left(\frac{z_{\alpha/2}CV_0\sqrt{CV_0^2 + \frac{1}{2}} + z_\beta CV_1\sqrt{CV_1^2 + \frac{1}{2}}}{CV_1 - CV_0}\right)^2. \tag{3.49}$$

Example 3.14 Determine the sample size required to reject $H_0 : CV = 0.5$ with 90% power when $CV = 0.8$.

Solution: With $CV_0 = 0.5$, $CV_1 = 0.8$, $\alpha = 0.05$, and $\beta = 0.10$ in Equation 3.49, the required sample size is

$$n = \left(\frac{1.96 \times 0.5\sqrt{(0.5)^2 + \frac{1}{2}} + 1.282 \times 0.8\sqrt{(0.8)^2 + \frac{1}{2}}}{0.8 - 0.5}\right)^2 = 42.$$

3.3.2.2 Test for Two Coefficients of Variation

The hypotheses to be tested are $H_0 : CV_1 = CV_2$ versus $H_A : CV_1 \neq CV_2$. The test is performed by drawing random samples of size n_1 and n_2 and calculating $\widehat{CV}_1$ and $\widehat{CV}_2$. If the sample sizes are sufficiently large and the populations being sampled are normal, then by the delta method (see Appendix G.3.14) the test statistic

$$z = \frac{\Delta\widehat{CV}}{\hat{\sigma}_{\Delta\widehat{CV}}} \tag{3.50}$$

$$= \frac{\widehat{CV}_1 - \widehat{CV}_2}{\sqrt{\hat{\sigma}^2_{\widehat{CV}_1} + \hat{\sigma}^2_{\widehat{CV}_2}}} \tag{3.51}$$

is approximately normal and the acceptance interval for H_0 is $-z_{\alpha/2} < z < z_{\alpha/2}$. Under $H_0 : CV_1 = CV_2$, $\hat{\sigma}_{\Delta\widehat{CV}}$ can be approximated with the pooled standard deviation

$$\hat{\sigma}_{\widehat{\Delta CV}} = \sqrt{\left(\frac{1}{n_1} + \frac{1}{n_2}\right) \widehat{CV}^2 \left(\widehat{CV}^2 + \frac{1}{2}\right)} \quad (3.52)$$

where

$$\widehat{CV} = \frac{n_1 \widehat{CV}_1 + n_2 \widehat{CV}_2}{n_1 + n_2}. \quad (3.53)$$

Under $H_A : CV_1 \neq CV_2$, the probability of rejecting H_0 is given by

$$\pi = P(-\infty < z < -z_\beta) \quad (3.54)$$

where

$$z_\beta = \frac{|CV_1 - CV_2| - z_{\alpha/2}\sigma_{CV_1}}{\sigma_{CV_2}}$$

$$= \frac{|CV_1 - CV_2| - z_{\alpha/2} CV_1 \sqrt{\frac{1}{n_1}\left(CV_1^2 + \frac{1}{2}\right)}}{CV_2 \sqrt{\frac{1}{n_2}\left(CV_2^2 + \frac{1}{2}\right)}}. \quad (3.55)$$

For a specified value of the sample size ratio n_1/n_2, the sample size n_1 required to obtain power $\pi = 1 - \beta$ is

$$n_1 = \left(\frac{z_{\alpha/2} CV_1 \sqrt{CV_1^2 + \frac{1}{2}} + z_\beta CV_2 \sqrt{\frac{n_1}{n_2}\left(CV_2^2 + \frac{1}{2}\right)}}{CV_1 - CV_2}\right)^2. \quad (3.56)$$

When the two sample sizes are equal, so $n_1/n_2 = 1$, the common sample size required is

$$n = \left(\frac{z_{\alpha/2} CV_1 \sqrt{CV_1^2 + \frac{1}{2}} + z_\beta CV_2 \sqrt{CV_2^2 + \frac{1}{2}}}{CV_1 - CV_2}\right)^2. \quad (3.57)$$

Example 3.15 Determine the sample size required to reject $H_0 : CV_1 = CV_2$ in favor of $H_A : CV_1 \neq CV_2$ with 90% power when $CV_1 = 0.3$ and $CV_2 = 0.5$.
Solution: With $CV_1 = 0.3$, $CV_2 = 0.5$, $\alpha = 0.05$, and $\beta = 0.10$ in Equation 3.57, the required sample size is

$$n = \left(\frac{1.96 \times 0.3 \sqrt{(0.3)^2 + \frac{1}{2}} + 1.282 \times 0.5 \sqrt{(0.5)^2 + \frac{1}{2}}}{0.3 - 0.5}\right)^2 = 26.$$

Chapter 4

Proportions

When a random sample of n units is drawn from a population and x of the units are judged to be successes, then the sample proportion

$$\widehat{p} = \frac{x}{n} \qquad (4.1)$$

is a point estimate for the true but unknown success proportion p of the population. The purpose of this chapter is to present methods for determining sample size and power for experiments to quantify or test the value of one or more proportions.

When a method is presented for a narrow range of p, it may be necessary to apply the method to the complement of the success probability, $p' = 1 - p$, instead of to p directly. For example, many quality engineering methods take the viewpoint of an inspector to whom a success is a defective unit, whereas the manufacturing manager might prefer to talk in terms of the process yield, which is the complement of the proportion defective.

4.1 One Proportion (Large Population)

When the population being sampled is very large compared to the sample size, the probability of obtaining x successes in n trials when the probability of a success on any trial is a constant value p is given by the binomial distribution:

$$b(x; n, p) = \binom{n}{x} p^x (1-p)^{n-x}. \qquad (4.2)$$

The binomial distribution provides the basis for constructing confidence intervals and performing hypothesis tests for proportions; however, the binomial distribution can be unwieldy and approximate methods are usually easier to use and still quite accurate. This section presents exact and approximate methods for finding sample size and power for problems involving one proportion under the binomial distribution.

4.1.1 Confidence Interval for One Proportion

Confidence interval calculations for proportions are presented for five methods: the exact binomial method, the exact F distribution method, the χ^2 approximation method, the normal approximation method, and the arcsine transformation approximation method. There is substantial overlap among the various methods, and some methods are easier to use than others in certain circumstances. Sample size calculations for some special but common cases are also presented.

4.1.1.1 Confidence Interval Formulas

Table 4.1 presents exact and approximate methods for determining one-sided upper, one-sided lower, and two-sided confidence limits for the population proportion p based on the observation of X successes in n trials. The *Exact Binomial* row gives the fundamental conditions that determine the exact confidence limits in terms of the binomial distribution. These equations are transcendental, however, so they must be solved by iteration. The *Exact F* row gives equations for the exact confidence bounds in terms of the F distribution. These are the formulas that are implemented in software.

The χ^2 and normal approximation confidence interval calculation methods in Table 4.1 are the most useful ones for determining sample size and power. The χ^2 approximation is justified for $X \ll n$ and the normal approximation is justified when both conditions $np > 5$ and $n(1-p) > 5$ are satisfied. In most cases at least one of these two approximation methods is available and provides a simple algorithm for calculating sample size and power for experiments to study one proportion. The arcsine transformation method is valid for a wider range of proportions than the normal approximation method; but, the confidence interval half-width is not expressed in terms that can be used to derive a practical sample size calculation.

4.1.1.2 One-sided Upper Confidence Interval for Small Proportions

When the population proportion p is expected to be very small and a one-sided upper confidence limit for p is required of the form

$$P(0 < p < p_U) = 1 - \alpha, \tag{4.3}$$

then the χ^2 approximation formula from Table 4.1 can be solved to obtain the following estimate for the sample size:

$$n \simeq \frac{\chi^2_{1-\alpha, 2(X+1)}}{2p_U} \tag{4.4}$$

where the χ^2 distribution has $2(X+1)$ degrees of freedom and $1-\alpha$ is the left tail area. This method requires that some value of X, the allowed number of successes in the sample, be selected before doing the sample size calculation. The choice of $X = 0$ gives the smallest sample size.

4.1. One Proportion (Large Population)

	$P(p_L < p < 1) = 1-\alpha$	$P(0 < p < p_U) = 1-\alpha$	$P(p_L < p < p_U) = 1-\alpha$
Exact Binomial	$\sum_{x=X}^{n} b(x; n, p_L) = \alpha$	$\sum_{x=0}^{X} b(x; n, p_U) = \alpha$	$\sum_{x=X}^{n} b(x; n, p_L) = \alpha/2$ $\sum_{x=0}^{X} b(x; n, p_U) = \alpha/2$
Exact F	$p_L = \dfrac{1}{1 + \frac{2(n-X+1)}{2X} F_{\alpha, 2X, 2(n-X+1)}}$	$p_U = \dfrac{1}{1 + \frac{2(n-X)}{2(X+1)} F_{1-\alpha, 2(X+1), 2(n-X)}}$	$p_L = \dfrac{1}{1 + \frac{2(n-X+1)}{2X} F_{\alpha/2, 2X, 2(n-X+1)}}$ $p_U = \dfrac{1}{1 + \frac{2(n-X)}{2(X+1)} F_{1-\alpha/2, 2(X+1), 2(n-X)}}$
χ^2 Approximation	$p_L = \dfrac{\chi^2_{\alpha, 2X}}{2n}$	$p_U = \dfrac{\chi^2_{1-\alpha, 2(X+1)}}{2n}$	$p_L = \dfrac{\chi^2_{\alpha/2, 2X}}{2n}$ $p_U = \dfrac{\chi^2_{1-\alpha/2, 2(X+1)}}{2n}$
Normal Approximation	$p_L = \widehat{p} - z_\alpha \sqrt{\dfrac{\widehat{p}(1-\widehat{p})}{n}}$	$p_U = \widehat{p} + z_\alpha \sqrt{\dfrac{\widehat{p}(1-\widehat{p})}{n}}$	$p_U/p_L = \widehat{p} \pm z_{\alpha/2} \sqrt{\dfrac{\widehat{p}(1-\widehat{p})}{n}}$
Arcsine Approximation	$p_L = 2\arcsin\left(\sqrt{\widehat{p}}\right) - \dfrac{z_\alpha}{\sqrt{n}}$	$p_U = 2\arcsin\left(\sqrt{\widehat{p}}\right) + \dfrac{z_\alpha}{\sqrt{n}}$	$p_U/p_L = 2\arcsin\left(\sqrt{\widehat{p}}\right) \pm \dfrac{z_{\alpha/2}}{\sqrt{n}}$

Table 4.1: Summary of confidence limit calculations for the population proportion.

For the special case of a 95% one-sided upper confidence interval for p, given by $P(0 < p < p_U) = 0.95$, without any successes found in the sample, so $X = 0$, the approximate sample size is given by

$$n \simeq \frac{\chi^2_{0.95,2}}{2p_U}$$

$$\simeq \frac{3}{p_U}. \qquad (4.5)$$

This result is often referred to as the *rule of three*.

The confidence interval for the success proportion p given by Equation 4.3 can be manipulated to give a confidence interval for the number of successes S in the population

$$P(0 < S \leq S_U) = 1 - \alpha \qquad (4.6)$$

where S_U is the $(1 - \alpha)\,100\%$ upper confidence limit on S. S_U and p_U are related by $p_U = S_U/N$, where N is the population size, so from Equation 4.4 the fraction of the population that must be sampled to demonstrate the confidence interval in Equation 4.6 is

$$\frac{n}{N} \simeq \frac{\chi^2_{1-\alpha,2(X+1)}}{2S_U}. \qquad (4.7)$$

Example 4.1 How large a random sample is required to demonstrate that the fraction defective of a process is less than 1% with 95% confidence?
Solution: The required confidence interval has the form

$$P(0 < p < 0.01) = 0.95$$

so $p_U = 0.01$ and $\alpha = 0.05$. If we assume that the sample size is small compared to the lot size, then Equation 4.4 can be used to approximate the sample size. However, because the number of defectives allowed in the sample was not specified, we must consider the possibility of different X values. For $X = 0$, by the rule of three (Equation 4.5), the sample size is

$$n \simeq \frac{3}{0.01}$$

$$\simeq 300.$$

For $X = 1$, by Equation 4.4

$$n \simeq \frac{\chi^2_{0.95,4}}{2(0.01)}$$

$$\simeq \frac{9.49}{2(0.01)}$$

$$\simeq 475.$$

The values of n can be found for other choices of X in a similar manner.

4.1. One Proportion (Large Population)

Example 4.2 What fraction of a large population must be inspected and found to be free of defectives to be 95% confident that the population contains no more than ten defectives?

Solution: The goal of the experiment is to demonstrate that the population defective count satisfies the confidence interval $P(0 < S \leq 10) = 0.95$. With $X = 0$ and $\alpha = 0.05$ in Equation 4.7, the fraction of the population that will need to be inspected is

$$\frac{n}{N} \simeq \frac{\chi^2_{0.95}}{2S_U} \tag{4.8}$$

$$\simeq \frac{3}{10} \tag{4.9}$$

$$\simeq 0.30.$$

This result violates the small-sample approximation requirement that $n \ll N$, but it provides a good starting point for iterations toward a more accurate result. When n becomes a substantial fraction of N, use the method shown in Section 10.4.1.2 instead. (This example is re-solved using that method in Example 10.21.)

4.1.1.3 Confidence Interval for Intermediate Proportions

When the sample size n is large and the expected population proportion p is intermediate in size, say $0.1 < p < 0.9$, then the normal approximation formulas from Table 4.1 for p_L and/or p_U may be solved to determine the sample size. Under this condition, if a two-sided confidence interval for p is required of the form

$$P(\hat{p} - \delta < p < \hat{p} + \delta) = 1 - \alpha, \tag{4.10}$$

then by comparison to the two-sided interval formulas for p_U and p_L from Table 4.1 we have

$$\delta = z_{\alpha/2}\sqrt{\frac{\hat{p}(1-\hat{p})}{n}}$$

which leads to

$$n \simeq \hat{p}(1-\hat{p})\left(\frac{z_{\alpha/2}}{\delta}\right)^2. \tag{4.11}$$

The sample size n is a maximum with respect to p when $p = \frac{1}{2}$, so when an initial estimate for p is not available $p = \frac{1}{2}$ gives a conservative sample size. See Section 4.1.1.4 for special consideration of this case.

4.1.1.4 Confidence Interval for $p \simeq \frac{1}{2}$

Often, the expected value of the proportion p is about 50%, so this condition deserves special attention. If the goal of an experiment is to determine a confidence interval for the population proportion p of the form

$$P(\hat{p} - \delta < p < \hat{p} + \delta) = 1 - \alpha \qquad (4.12)$$

where $p \simeq \frac{1}{2}$ is expected, then under the assumption that the binomial distribution of $\hat{p}$ may be approximated with the normal distribution with standard deviation $\sigma_{\hat{p}} = \sqrt{p(1-p)/n}$, we have

$$\begin{aligned} \delta &= z_{\alpha/2} \sqrt{\frac{p(1-p)}{n}} \\ &\simeq \frac{z_{\alpha/2}}{2\sqrt{n}} \end{aligned} \qquad (4.13)$$

from which we obtain the following sample size requirement:

$$n = \frac{1}{4}\left(\frac{z_{\alpha/2}}{\delta}\right)^2. \qquad (4.14)$$

For the special case of $\alpha = 0.05$, because $z_{0.025} \simeq 2$, this simplifies to

$$n \simeq \frac{1}{\delta^2}. \qquad (4.15)$$

Example 4.3 How many people should be polled to estimate voter preference for two candidates in a close election if the poll result must be within 2% of the truth with 95% confidence?

Solution: From Equation 4.15 with confidence interval half-width $\delta = 0.02$ the required sample size is

$$n = \frac{1}{(0.02)^2} = 2500.$$

4.1.1.5 Correction for Small Populations

If the sample size n calculated by a method from this section is greater than about 10% of the population size, then the sample size should be corrected for the relatively small population using

$$n' = \frac{n}{1 + \frac{n-1}{N}}. \qquad (4.16)$$

Section 4.2 presents more accurate methods for estimating proportions in small populations.

4.1.2 Tests for One Proportion

The hypotheses to be tested are $H_0 : p = p_0$ versus one of the alternatives: $H_A : p \neq p_0$, $H_A : p < p_0$, or $H_A : p > p_0$. This test may be performed using the

4.1. One Proportion (Large Population)

exact binomial distribution or using the normal approximation when the large-sample conditions are satisfied.[1] The sample size calculation by the binomial method is calculation intensive, so before appropriate software was available the approximation methods were used exclusively. Because exact and approximate methods are both still in use, they are both presented here. In the normal approximation method, the continuity correction is ignored here because it has a relatively small effect on the power and sample size.

When the units being inspected are judged to be defective or not defective based on comparing a measured response to an upper and/or lower quantitative specification limit, the variables inspection method gives much smaller sample size than the attribute inspection method for the same protection against type I and II errors. (Section 10.4.3.1 provides variables inspection sample size calculations analogous to the attributes inspection sample size calculations presented here.)

4.1.2.1 Test for One Proportion (Normal Approximation)

The two-sided test of $H_0 : p = p_0$ versus $H_A : p \neq p_0$ is performed by drawing a random sample of size n and counting the number of successes (x) in the sample. If the sample size is sufficiently large so that the binomial distribution of x may be approximated with the normal distribution with $\mu_0 = np_0$ and $\sigma_0 = \sqrt{np_0(1-p_0)}$, the decision to reject H_0 or not is based on the standard normal z statistic

$$z = \frac{x - np_0}{\sqrt{np_0(1-p_0)}} \quad (4.17)$$

where the acceptance interval for H_0 is given by

$$\Phi\left(-z_{\alpha/2} < z < z_{\alpha/2}\right) = 1 - \alpha. \quad (4.18)$$

Under H_A with $p = p_1$, the binomial distribution of x will again be approximately normal with $\mu_1 = np_1$ and $\sigma_1 = \sqrt{np_1(1-p_1)}$. Figure 4.1 shows the x distributions under H_0 and H_A. The effect size, that is, the difference between p_1 and p_0, that can be detected with specified sample size and power $\pi = 1 - \beta$ is

$$\begin{aligned}
\delta &= p_1 - p_0 \\
&= \frac{z_{\alpha/2}\sigma_0 + z_\beta \sigma_1}{n} \\
&= \frac{z_{\alpha/2}\sqrt{np_0(1-p_0)} + z_\beta \sqrt{np_1(1-p_1)}}{n} \\
&= z_{\alpha/2}\sqrt{\frac{p_0(1-p_0)}{n}} + z_\beta \sqrt{\frac{p_1(1-p_1)}{n}}. \quad (4.19)
\end{aligned}$$

---

[1] The normal approximation to the binomial distribution is valid when $np > 5$ and $n(1-p) > 5$.

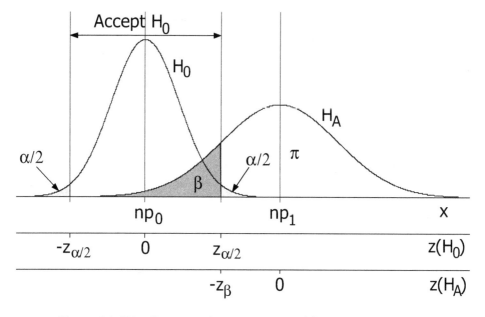

Figure 4.1: Distributions of success counts (x) under H_0 and H_A.

The power, that is, the probability of rejecting H_0 when $p = p_1$, for specified sample size and effect size is given by

$$\pi = 1 - \Phi \left(-\infty < z < z_\beta \right) \tag{4.20}$$

where

$$\begin{aligned} z_\beta &= \frac{n \left| p_1 - p_0 \right| - z_{\alpha/2} \sqrt{n p_0 \left(1 - p_0 \right)}}{\sqrt{n p_1 \left(1 - p_1 \right)}} \\ &= \frac{\sqrt{n} \left| p_1 - p_0 \right| - z_{\alpha/2} \sqrt{p_0 \left(1 - p_0 \right)}}{\sqrt{p_1 \left(1 - p_1 \right)}}. \end{aligned} \tag{4.21}$$

By solving Equation 4.21 for n for specified values of p_0, p_1, α, and β, the required sample size is

$$n = \left(\frac{z_{\alpha/2} \sqrt{p_0 \left(1 - p_0 \right)} + z_\beta \sqrt{p_1 \left(1 - p_1 \right)}}{p_1 - p_0} \right)^2. \tag{4.22}$$

When $p_0 \simeq p_1$, then Equation 4.22 may be approximated by

$$n = \bar{p} \left(1 - \bar{p} \right) \left(\frac{z_{\alpha/2} + z_\beta}{\Delta p} \right)^2 \tag{4.23}$$

where $\bar{p} = (p_0 + p_1)/2$ and $\Delta p = p_1 - p_0$.

4.1. One Proportion (Large Population)

Example 4.4 Find the power to reject $H_0 : p = 0.1$ when in fact $p = 0.2$ and the sample will be of size $n = 200$.

Solution: Under both H_0 and H_A the sample size is sufficiently large to justify the use of normal approximations to the binomial distributions. From Equation 4.21 with $\alpha = 0.05$ we have

$$z_\beta = \frac{\sqrt{200}\,|0.2 - 0.1| - z_{0.025}\sqrt{(0.1)(1-0.1)}}{\sqrt{(0.2)(1-0.2)}} = 2.385,$$

so the power is

$$\pi = 1 - \Phi(-\infty < z < 2.385)$$
$$= 0.9915.$$

Example 4.5 What sample size is required to reject $H_0 : p = 0.05$ when in fact $p = 0.10$ using a two-sided test with 90% power?

Solution: Assuming that the sample size will be sufficiently large to justify the normal approximation method, from Equation 4.22 the required sample size is

$$n = \left(\frac{1.96\sqrt{(0.05)(1-0.05)} + 1.282\sqrt{(0.10)(1-0.10)}}{0.10 - 0.05}\right)^2$$
$$= 264.$$

4.1.2.2 Test for One Proportion (Exact Binomial Method, $p > p_0$)

The one-sided test of $H_0 : p = p_0$ versus $H_A : p > p_0$ is performed by drawing a random sample of size n and counting the number of successes x in the sample. We will reject H_0 if x is greater than a critical value c which is chosen to limit the type I error rate α, where these parameters are related by

$$\sum_{x=0}^{c} b(x; n, p_0) \geq 1 - \alpha. \tag{4.24}$$

To reject H_0 with power $\pi = 1 - \beta$ when $p = p_1$ is some specified value greater than p_0, the sampling plan must meet the condition

$$\sum_{x=0}^{c} b(x; n, p_1) \leq \beta. \tag{4.25}$$

Then for specified values of $(p, P_A(H_0)) = (p_0, 1 - \alpha)$ and (p_1, β), which correspond to two points on an operating characteristic (OC) curve, there are unique values of n and c which simultaneously minimize n and satisfy both Equation 4.24 and Equation 4.25. This (n, c) pair can be found using one of the following methods:

1. By Larson's nomogram for the cumulative binomial distribution. Although obtaining (n, c) is very simple by Larson's nomogram, the values obtained can be imprecise, but they are usually accurate enough for planning experiments.

2. By the χ^2 form of the Poisson distribution when Poisson approximations to the two binomial distributions are justified. This method is described in detail in Section 10.4.1.1. It provides only an approximate sample size and under somewhat limited conditions, but it is very easy to use and is usually accurate enough for planning experiments

3. By manual iterative calculation or using an appropriate software package to find the exact solution.

Example 4.6 What sample size is required to reject $H_0 : p = 0.01$ with 90% power when in fact $p = 0.03$?

Solution: The hypotheses to be tested are $H_0 : p = 0.01$ versus $H_A : p > 0.01$ and the two points on the OC curve are $(p_0, 1 - \alpha) = (0.01, 0.95)$ and $(p_1, \beta) = (0.03, 0.10)$. The exact simultaneous solution to Equations 4.24 and 4.25, obtained using Larson's nomogram and then iterating to the exact solution using a binomial calculator, is $(n, c) = (390, 7)$. The distributions of the success counts under H_0 and H_A are shown in Figure 4.2.

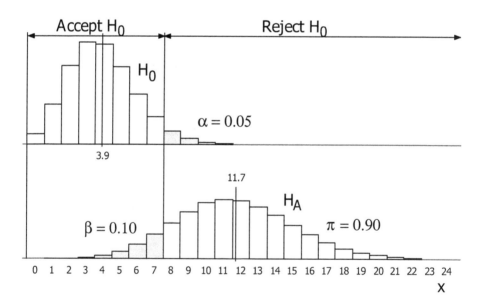

Figure 4.2: Distributions of success counts (x) under H_0 and H_A.

4.1. One Proportion (Large Population)

Example 4.7 Use Larson's nomogram to find n and c for the sampling plan for defectives that will accept 95% of lots with 2% defectives and 10% of lots with 8% defectives. Draw the OC curve.

Solution: Figure 4.3 shows the solution using Larson's nomogram with the two specified points on the OC curve at $(p, P_A(H_0)) = (0.02, 0.95)$ and $(0.08, 0.10)$. The required sampling plan is $n = 100$ and $c = 4$. The OC curve is shown in Figure 4.4. Points on the OC were obtained by rocking a line about the point at $n = 100$ and $c = 4$ in the nomogram and reading off p and P_A values.

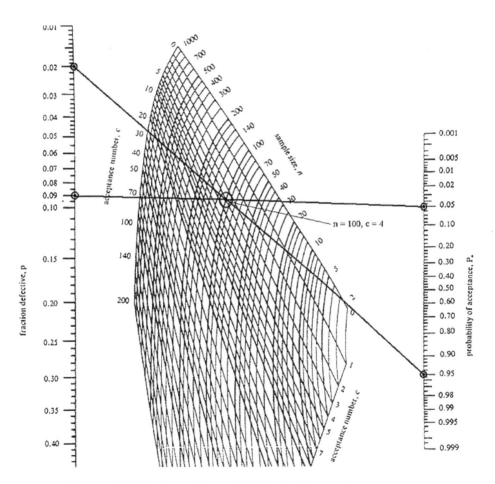

Figure 4.3: Solution to Example 4.7 using Larson's nomogram.

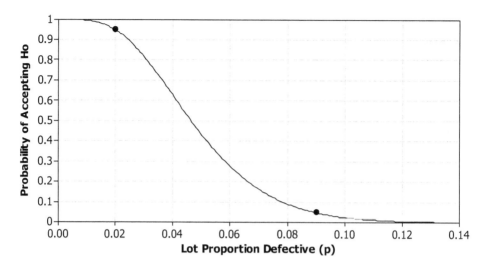

Figure 4.4: OC curve for the sampling plan with $n = 100, c = 4$.

4.1.2.3 Test for One Proportion (Exact Binomial Method, $p < p_0$)

Using arguments similar to those in the preceding section for the one-sided test of $H_0 : p = p_0$ versus $H_A : p < p_0$, we will reject H_0 if x is less than or equal to some critical value r which satisfies

$$\sum_{x=0}^{r} b(x; n, p_0) \leq \alpha. \tag{4.26}$$

Under H_A, where $p = p_1$ is some specified value less than p_0, the type II error rate must meet the condition

$$\sum_{x=0}^{r} b(x; n, p_1) \geq 1 - \beta. \tag{4.27}$$

Then, for specified values of $(p_0, 1 - \alpha)$ and (p_1, β), the simultaneous solution to Equations 4.26 and 4.27 gives the unique values of n and c.

Example 4.8 What sample size is required to reject $H_0 : p = 0.03$ with 90% power when in fact $p = 0.01$?

Solution: The hypotheses to be tested are $H_0 : p = p_0$ versus $H_A : p < p_0$ and the two points on the OC curve are $(p_0, 1 - \alpha) = (0.03, 0.95)$ and $(p_1, \beta) = (0.01, 0.10)$. The exact simultaneous solution to Equations 4.26 and 4.27, determined using Larson's nomogram followed by manual iterations with a binomial calculator, is $(n, r) = (436, 7)$.

4.2 One Proportion (Small Population)

The preceding section was dedicated to sampling from a single population that was infinite or practically infinite in size. When the population size is small or when the sample size is relatively large compared to the population size, the binomial distribution is inappropriate and the hypergeometric distribution must be used instead. This section is dedicated to the treatment of power and sample size calculations for small populations when the hypergeometric distribution is necessary. In many practical situations the hypergeometric distribution may be approximated with the binomial distribution. Because hypergeometric probability calculations can be very difficult, even with software, these approximations become very important in power and sample size calculations, even if they are used only for the first iteration of an exact solution.

The probability of obtaining x successes in n trials when observations are drawn without replacement from a finite population of N units of which S are successes is given by the hypergeometric probability distribution

$$h(x; S, N, n) = \frac{\binom{S}{x}\binom{N-S}{n-x}}{\binom{N}{n}} \qquad (4.28)$$

where the values of x are constrained by $0 \leq x \leq min(S, n)$. The proportion of successes in the population is $p = S/N$.

Another method of calculating hypergeometric probabilities from the same input parameters is

$$h(x; S, N, n) = \frac{\binom{n}{x}\binom{N-n}{S-x}}{\binom{N}{S}}. \qquad (4.29)$$

This method is not as well known as the first one, but it has practical applications that fall outside the scope of the first method.

4.2.1 Approximations to the Hypergeometric Distribution

Hypergeometric probability calculations can exceed the computational abilities of the calculator or software being used. However, there are two situations in which the hypergeometric distribution may be approximated with the binomial distribution. These approximations are called the *small-sample* and the *rare-event* approximations for the hypergeometric distribution.[2] As their names imply, the small-sample approximation is valid when the sample size is less than about 10% of the population size and the rare-event approximation is valid when the number of successes in the lot is less than about 10% of the population size. The two approximations converge when both conditions are satisfied.

---

[2]These two approximations are also referred to as the p-binomial and f-binomial approximations, respectively. These names aren't very descriptive, so they won't be used here.

The small-sample binomial approximation for the hypergeometric distribution is given by

$$h(x; S, N, n) \simeq b(x; n, p = S/N)$$
$$\simeq \binom{n}{x} \left(\frac{S}{N}\right)^x \left(1 - \frac{S}{N}\right)^{n-x}. \tag{4.30}$$

The rare-event binomial approximation for the hypergeometric distribution is given by

$$h(x; S, N, n) \simeq b(x; S, p = n/N)$$
$$\simeq \binom{S}{x} \left(\frac{n}{N}\right)^x \left(1 - \frac{n}{N}\right)^{S-x}. \tag{4.31}$$

4.2.2 Confidence Intervals for One Proportion

This section describes sample size calculations for experiments to determine confidence intervals for the number of successes or the proportion of successes in small populations. The special case of zero successes in the sample is also presented.

4.2.2.1 General Case

When X successes are obtained in n trials drawn from a population of size N containing S successes, confidence intervals may be written for the unknown value of S and for the fraction of successes in the population $p = S/N$. The one-sided upper $(1 - \alpha) \, 100\%$ confidence interval for S is given by

$$P(S \leq S_U) \geq 1 - \alpha \tag{4.32}$$

or, in terms of the population's success fraction p,

$$P\left(p \leq \frac{S_U}{N}\right) \geq 1 - \alpha \tag{4.33}$$

where S_U is the smallest value of S which satisfies

$$\sum_{x=0}^{X} h(x; S_U, N, n) \leq \alpha. \tag{4.34}$$

Analogously, for the one-sided lower $(1 - \alpha) \, 100\%$ confidence interval for S we have

$$P(S \geq S_L) \geq 1 - \alpha \tag{4.35}$$

or, in terms of the population's success fraction p,

$$P\left(p \geq \frac{S_L}{N}\right) \geq 1 - \alpha \tag{4.36}$$

4.2. One Proportion (Small Population)

where S_L is the largest value of S which satisfies

$$\sum_{x=X}^{\min(S_L,n)} h(x; S_L, N, n) \le \alpha. \tag{4.37}$$

Example 4.9 Suppose that a sample of size $n = 20$ drawn from a population of $N = 100$ units was found to have $X = 2$ defective units. Determine the one-sided upper confidence limit for the population fraction defective.
Solution: From the following hypergeometric probabilities:

$$h(0 \le x \le 2; S = 26, N = 100, n = 20) = 0.0555$$
$$h(0 \le x \le 2; S = 27, N = 100, n = 20) = 0.0448$$

the smallest value of S that satisfies the inequality in Equation 4.34 is $S = 27$, so the 95% one-sided upper confidence limit for S is $S_U = 27$ or

$$P(S \le 27) \ge 0.95.$$

Without software, sample size calculations for estimating the number of successes or the proportion of successes in a small population are easiest to perform by first estimating the sample size using one of the binomial approximations and then fine-tuning the sample size using exact hypergeometric calculations. Table 4.2 summarizes the exact and approximate formulas for one- and two-sided confidence intervals for the number of successes in the population.

Example 4.10 A hospital is asked by an auditor to confirm that its billing error rate is less than 10% for a day chosen randomly by the auditor. However, it is impractical to inspect all 120 bills issued on that day. How many of the bills must be inspected to demonstrate, with 95% confidence, that the billing error rate is less than 10%?
Solution: The goal of the analysis is to demonstrate that the one-sided upper 95% confidence limit on the billing error rate p is 10% or

$$P(p \le 0.10) = 0.95.$$

Under the assumption that the auditor will accept a zero defectives sampling plan, by the rule of three (Equation 4.5) the approximate sample size must be

$$n \simeq \frac{3}{p} = \frac{3}{0.10} = 30.$$

Because $n = 30$ is large compared to $N = 120$, the finite population correction

Interval	Exact	Binomial Approximations	
		Small-Sample ($n \ll N$)	Rare Event ($S \ll N$)
$P(0 < S < S_U) \leq 1 - \alpha$	$\sum_{x=0}^{X} h(x; S_U, N, n) \leq \alpha$	$\sum_{x=0}^{X} b(x; n, p_U) = \alpha$	$\sum_{x=0}^{X} b\left(x; S_U, \frac{n}{N}\right) \leq \alpha$
$P(S_L < S \leq n) \leq 1 - \alpha$	$\sum_{x=X}^{min(S_L, n)} h(x; S_L, N, n) \leq \alpha$	$\sum_{x=X}^{n} b(x; n, p_L) = \alpha$	$\sum_{x=X}^{min(S_L, n)} b\left(x; S_L, \frac{n}{N}\right) \leq \alpha$
$P(S_L < S < S_U) \leq 1 - \alpha$	$\sum_{x=0}^{X} h(x; S_U, N, n) \leq \alpha/2$ $\sum_{x=X}^{min(S_L, n)} h(x; S_L, N, n) \leq \alpha/2$	$\sum_{x=0}^{X} b(x; n, p_U) = \alpha/2$ $\sum_{x=X}^{n} b(x; n, p_L) = \alpha/2$	$\sum_{x=0}^{X} b\left(x; S_U, \frac{n}{N}\right) \leq \alpha/2$ $\sum_{x=X}^{min(S_L, n)} b\left(x; S_L, \frac{n}{N}\right) \leq \alpha/2$

Table 4.2: Confidence interval formulas for the hypergeometric success count parameter S.

4.2. One Proportion (Small Population)

factor (Equation 4.16) should be used and gives

$$n' = \frac{30}{1 + \frac{30-1}{120}}$$
$$= 25.$$

Iterations with a hypergeometric probability calculator show that $n = 26$ is the smallest sample size that gives 95% confidence that the billing error rate is less than 10%.

4.2.2.2 Special Case: Zero Successes

For the important special case of demonstrating Equation 4.32 with $X = 0$ successes observed in the sample, Equation 4.34 becomes

$$h(0; S_U, N, n) \leq \alpha. \quad (4.38)$$

There are two sample size approximations for such situations, one applicable when n is small compared to N and the other applicable when S is small compared to N.

When the small-sample condition is satisfied ($n/N < 0.10$), then Equation 4.38 can be approximated by

$$h(0; S_U, N, n) \simeq b\left(0; n, p = \frac{S_U}{N}\right)$$
$$\simeq \left(1 - \frac{S_U}{N}\right)^n, \quad (4.39)$$

which leads to the following sample size requirement for the $(1 - \alpha)\,100\%$ one-sided upper confidence interval for S:

$$n \geq \frac{\ln(\alpha)}{\ln\left(1 - \frac{S_U}{N}\right)}. \quad (4.40)$$

When the rare-event condition is satisfied ($S/N < 0.10$), then Equation 4.38 may be approximated by

$$h(0; S_U, N, n) \simeq b\left(0; S_U, p = \frac{n}{N}\right)$$
$$\simeq \left(1 - \frac{n}{N}\right)^{S_U}, \quad (4.41)$$

which leads to the following sample size requirement for the $(1 - \alpha)\,100\%$ one-sided upper confidence interval for S:

$$n \geq N\left(1 - \alpha^{1/S_U}\right). \quad (4.42)$$

This result is discussed again in the context of acceptance sampling in Section 10.4.1.2.

Because the validity of the approximate methods is uncertain before the sample size calculations are performed, it might be necessary to perform both of them.

Example 4.11 What sample size n must be drawn from a population of size $N = 200$ and found to be free of defectives if we need to demonstrate, with 95% confidence, that there are no more than four defectives in the population?
Solution: The goal of the experiment is to demonstrate the confidence interval

$$P(0 \leq S \leq 4) \geq 0.95$$

using a zero-successes ($X = 0$) sampling plan. By the small-sample binomial approximation with $S_U = 4$ and $\alpha = 0.05$, the required sample size by Equation 4.40 is given by

$$n = \frac{\ln(0.05)}{\ln\left(1 - \frac{4}{200}\right)} = 149,$$

which violates the small-sample assumption. By Equation 4.42, the rare-event binomial approximation gives

$$\begin{aligned} n &\geq N\left(1 - \alpha^{1/S_U}\right) \\ &\geq 200\left(1 - 0.05^{1/4}\right) \\ &\geq 106. \end{aligned}$$

This solution meets the requirements of the rare-event approximation method, but just to check this result, the corresponding exact hypergeometric probability is $h(0; 4, 200, 106) = 0.047$ which is less than $\alpha = 0.05$ as required, however, because $h(0, 4, 200, 105) = 0.049$, the sample size $n = 105$ is the exact solution to the problem.

4.2.3 Tests for One Proportion

The hypotheses to be tested are $H_0 : p = p_0$ versus one of the alternatives $H_A : p \neq p_0$, $H_A : p < p_0$, or $H_A : p > p_0$ where $p = S/N$. These hypotheses may also be written in terms of the success count $S = Np$ when the population size N is specified. See Section 10.4.1.2 for applications of these methods in acceptance sampling.

4.2.3.1 Test for One Proportion ($p > p_0$)

The one-sided test of $H_0 : p = p_0$ versus $H_A : p > p_0$ (or, in terms of the number of successes $S = nP$, $H_0 : S = S_0$ versus $H_A : S > S_0$) is performed by drawing

4.2. One Proportion (Small Population)

a random sample of size n and counting the number of successes x in the sample. We reject H_0 if x is greater than the critical value c which is chosen to limit the type I error rate α, where these parameters are related by

$$\sum_{x=0}^{c} h(x; S_0, N, n) \geq 1 - \alpha. \tag{4.43}$$

To reject H_0 with power $\pi = 1 - \beta$ when $p = p_1$ (or $S = S_1 = Np_1$) is some specified value greater than p_0 (or $S_0 = Np_0$), the sampling plan must meet the condition

$$\sum_{x=0}^{c} h(x; S_1, N, n) \leq \beta. \tag{4.44}$$

The values of n and c are easiest to find by using a binomial approximation to obtain an approximate solution and then iterating to the exact solution using a hypergeometric calculator. It may take many iterations to find the solution that minimizes the sample size.

Example 4.12 A biologist needs to test the fraction of female frogs in a single brood, but the sex of the frog tadpoles is difficult to determine. The hypotheses to be tested are $H_0 : p = 0.5$ versus $H_A : p > 0.5$ where p is the fraction of the frogs that are female. If there are $N = 212$ viable frogs in the brood, how many of them must she sample to reject H_0 with 90% power when $p = 0.65$?

Solution: The exact sample size (n) and acceptance number (c) have to be determined by iteration. The approximate sample size given by the large-sample binomial approximation method in Equation 4.22 with $p_0 = 0.5$, $\alpha = 0.05$, $p_1 = 0.65$, and $\beta = 0.10$ is

$$n = \left(\frac{1.645\sqrt{0.5(1-0.5)} + 1.282\sqrt{0.65(1-0.65)}}{0.65 - 0.5} \right)^2 = 92.$$

However, this sample size is large compared to the population size, so the finite population correction factor (Equation 4.16) must be used, which gives

$$n' = \frac{92}{1 + \frac{92-1}{212}}$$
$$= 65.$$

The exact values of n and c are determined from the simultaneous solution of Equations 4.43 and Equation 4.44 with $S_0 = Np_0 = 106$ and $S_1 = Np_1 = 138$, which gives

$$\sum_{x=0}^{c} h(x; S = 106, N = 212, n) \geq 0.95$$

$$\sum_{x=0}^{c} h(x; S = 138, N = 212, n) \leq 0.10.$$

Using a hypergeometric calculator with $n = 65$ we find

$$\sum_{x=0}^{38} h(x; S = 106, N = 212, n = 65) = 0.963$$

$$\sum_{x=0}^{38} h(x; S = 138, N = 212, n = 65) = 0.117,$$

which satisfies the $1 - \alpha \geq 0.95$ requirement but does not satisfy the $\beta \leq 0.10$ requirement. A few more iterations determine that $n = 69$ and $c = 40$ gives $\alpha = 0.039$ and $\pi = 0.912$ which meets both requirements. This means that the biologist must sample $n = 69$ frogs and can reject H_0 if $x > 40$.

4.2.3.2 Test for One Proportion ($p < p_0$)

The one-sided test of $H_0 : p = p_0$ versus $H_A : p < p_0$ or $H_0 : S = S_0$ versus $H_A : S < S_0$ is performed by drawing a random sample of size n and counting the number of successes x in the sample. We reject H_0 if x is less than or equal to the critical value c which is chosen to limit the type I error rate α, where these parameters are related by

$$\sum_{x=0}^{c} h(x; S_0, N, n) < \alpha. \tag{4.45}$$

To reject H_0 with power $\pi = 1 - \beta$ when $p = p_1$ (or $S = S_1 = Np_1$) is some specified value less than p_0 (or $S_0 = Np_0$), the sampling plan must meet the condition

$$\sum_{x=0}^{c} h(x; S_1, N, n) \geq \pi. \tag{4.46}$$

4.3 Two Proportions

Sample size and power calculations for problems involving two independent proportions are presented here for several methods: Fisher's exact test, the normal approximation method, the *arcsine* transformation method, the relative risk or risk ratio method, and the odds ratio method. A sixth method, the χ^2 test for 2×2 contingency tables, is presented in the discussion of methods for contingency tables in Section 4.5.

4.3.1 Confidence Intervals for Two Proportions

Methods are presented for calculating sample sizes for experiments to estimate the difference between two proportions, the risk ratio, and the odds ratio.

4.3.1.1 Confidence Interval for the Difference Between Two Proportions

When two independent binomial distributions with parameters p_1 and p_2 may be approximated with normal distributions, then the sampling distribution of

$$\Delta \widehat{p} = \widehat{p}_1 - \widehat{p}_2 \tag{4.47}$$

is approximately normal with mean

$$\mu_{\Delta \widehat{p}} = p_1 - p_2 \tag{4.48}$$

and standard deviation

$$\begin{aligned}\sigma_{\Delta \widehat{p}} &= \sqrt{\sigma_{\widehat{p}_1}^2 + \sigma_{\widehat{p}_2}^2} \\ &= \sqrt{\frac{p_1(1-p_1)}{n_1} + \frac{p_2(1-p_2)}{n_2}} \tag{4.49} \\ &= \sqrt{\frac{1}{n_1}\left(p_1(1-p_1) + p_2(1-p_2)\left(\frac{n_1}{n_2}\right)\right)}. \tag{4.50}\end{aligned}$$

The two-sided confidence interval for Δp has the form

$$P(\Delta \widehat{p} - \delta < \Delta p < \Delta \widehat{p} + \delta) = 1 - \alpha \tag{4.51}$$

where the confidence interval half-width is

$$\delta = z_{\alpha/2} \widehat{\sigma}_{\Delta \widehat{p}}. \tag{4.52}$$

Then the sample size n_1 required to obtain the desired confidence interval half-width δ with sample size ratio n_1/n_2 is

$$n_1 = \left(\frac{z_{\alpha/2}}{\delta}\right)^2 \left(p_1(1-p_1) + p_2(1-p_2)\left(\frac{n_1}{n_2}\right)\right). \tag{4.53}$$

If p_1 and p_2 are expected to be approximately equal so that they can both be estimated by a nominal value p, then

$$n_1 = \left(\frac{z_{\alpha/2}}{\delta}\right)^2 \left(1 + \frac{n_1}{n_2}\right) p(1-p). \tag{4.54}$$

Example 4.13 Determine the sample size required to estimate the difference between two proportions to within 0.03 with 95% confidence if both proportions are expected to be about 0.45. Assume that the two sample sizes will be equal.
Solution: From Equation 4.54 with $\delta = 0.03$, $\bar{p} = 0.45$, $n_1/n_2 = 1$, and $\alpha = 0.05$, the required sample size is

$$\begin{aligned}n_1 = n_2 &= \left(\frac{1.96}{0.03}\right)^2 (2 \times 0.45 \times (1 - 0.45)) \\ &= 2113.\end{aligned}$$

4.3.1.2 Confidence Interval for the Risk Ratio

The risk ratio, or relative risk RR, is the ratio of two independent binomial proportion parameters p_1 and p_2:

$$RR = \frac{p_1}{p_2}. \tag{4.55}$$

A confidence interval is required for the risk ratio of the form

$$P\left(\widehat{RR}(1-\delta) < RR < \widehat{RR}(1+\delta)\right) = 1 - \alpha. \tag{4.56}$$

For large samples, the distribution of $\ln\left(\widehat{RR}\right)$ is approximately normal and the confidence interval for $\ln(RR)$ has the form

$$P\left(\ln\left(\widehat{RR}(1-\delta)\right) < \ln(RR) < \ln\left(\widehat{RR}(1+\delta)\right)\right) = 1 - \alpha. \tag{4.57}$$

Because $\ln(1+\delta) \simeq \delta$ when $|\delta| \ll 1$, a relatively narrow confidence interval may be approximated by

$$P\left(\ln\left(\widehat{RR}\right) - \delta < \ln(RR) < \ln\left(\widehat{RR}\right) + \delta\right) \simeq 1 - \alpha. \tag{4.58}$$

By the delta method (see Appendix G.3.4), the approximate standard deviation of the sampling distribution of $\ln\left(\widehat{RR}\right)$ is

$$\begin{aligned}
\hat{\sigma}_{\ln(\widehat{RR})} &= \sqrt{\frac{1-p_1}{n_1 p_1} + \frac{1-p_2}{n_2 p_2}} \\
&= \sqrt{\frac{1}{n_1}\left(\frac{1-p_1}{p_1} + \frac{1-p_2}{p_2}\left(\frac{n_1}{n_2}\right)\right)},
\end{aligned} \tag{4.59}$$

so the confidence interval half-width δ is

$$\delta = z_{\alpha/2}\hat{\sigma}_{\ln(\widehat{RR})}. \tag{4.60}$$

Then the sample size n_1 required to obtain confidence interval half-width δ with specified sample size ratio n_1/n_2 is

$$n_1 = \left(\frac{z_{\alpha/2}}{\delta}\right)^2 \left(\frac{1-p_1}{p_1} + \frac{1-p_2}{p_2}\left(\frac{n_1}{n_2}\right)\right). \tag{4.61}$$

It can be shown that when the total number of units is constrained, the optimal sample size allocation ratio is

$$\frac{n_1}{n_2} = \sqrt{\frac{p_2/(1-p_2)}{p_1/(1-p_1)}} = \frac{1}{\sqrt{OR}} \tag{4.62}$$

where OR is the odds ratio.

4.3. Two Proportions

Example 4.14 An experiment is planned to estimate the risk ratio. The two proportions are expected to be $p_1 \simeq 0.2$ and $p_2 \simeq 0.05$. Determine the optimal allocation ratio and the sample size required to determine the risk ratio to within 20% of its true value with 95% confidence?

Solution: A 95% confidence interval for the risk ratio is required of the form in Equation 4.56. With $p_1 = 0.2$ and $p_2 = 0.05$, the anticipated value of the risk ratio is $RR \simeq 0.2/0.05 = 4$ and from Equation 4.62 the optimal sample size allocation ratio is

$$\frac{n_1}{n_2} = \sqrt{\frac{0.05/0.95}{0.2/0.8}} = 0.4588.$$

Then with $\delta = 0.2$ and $\alpha = 0.05$ in Equation 4.61, the required sample size n_1 is

$$n_1 = \left(\frac{1.96}{0.2}\right)^2 \left(\frac{1 - 0.2}{0.2} + \frac{1 - 0.05}{0.05}(0.4588)\right)$$
$$= 1222$$

and the sample size n_2 is

$$n_2 = \frac{n_1}{\left(\frac{n_1}{n_2}\right)} = \frac{1222}{0.4588} = 2664.$$

These sample sizes minimize the total number of samples required for the experiment.

4.3.1.3 Confidence Interval for the Odds Ratio

The odds ratio OR is given by the ratio of two independent odds:

$$OR = \frac{O_1}{O_2} = \frac{\left(\frac{p_1}{1-p_1}\right)}{\left(\frac{p_2}{1-p_2}\right)} = \frac{p_1(1-p_2)}{p_2(1-p_1)} \qquad (4.63)$$

where p_1 and p_2 are binomial proportion parameters. A confidence interval is required for the odds ratio of the form

$$P\left(\widehat{OR}(1-\delta) < OR < \widehat{OR}(1+\delta)\right) = 1 - \alpha. \qquad (4.64)$$

For large samples, the sampling distribution of $\ln\left(\widehat{OR}\right)$ is approximately normal and the confidence interval for $\ln(OR)$ has the form

$$P\left(\ln\left(\widehat{OR}(1-\delta)\right) < \ln(OR) < \ln\left(\widehat{OR}(1+\delta)\right)\right) = 1 - \alpha. \qquad (4.65)$$

When $|\delta| \ll 1$, the confidence interval may be approximated by

$$P\left(\ln\left(\widehat{OR}\right) - \delta < \ln(OR) < \ln\left(\widehat{OR}\right) + \delta\right) \simeq 1 - \alpha. \qquad (4.66)$$

By the delta method (see Appendix G.3.5), the approximate standard deviation of $\ln\left(\widehat{OR}\right)$ is

$$\begin{aligned}
\widehat{\sigma}_{\ln(\widehat{OR})} &= \sqrt{\frac{1}{n_1 p_1 (1 - p_1)} + \frac{1}{n_2 p_2 (1 - p_2)}} \\
&= \sqrt{\frac{1}{n_1}\left(\frac{1}{p_1(1-p_1)} + \frac{1}{p_2(1-p_2)}\left(\frac{n_1}{n_2}\right)\right)}, \qquad (4.67)
\end{aligned}$$

so the confidence interval half-width δ is

$$\delta = z_{\alpha/2}\widehat{\sigma}_{\ln(\widehat{OR})}. \qquad (4.68)$$

Then the sample size n_1 required to obtain confidence interval half-width δ with sample size ratio n_1/n_2 must be

$$n_1 = \left(\frac{z_{\alpha/2}}{\delta}\right)^2 \left(\frac{1}{p_1(1-p_1)} + \frac{1}{p_2(1-p_2)}\left(\frac{n_1}{n_2}\right)\right). \qquad (4.69)$$

It can be shown that when the total number of units is constrained, the optimal sample size allocation ratio is

$$\frac{n_1}{n_2} = \sqrt{\frac{p_2(1-p_2)}{p_1(1-p_1)}}. \qquad (4.70)$$

Example 4.15 An experiment is planned to estimate the odds ratio. The two proportions are expected to be $p_1 \simeq 0.5$ and $p_2 \simeq 0.25$. Determine the optimal allocation ratio and the sample size required to determine, with 90% confidence, the odds ratio to within 20% of its true value?

Solution: The desired confidence interval has the form given by Equation 4.64 with $\delta = 0.2$. With $p_1 = 0.5$ and $p_2 = 0.25$, the anticipated value of the odds ratio is $OR = \frac{0.5/0.5}{0.25/0.75} = 3$ and from Equation 4.70 the optimal sample size allocation ratio is

$$\frac{n_1}{n_2} = \sqrt{\frac{0.25 \times 0.75}{0.5 \times 0.5}} = 0.866.$$

Then with $\delta = 0.2$ and $\alpha = 0.10$ in Equation 4.69, the required sample size n_1 is

$$\begin{aligned}
n_1 &= \left(\frac{1.645}{0.2}\right)^2 \left(\frac{1}{0.5 \times 0.5} + \frac{1}{0.25 \times 0.75}(0.866)\right) \\
&= 584
\end{aligned}$$

4.3. Two Proportions

and the sample size n_2 is

$$n_2 = \frac{n_1}{\left(\frac{n_1}{n_2}\right)} = \frac{584}{0.866} = 675.$$

4.3.2 Tests for Two Proportions

Sample size and power calculations for tests for two independent proportions are presented using Fisher's exact test, the normal approximation method, the *arcsine* transformation method, the risk ratio method, and the log odds ratio method. The power calculations for McNemar's test for two related proportions is also presented.

4.3.2.1 Fisher's Exact Test

Fisher's exact test is used to test two independent populations for a difference in proportions.[3] It is usually suspected that one of the populations has a smaller proportion than the other population, so the test is most often used in its one-sided form. The hypotheses to be tested are $H_0 : p_1 = p_2$ versus $H_A : p_1 < p_2$. The experimental data consist of two random samples of size n_1 and n_2 that are inspected for successes. The number of successes found in the samples are x_1 and x_2, respectively. The decision to reject or not reject the null hypothesis is made on the basis of the p value for Fisher's test given by

$$\begin{aligned} p &= \sum_{x=0}^{x_1} h\left(x; n_1, n_2, x_1 + x_2\right) \\ &= \sum_{x=0}^{x_1} \frac{\binom{n_1}{x}\binom{n_2}{x_1+x_2-x}}{\binom{n_1+n_2}{x_1+x_2}} \end{aligned} \quad (4.71)$$

where $h(x)$ is the hypergeometric probability distribution. If the p value meets the condition $p \leq \alpha$, then H_0 is rejected.

The probabilities of obtaining a specific set of x_1 and x_2 values when samples are drawn of size n_1 and n_2, respectively, from populations with proportions p_1 and p_2, respectively, are given by the binomial probabilities

$$b\left(x_1; n_1, p_1\right) = \binom{n_1}{x_1} p_1^{x_1} (1-p_1)^{n_1-x_1} \quad (4.72)$$

and

---

[3] Fisher's exact test is used under more general circumstances than only the two-proportions test presented here, but the power calculations for those other methods are performed the same way.

		x_2								
		0	1	2	3	4	5	6	7	8
x_1	0	1	0.500	0.233	0.100	**0.038**	**0.013**	**0.003**	**0.001**	**0.000**
	1	1	0.767	0.500	0.285	0.141	0.059	**0.020**	**0.005**	**0.001**
	2	1	0.900	0.715	0.500	0.304	0.157	0.066	**0.020**	**0.003**
	3	1	0.962	0.859	0.696	0.500	0.310	0.157	0.059	**0.013**
	4	1	0.987	0.941	0.843	0.690	0.500	0.304	0.141	**0.038**
	5	1	0.997	0.980	0.934	0.843	0.696	0.500	0.100	0.100
	6	1	0.999	0.995	0.980	0.941	0.859	0.715	0.500	0.233
	7	1	1	0.999	0.997	0.987	0.962	0.900	0.767	0.500
	8	1	1	1	1	1	1	1	1	1

Table 4.3: Table of Fisher's test p values.

$$b(x_2; n_2, p_2) = \binom{n_2}{x_2} p_2^{x_2} (1 - p_2)^{n_2 - x_2}. \quad (4.73)$$

The two events are independent, so the probability of their both happening is the product of the two probabilities. Then the power of Fisher's exact test is given by

$$\pi = \sum_{x_1=0}^{n_1} \sum_{x_2=0}^{n_2} \delta(p \leq \alpha) b(x_1; n_1, p_1) b(x_2; n_2, p_2) \quad (4.74)$$

where $\delta(p \leq \alpha)$ is a Boolean function that is equal to 1 when the condition $p \leq \alpha$ is true and 0 otherwise.

Because of the complexity of Equation 4.74, the only practical ways to obtain the exact power for Fisher's test is from a published table of values or with software. When those are not available, one of the following approximate methods is usually sufficient for experiment planning purposes.

Example 4.16 presents a simple problem to demonstrate the power calculation for Fisher's test.

Example 4.16 Determine the power for Fisher's test to reject $H_0 : p_1 = p_2$ in favor of $H_A : p_1 < p_2$ when $p_1 = 0.01$, $p_2 = 0.50$, and $n_1 = n_2 = 8$.
Solution: The Fisher's test p values for all possible combinations of x_1 and x_2 were calculated using Equation 4.71 and are shown in Table 4.3. The few cases that are statistically significant, where $p \leq 0.05$, are shown in a bold font in the upper right corner of the table. Table 4.4 shows the contributions to the power given by the product of the two binomial probabilities in Equation 4.74. The sum of the individual contributions, that is, the power of Fisher's test, is $\pi = 0.69$.

4.3. Two Proportions

		x_2								
		0	1	2	3	4	5	6	7	8
x_1	0					0.252	0.202	0.101	0.029	0.004
	1							0.008	0.002	0.000
	2								0.000	0.000
	3									0.000
	4									0.000
	5									
	6									
	7									
	8									

Table 4.4: Contributions to the power for Fisher's test.

4.3.2.2 Test for a Difference Between Two Proportions (Normal Approximation)

The hypotheses to be tested are $H_0 : p_1 = p_2$ versus $H_A : p_1 \neq p_2$. When the two binomial distributions may be approximated with the normal distribution, the two-sample test for proportions may be performed using the standard normal test statistic

$$z = \frac{\Delta \widehat{p}}{\sqrt{\widehat{p}(1-\widehat{p})\left(\frac{1}{n_1}+\frac{1}{n_2}\right)}} \tag{4.75}$$

where $\widehat{p}_1$ and $\widehat{p}_2$ are the two sample proportions determined from samples of size n_1 and n_2, respectively, $\Delta \widehat{p} = |\widehat{p}_1 - \widehat{p}_2|$, and the grand mean proportion is

$$\widehat{p} = \frac{x_1 + x_2}{n_1 + n_2} = \frac{n_1 \widehat{p}_1 + n_2 \widehat{p}_2}{n_1 + n_2}. \tag{4.76}$$

For power and sample size calculations, if there is no prior knowledge about the difference between p_1 and p_2, then the optimal allocation of observations is $n = n_1 = n_2$, so the z statistic simplifies to

$$z = \frac{\Delta \widehat{p}}{\sqrt{\frac{2\widehat{p}(1-\widehat{p})}{n}}} \tag{4.77}$$

where $\widehat{p} = \frac{1}{2}(\widehat{p}_1 + \widehat{p}_2)$. Then the power of the test is

$$\pi = \Phi(-\infty < z < z_\beta) \tag{4.78}$$

where

$$z_\beta = \frac{\Delta \widehat{p}}{\sqrt{\frac{2\widehat{p}(1-\widehat{p})}{n}}} - z_{\alpha/2}. \tag{4.79}$$

When the sample size is required, this equation may be solved for n to obtain

$$n = \frac{2\hat{p}(1-\hat{p})}{(\Delta\hat{p})^2}\left(z_\beta + z_{\alpha/2}\right)^2. \tag{4.80}$$

Example 4.17 Determine the power for the test of $H_0 : p_1 = p_2$ versus $H_A : p_1 \neq p_2$ when $n_1 = n_2 = 200$, $p_1 = 0.10$, and $p_2 = 0.20$. Use a two-tailed test with $\alpha = 0.05$.

Solution: The normal approximation to the binomial distribution is justified for both samples, so with $\hat{p} = 0.15$ and $\Delta\hat{p} = 0.10$ in Equations 4.78 and 4.79, the power is

$$\pi = \Phi\left(-\infty < z < \frac{0.10}{\sqrt{\frac{2(0.15)(1-0.15)}{200}}} - 1.96\right)$$
$$= \Phi(-\infty < z < 0.84)$$
$$= 0.80.$$

Example 4.18 What common sample size is required to resolve the difference between two proportions with 90% power using a two-sided test when $p_1 = 0.10$ and $p_2 = 0.20$ is expected?

Solution: From Equation 4.80 with $\hat{p} = 0.15$ and $\Delta\hat{p} = 0.10$ the required sample size is

$$n = \frac{2 \times 0.15 \times 0.85}{(0.10)^2}(1.28 + 1.96)^2$$
$$= 268.$$

4.3.2.3 Test for Two Proportions (Arcsine Transform Method)

The normal approximation method presented in Section 4.3.2.2 suffers slightly because the binomial distribution of $\hat{p}$ is skewed. The *arcsine*-transformed proportion p' given by

$$p' = 2arcsin\left(\sqrt{p}\right) \tag{4.81}$$

where the sampling distribution of p', expressed in radians, is more closely normal except near the ends of the p interval $[0, 1]$. There are variations on the form of this transformation that are more effective when the sample size for estimating $\hat{p}_i$ is small [70]. However, all of these forms approach Equation 4.81 in the large-sample limit. For most situations with moderate p, Equation 4.81 is accurate enough for approximating sample sizes and power.

Another advantage of the *arcsine* transformation is that while the variance of the $\hat{p}$ distribution depends on p and n, the variance of $\hat{p}' = 2arcsin\left(\sqrt{\hat{p}}\right)$

4.3. Two Proportions

is approximately constant with respect to p. The approximate variance of $\widehat{p}'$, estimated by the delta method (see Appendix G.3.1), is

$$\widehat{\sigma}^2_{\widehat{p}'} = \frac{1}{n}. \tag{4.82}$$

Consequently, the test of $H_0 : p_1 = p_2$ versus $H_A : p_1 \neq p_2$ may be performed using the approximately normal statistic

$$z = \frac{\widehat{p}'_1 - \widehat{p}'_2}{\sqrt{\widehat{\sigma}^2_{\widehat{p}'_1} + \widehat{\sigma}^2_{\widehat{p}'_2}}} = \frac{2arcsin\left(\sqrt{\widehat{p}_1}\right) - 2arcsin\left(\sqrt{\widehat{p}_2}\right)}{\sqrt{\frac{1}{n_1} + \frac{1}{n_2}}}. \tag{4.83}$$

Then, for specified values of p_1, p_2, n_1, and n_2, the power of the test is

$$\pi = \Phi\left(-\infty < z < z_\beta\right) \tag{4.84}$$

where

$$z_\beta = \frac{\left|2arcsin\left(\sqrt{p_1}\right) - 2arcsin\left(\sqrt{p_2}\right)\right|}{\sqrt{\frac{1}{n_1} + \frac{1}{n_2}}} - z_{\alpha/2}. \tag{4.85}$$

In the equal-sample-size case ($n = n_1 = n_2$), the sample size required to obtain specified power $\pi = 1 - \beta$ is

$$n = \frac{1}{2}\left(\frac{z_\beta + z_{\alpha/2}}{arcsin\left(\sqrt{p_1}\right) - arcsin\left(\sqrt{p_2}\right)}\right)^2. \tag{4.86}$$

Example 4.19 Repeat the calculation of the sample size for Example 4.18.
Solution: From Equation 4.86 the required sample size is

$$n = \frac{1}{2}\left(\frac{1.28 + 1.96}{arcsin\sqrt{0.10} - arcsin\sqrt{0.20}}\right)^2 = 261,$$

which is in excellent agreement with the sample size determined by the normal approximation method.

4.3.2.4 Test for Two Proportions (Risk Ratio Method)

In terms of the risk ratio $RR = p_1/p_2$, the hypotheses $H_0 : p_1 = p_2$ versus $H_A : p_1 \neq p_2$ transform to $H_0 : RR = 1$ versus $H_A : RR \neq 1$. For large sample size, the sampling distribution of $\ln\left(\widehat{RR}\right)$ is approximately normal with approximate standard deviation given by the delta method (see Appendix G.3.4):

$$\widehat{\sigma}_{\ln(\widehat{RR})} = \sqrt{\frac{1 - p_1}{n_1 p_1} + \frac{1 - p_2}{n_2 p_2}}$$

$$= \sqrt{\frac{1}{n_1}\left(\frac{1 - p_1}{p_1} + \frac{1 - p_2}{p_2}\left(\frac{n_1}{n_2}\right)\right)}. \tag{4.87}$$

Then the power to reject H_0 for a specified value of RR under H_A is given by

$$\pi = \Phi\left(-\infty < z < z_\beta\right) \qquad (4.88)$$

where

$$\begin{aligned}
z_\beta &= \frac{|\ln(RR)|}{\widehat{\sigma}_{\ln(\widehat{RR})}} - z_{\alpha/2} \\
&= \frac{|\ln(RR)|}{\sqrt{\frac{1}{n_1}\left(\frac{1-p_1}{p_1} + \frac{1-p_2}{p_2}\left(\frac{n_1}{n_2}\right)\right)}} - z_{\alpha/2}.
\end{aligned} \qquad (4.89)$$

Then the sample sizes required to reject H_0 with specified power and sample size ratio are given by

$$n_1 = \left(\frac{z_{\alpha/2} + z_\beta}{\ln(RR)}\right)^2 \left(\frac{1-p_1}{p_1} + \frac{1-p_2}{p_2}\left(\frac{n_1}{n_2}\right)\right) \qquad (4.90)$$

and

$$n_2 = \frac{n_1}{\left(\frac{n_1}{n_2}\right)}. \qquad (4.91)$$

The optimal sample size allocation is

$$\frac{n_1}{n_2} = \sqrt{\frac{p_2/(1-p_2)}{p_1/(1-p_1)}} = \frac{1}{\sqrt{OR}} \qquad (4.92)$$

where OR is the odds ratio.

Example 4.20 Repeat the calculation of the sample size for Example 4.18 using the log risk ratio method.
Solution: With $RR = 0.1/0.2 = 0.5$, $\alpha = 0.05$, $\beta = 0.10$, and $n_1/n_2 = 1$ in Equation 4.90, the required sample size is

$$n_1 = n_2 = \left(\frac{1.96 + 1.282}{\ln(0.5)}\right)^2 \left(\frac{1-0.1}{0.1} + \frac{1-0.2}{0.2}\right) = 285,$$

which is in excellent agreement with the sample size obtained by the normal approximation method.

4.3.2.5 Test for Two Proportions (Log Odds Ratio Method)

In terms of the odds ratio OR given by

$$OR = \frac{p_1/(1-p_1)}{p_2/(1-p_2)}, \qquad (4.93)$$

4.3. Two Proportions

the hypotheses $H_0 : p_1 = p_2$ versus $H_A : p_1 \neq p_2$ transform to $H_0 : OR = 1$ versus $H_A : OR \neq 1$. For large sample size, the sampling distribution of $\ln(OR)$ is approximately normal with approximate standard deviation given by the delta method (see Appendix G.3.5):

$$\sigma_{\ln(\widehat{OR})} = \sqrt{\frac{1}{n_1 p_1 (1 - p_1)} + \frac{1}{n_2 p_2 (1 - p_2)}}$$

$$= \sqrt{\frac{1}{n_1} \left(\frac{1}{p_1 (1 - p_1)} + \frac{1}{p_2 (1 - p_2)} \left(\frac{n_1}{n_2} \right) \right)}. \qquad (4.94)$$

Then the power to reject H_0 for a specified value of OR under H_A is given by

$$\pi = \Phi(-\infty < z < z_\beta) \qquad (4.95)$$

where

$$z_\beta = \frac{|\ln(OR)|}{\widehat{\sigma}_{\ln(\widehat{OR})}} - z_{\alpha/2}$$

$$= \frac{|\ln(OR)|}{\sqrt{\frac{1}{n_1} \left(\frac{1}{p_1(1-p_1)} + \frac{1}{p_2(1-p_2)} \left(\frac{n_1}{n_2} \right) \right)}} - z_{\alpha/2}. \qquad (4.96)$$

The sample sizes required to reject H_0 with specified power and sample size ratio are

$$n_1 = \left(\frac{z_{\alpha/2} + z_\beta}{\ln(OR)} \right)^2 \left(\frac{1}{p_1(1-p_1)} + \frac{1}{p_2(1-p_2)} \left(\frac{n_1}{n_2} \right) \right) \qquad (4.97)$$

and

$$n_2 = \frac{n_1}{\left(\frac{n_1}{n_2} \right)}. \qquad (4.98)$$

The optimal sample size allocation is

$$\frac{n_1}{n_2} = \sqrt{\frac{p_2(1-p_2)}{p_1(1-p_1)}}. \qquad (4.99)$$

Example 4.21 Repeat the calculation of the sample size for Example 4.18 using the log odds ratio method.

Solution: With $n_1/n_2 = 1$ in Equation 4.97, the required common sample size is

$$n = \left(\frac{1.96 + 1.282}{\ln\left(\frac{0.10/0.90}{0.20/0.80} \right)} \right)^2 \left(\frac{1}{0.10(0.90)} + \frac{1}{0.20(0.80)} \right) = 278,$$

which is in excellent agreement with the sample size obtained by the normal approximation method.

		$X = 2$	
		$Y = 1$	$Y = 2$
$X = 1$	$Y = 1$	f_{11}	f_{12}
	$Y = 2$	f_{21}	f_{22}

Table 4.5: Frequencies in McNemar's test table.

4.3.2.6 McNemar's Test for Correlated Proportions

McNemar's test is used to analyze a dichotomous response (Y) as a function of a dichotomous predictor (X) using subjects that are observed under both levels of the predictor. The experimental data can be summarized in a two-by-two table of frequencies as shown in Table 4.5.

The McNemar's test hypotheses can be stated H_0: *The response is the same for the two levels of the predictor* versus H_A: *The response is different for the two levels of the predictor*. In terms of the risk ratio, $RR = p_{12}/p_{21}$, the hypotheses are $H_0 : RR = 1$ versus $H_A : RR \neq 1$, where p_{ij} is the population proportion associated with table cell ij.

There are three ways to determine the significance of McNemar's test. Under H_0, the test statistic given by

$$z_M = \frac{|f_{12} - f_{21}|}{\sqrt{f_{12} + f_{21}}} \qquad (4.100)$$

is approximately standard normal, so the acceptance region for the null hypothesis is $-z_{\alpha/2} < z_M < z_{\alpha/2}$. An equivalent form of the McNemar's test statistic is given by $\chi^2 = z_M^2$, where the χ^2 distribution has $df = 1$ degree of freedom and the acceptance region for H_0 is $P\left(0 \leq \chi^2 \leq \chi^2_{1-\alpha}\right) = 1 - \alpha$. A third method for determining the significance of McNemar's test is to calculate the p value of the test using the nonparametric sign test method

$$p = 2 \sum_{x=0}^{min(f_{12},f_{21})} b\left(x; n = f_{12} + f_{21}, p = \frac{1}{2}\right), \qquad (4.101)$$

where $min(f_{12}, f_{21})$ is a function that returns the smallest value in the list of arguments and $b(x; n, p)$ is the binomial distribution. These three methods give identical or similar results.

Equation 4.101 shows that the McNemar's test p value depends only on those subjects that change response states; however, for the purpose of power and sample size calculations, it is also necessary to consider the rate that these events occur in the population of test subjects. This rate, called the *rate of discordant observations*, is usually estimated from a preliminary study using

$$\widehat{p}_D = \frac{f_{12} + f_{21}}{\sum_i \sum_j f_{ij}}. \qquad (4.102)$$

4.3. Two Proportions

The total number of subjects required for a study to be analyzed by McNemar's test is determined by finding the number of discordant observations required to detect a specified difference between p_{12} and p_{21} with desired power, and then increasing that number by a factor of $1/p_D$.

The power of the two-sided McNemar's test for a specified risk ratio under H_A is

$$\pi = P\left(-\infty < z < z_\beta\right) \qquad (4.103)$$

where

$$z_\beta = \widehat{z}_M - z_{\alpha/2}. \qquad (4.104)$$

For specified values of $\pi = 1 - \beta$, RR, p_D, and α, this system of equations can be solved to estimate the total number of subjects required for the study:

$$\sum_i \sum_j \widehat{f}_{ij} \simeq \frac{(z_{\alpha/2} + z_\beta)^2}{p_D} \left(\frac{RR+1}{RR-1}\right)^2. \qquad (4.105)$$

Example 4.22 Determine the number of subjects required for McNemar's test to reject $H_0 : RR = 1$ with 90% power when $RR = 2$ and the rate of discordant observations is estimated to be $p_D = 0.2$ from a preliminary study.

Solution: With $\beta = 0.10$ and $\alpha = 0.05$ in Equation 4.105, the approximate number of subjects required for the study is

$$\sum_i \sum_j \widehat{f}_{ij} \simeq \frac{(1.282 + 1.96)^2}{0.20} \left(\frac{2+1}{2-1}\right)^2$$

$$\simeq 473.$$

Example 4.23 Determine the McNemar's test power to reject $H_0 : RR = 1$ in favor of $H_A : RR \neq 1$ for a study with 200 subjects when in fact $RR = 3$ using $p_D = 0.3$.

Solution: With 200 subjects in the study, the expected number of discordant pairs is

$$\widehat{f}_{12} + \widehat{f}_{21} = p_D \sum_i \sum_j \widehat{f}_{ij} = 0.3 \times 200 = 60.$$

Under H_A with $RR = 3$, we have $\widehat{f}_{21} = 15$ and $\widehat{f}_{12} = 45$, so the expected value of the McNemar's z_M statistic is

$$z_M = \frac{|45 - 15|}{\sqrt{45 + 15}} = 3.87$$

and the approximate power is

$$\begin{aligned}
\pi &= P(-\infty < z < z_\beta) \\
&= P(-\infty < z < (z_M - z_{\alpha/2})) \\
&= P(-\infty < z < (3.87 - 1.96)) \\
&= P(-\infty < z < 1.91) \\
&= 0.972.
\end{aligned}$$

4.4 Equivalence Tests

Power and sample size calculations are presented for one- and two-sample equivalence tests for proportions. The purpose of a one-sample equivalence test is to demonstrate that a population proportion is practically equivalent to a specified value. The purpose of a two-sample equivalence test is to demonstrate that two independent population proportions are practically equivalent to each other.

4.4.1 Equivalence Test for One Proportion

The hypotheses to be tested are $H_{01} : p < p_1$ or $H_{02} : p > p_2$ versus $H_A : p_1 < p < p_2$ where p is the binomial proportion and p_1 and p_2 are chosen to be so close together that they are practically equivalent to each other.

Figure 4.5a shows the distributions of binomial counts x when $p = p_1$ and $p = p_2$. For a given sample size n, the rejection region for H_0 is given by $x_1 \leq x \leq x_2$ where x_1 and x_2 are the smallest and largest values, respectively, that satisfy the following conditions to control the type I error rate:

$$\sum_{x=0}^{x_1-1} b(x; n, p_1) > 1 - \alpha \qquad (4.106)$$

and

$$\sum_{x=0}^{x_2} b(x; n, p_2) < \alpha. \qquad (4.107)$$

Then for a specified value of p under H_A, as in Figure 4.5b, the power of the test is given by

$$\pi = \sum_{x_1}^{x_2} b(x; n, p). \qquad (4.108)$$

When $n, p_1, p_2,$ and p are chosen so that all three x distributions under H_0 and H_A are approximately normal, H_0 is rejected if both of the following conditions are satisfied:

$$\frac{\widehat{p} - p_1}{\sqrt{\frac{\widehat{p}(1-\widehat{p})}{n}}} > z_\alpha \qquad (4.109)$$

4.4. Equivalence Tests

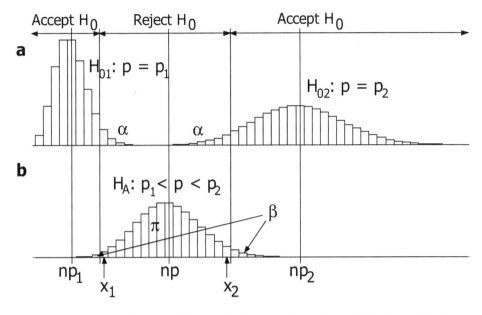

Figure 4.5: Binomial count (x) distributions under H_0 and H_A for equivalence test for one proportion.

and
$$\frac{\widehat{p} - p_2}{\sqrt{\frac{\widehat{p}(1-\widehat{p})}{n}}} < -z_\alpha. \tag{4.110}$$

Then the approximate power for a specified value of p is given by

$$\pi = \Phi\left(\frac{p_1 - p}{\sqrt{\frac{p(1-p)}{n}}} + z_\alpha < z < \frac{p_2 - p}{\sqrt{\frac{p(1-p)}{n}}} - z_\alpha\right). \tag{4.111}$$

Equation 4.111 can be solved to estimate the sample size. If p is chosen to be midway between p_1 and p_2, that is, $p = (p_1 + p_2)/2$, then the sample size to obtain power $\pi = 1 - \beta$ is given by

$$n \simeq (z_\alpha + z_{\beta/2})^2 \frac{(p_1 + p_2)(2 - p_1 - p_2)}{(p_2 - p_1)^2}. \tag{4.112}$$

If p is chosen to be closer to p_1 or p_2 so that the power is substantially determined by one tail of the normal distribution, then if p_i is the closer of p_1 and p_2 to p, the sample size is given by

$$n \simeq (z_\alpha + z_\beta)^2 \frac{p(1-p)}{(p_i - p)^2}. \tag{4.113}$$

Example 4.24 For the test of $H_0 : p < 0.45$ or $p > 0.55$ versus $H_A : 0.45 < p < 0.55$, calculate the exact and approximate power when $p = 0.5$ assuming that the sample size is $n = 800$ and $\alpha = 0.05$.
Solution: The value of x_1, determined from Equation 4.106, is $x_1 = 384$ because

$$\sum_{x=0}^{383} b(x; n = 800, p_1 = 0.45) = 0.952.$$

The value of x_2, determined from Equation 4.107, is $x_2 = 416$ because

$$\sum_{x=0}^{416} b(x; n = 800, p_2 = 0.55) = 0.048.$$

Then the power when $p = 0.5$ is given by Equation 4.108:

$$\begin{aligned} \pi &= \sum_{x=383}^{416} b(x; n = 800, p_2 = 0.50) \\ &= 0.757. \end{aligned}$$

The approximate power by the normal approximation method, given by Equation 4.111, is

$$\begin{aligned} \pi &= \Phi\left(\frac{0.45 - 0.5}{\sqrt{\frac{0.5(1-0.5)}{800}}} + 1.645 < z < \frac{0.55 - 0.5}{\sqrt{\frac{0.5(1-0.5)}{800}}} - 1.645 \right) \\ &= \Phi(-1.183 < z < 1.183) \\ &= 0.763, \end{aligned}$$

which is in good agreement with the exact solution.

4.4.2 Equivalence Test for Two Independent Proportions

The calculation of power and sample size for the equivalence test of two proportions by the exact binomial method is not practical without software, so the normal approximation method is the only one presented here.

The hypotheses to be tested are $H_0 : \Delta p < -\delta$ or $\Delta p > \delta$ versus $H_A : -\delta < \Delta p < \delta$ where $\Delta p = p_1 - p_2$ is the difference between the two binomial proportions p_1 and p_2, $\delta > 0$, and δ is chosen to be so small that p_1 and p_2 are considered to be practically equivalent to each other under H_A.

The sampling distribution of Δp is approximately normal with mean

$$\mu_{\Delta \hat{p}} = \Delta p$$

4.4. Equivalence Tests

and standard deviation

$$\sigma_{\Delta\hat{p}} = \sqrt{\frac{p_1(1-p_1)}{n_1} + \frac{p_2(1-p_2)}{n_2}}. \tag{4.114}$$

H_0 is rejected if both of the following conditions are satisfied:

$$\frac{\Delta\hat{p} + \delta}{\sigma_{\Delta\hat{p}}} > z_\alpha \tag{4.115}$$

and

$$\frac{\Delta\hat{p} - \delta}{\sigma_{\Delta\hat{p}}} < -z_\alpha. \tag{4.116}$$

Then the approximate power for specified values of p_1, p_2, δ, n, and α is given by

$$\pi = \Phi\left(\frac{-\delta + \Delta p}{\sigma_{\Delta\hat{p}}} + z_\alpha < z < \frac{\delta - \Delta p}{\sigma_{\Delta\hat{p}}} - z_\alpha\right). \tag{4.117}$$

Equation 4.117 can be solved to estimate the sample size. For $\Delta p = 0$ and specified sample size ratio n_1/n_2, the sample size n_1 to obtain power $\pi = 1 - \beta$ is given by

$$n_1 \simeq (z_\alpha + z_{\beta/2})^2 \left(1 + \frac{n_1}{n_2}\right) \frac{p(1-p)}{\delta^2} \tag{4.118}$$

where $p = p_1 = p_2$. Because equivalence is expected, the optimal sample size allocation is $n_1 = n_2$.

For $\Delta p \neq 0$ such that the power is substantially determined by one tail of the normal distribution, the sample size n_1 to obtain power $\pi = 1 - \beta$ is given by

$$n_1 \simeq (z_\alpha + z_\beta)^2 \frac{p_1(1-p_1) + \frac{n_1}{n_2}p_2(1-p_2)}{(\delta - |\Delta p|)^2}. \tag{4.119}$$

Example 4.25 An experiment is to be performed to test the hypotheses $H_0: p_1 \neq p_2$ versus $H_A: p_1 = p_2$. The two proportions are expected to be $p \simeq 0.12$ and the limit of practical equivalence is $\delta = 0.02$. What sample size is required to reject H_0 when $p_1 = p_2$ with 80% power?
Solution: With $\alpha = 0.05$ and $n_1/n_2 = 1$ in Equation 4.118, the sample size $n = n_1 = n_2$ is

$$\begin{aligned}
n &= 2(z_{0.05} + z_{0.10})^2 \frac{0.12(1 - 0.12)}{(0.02)^2} \\
&= 2(1.645 + 1.282)^2 \frac{0.12(1 - 0.12)}{(0.02)^2} \\
&= 4524.
\end{aligned}$$

Example 4.26 What sample size is required if the true difference between the two proportions in Example 4.25 is $\Delta p = 0.01$?

Solution: p_1 and p_2 are not specified, but they are both approximately $p = 0.12$, so from Equation 4.119 the sample size must be

$$n_1 \simeq 2(z_\alpha + z_\beta)^2 \frac{p(1-p)}{(\delta - |\Delta p|)^2}$$

$$\simeq 2(1.645 + 0.842)^2 \frac{0.12(1-0.12)}{(0.02 - 0.01)^2}$$

$$\simeq 13063.$$

4.5 Chi-square Tests

Power calculations for one- and two-dimensional tables by the χ^2 test method are presented. Two-dimensional tables appear in problems of association involving two responses or one response and one predictor. One-dimensional tables appear in distribution goodness-of-fit problems.

4.5.1 General Case

For a one- or two-dimensional table of counts with a total of k cells, where the observed cell frequencies are f_i and the expected cell frequencies under H_0 are $\hat{f}_i$, the χ^2 statistic to test for agreement between the observed and expected frequencies (H_0) versus disagreement between the observed and expected frequencies (H_A) is

$$\chi^2 = \sum_{i=1}^{k} \frac{\left(f_i - \hat{f}_i\right)^2}{\hat{f}_i}. \qquad (4.120)$$

The decision to reject H_0 or not is made by comparing the χ^2 statistic to the critical value $\chi^2_{1-\alpha}$ with appropriate degrees of freedom for the situation.

To determine the power associated with a χ^2 test, it is necessary to specify expected proportions for all cells of the table under H_0 and H_A. If these proportions are $(p_i)_0$ and $(p_i)_A$, respectively, where in both cases $\sum p_i = 1$, then the χ^2 distribution noncentrality parameter is

$$\phi = n \sum_{i=1}^{k} \frac{\left((p_i)_A - (p_i)_0\right)^2}{(p_i)_0} \qquad (4.121)$$

where n is the number of observations in the experiment and the power of the χ^2 test (π) is determined by the condition

$$\chi^2_{1-\alpha} = \chi^2_{1-\pi,\phi}. \qquad (4.122)$$

4.5. Chi-square Tests

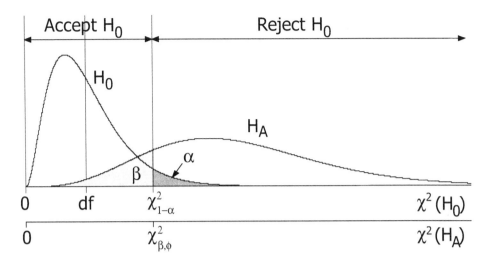

Figure 4.6: Distributions of χ^2 under H_0 and H_A.

The distributions of χ^2 under H_0 and H_A are shown in Figure 4.6.

If the sample size must be calculated to meet a specified power requirement, it must be determined by iterating over values of n until the desired power is obtained.

4.5.2 Test for Independence in a Two-way Contingency Table

There are several ways of phrasing the hypotheses to be tested for contingency tables; however, all of them are performed using the same χ^2 test method.

After the experimental data have been collected and the observed frequencies are presented in tabular form, the table row and column sums or *marginals* are calculated. If f_{ij} is the observed frequency in the ith of r rows and jth of c columns, then the corresponding expected frequency for that cell is given by

$$\widehat{f}_{ij} = \frac{\sum_{j=1}^{c} f_{ij} \sum_{i=1}^{r} f_{ij}}{\sum_{i=1}^{r}\sum_{j=1}^{c} f_{ij}} \tag{4.123}$$

where the two sums in the numerator are the row and column marginals, respectively. The χ^2 test statistic is calculated using Equation 4.120 and is compared to $\chi^2_{1-\alpha}$ with $df = (r-1)(c-1)$ degrees of freedom. The power of the test is calculated using the method of Section 4.5.1.

For the special case of 2×2 tables, especially when some of the $\widehat{f}_{ij}$ are small, Yates' continuity correction should be applied to Equation 4.120. This case and method is just an approximation to Fisher's exact test for which exact power calculations may be performed as described in Section 4.3.2.1.

Example 4.27 Confirm the sample size for Example 4.18 using the χ^2 test method for a 2 × 2 table.

Solution: Under H_A with $p_1 = 0.10$ and $p_2 = 0.20$ the expected proportion of observations in each cell of the 2 × 2 table is

$$(p_{ij})_A = \frac{1}{2}\left\{\begin{array}{cc} 0.1 & 0.9 \\ 0.2 & 0.8 \end{array}\right\} = \left\{\begin{array}{cc} 0.05 & 0.45 \\ 0.1 & 0.4 \end{array}\right\}.$$

Under H_0 with $p_1 = p_2 = (0.1 + 0.2)/2 = 0.15$ the expected distribution of observations is

$$(p_{ij})_0 = \frac{1}{2}\left\{\begin{array}{cc} 0.15 & 0.85 \\ 0.15 & 0.85 \end{array}\right\} = \left\{\begin{array}{cc} 0.075 & 0.425 \\ 0.075 & 0.425 \end{array}\right\}.$$

From Equation 4.121 with a total of $2 \times 268 = 536$ observations the noncentrality parameter is

$$\phi = 536\left(\frac{(0.05 - 0.075)^2}{0.075} + \frac{(0.1 - 0.075)^2}{0.075} + \frac{(0.45 - 0.425)^2}{0.425}\right.$$
$$\left. + \frac{(0.4 - 0.425)^2}{0.425}\right)$$
$$= 10.51.$$

Then, with $\alpha = 0.05$ and $df = 1$ degree of freedom in Equation 4.122,

$$\chi^2_{0.95} = 3.8415 = \chi^2_{\beta, 10.51}$$

which is satisfied by $\beta = 0.10$, so the power is $\pi = 1 - \beta = 0.90$ and is consistent with the original example problem solution.

Example 4.28 A large school district intends to perform pass/fail testing of students from four large schools to test for performance differences among schools. If 50 students are chosen randomly from each school, what is the power of the χ^2 test to reject the null hypothesis of homogeneity when the student failure rates at the four schools are in fact 10%, 10%, 10%, and 30%?

Solution: To calculate the power of the χ^2 test we must specify the two 2 × 4 tables (result by school) associated with $(p_{ij})_0$ and $(p_{ij})_A$. From the problem

4.5. Chi-square Tests

statement, under H_A with $(p_{1j})_A = \{0.1, 0.1, 0.1, 0.3\}$, the table of $(p_{ij})_A$ is

$$(p_{ij})_A = \frac{1}{4}\left\{\begin{array}{cccc} 0.1 & 0.1 & 0.1 & 0.3 \\ 0.9 & 0.9 & 0.9 & 0.7 \end{array}\right\}$$

$$= \left\{\begin{array}{cccc} 0.025 & 0.025 & 0.025 & 0.075 \\ 0.225 & 0.225 & 0.225 & 0.175 \end{array}\right\}.$$

The mean failure rate of all four schools is $(3(0.1) + 0.3)/4 = 0.15$ under H_0, so the corresponding table of $(p_{ij})_0$ is

$$(p_{ij})_0 = \frac{1}{4}\left\{\begin{array}{cccc} 0.15 & 0.15 & 0.15 & 0.15 \\ 0.85 & 0.85 & 0.85 & 0.85 \end{array}\right\}$$

$$= \left\{\begin{array}{cccc} 0.0375 & 0.0375 & 0.0375 & 0.0375 \\ 0.2125 & 0.2125 & 0.2125 & 0.2125 \end{array}\right\}.$$

Under these definitions, the χ^2 distribution noncentrality parameter is

$$\phi = 200\left[3\left(\frac{(0.025 - 0.0375)^2}{0.0375}\right) + \frac{(0.075 - 0.0375)^2}{0.0375}\right.$$

$$\left. + 3\left(\frac{(0.225 - 0.2125)^2}{0.2125}\right) + \frac{(0.175 - 0.2125)^2}{0.2125}\right]$$

$$= 11.77.$$

The χ^2 test statistic will have $df = (2-1)(4-1) = 3$ degrees of freedom, so the critical value of the test statistic is $\chi^2_{0.95,3} = 7.81$. The power of the test determined from the condition

$$\chi^2_{0.95} = 7.81 = \chi^2_{1-\pi, 11.77}$$

is $\pi = 0.833$.

4.5.3 Chi-square Goodness of Fit Test

In a χ^2 goodness of fit test, the hypotheses to be tested are H_0: *the observed distribution follows the specified distribution* versus H_A: *the observed distribution does not follow the specified distribution*. For the power calculation, the $(p_i)_0$ are calculated directly from the specified distribution under H_0 and the $(p_i)_A$ under H_A must include an appropriate deviation from the H_0 distribution that would cause rejection of H_0. The number of degrees of freedom for the χ^2 test is given by

$$df = k - 1 - m \tag{4.124}$$

where k is the number of classes that the observations are sorted into and m is the number of parameters that are estimated from the sample data.

Example 4.29 What is the power to reject the claim that a die is balanced (H_0 : $\theta_i = \frac{1}{6}$ for $i = 1$ to 6) when it is in fact slightly biased toward one die face ($H_A : \theta_i = \{0.16, 0.16, 0.16, 0.16, 0.16, 0.20\}$) based on 100 rolls of the die?

Solution: The table of observations will have six cells and there will be no parameters estimated from the sample data, so the χ^2 test will have $df = 6-1 = 5$ degrees of freedom. From Equation 4.121 the noncentrality parameter will be

$$\phi = 100 \left[5 \left(\frac{(0.16 - \frac{1}{6})^2}{\frac{1}{6}} \right) + \frac{(0.20 - \frac{1}{6})}{\frac{1}{6}} \right] = 20.13.$$

With $\alpha = 0.05$ we have $\chi^2_{0.95} = 11.07$, so the power to reject H_0 is determined from the condition

$$\chi^2_{0.95} = 11.07 = \chi^2_{1-\pi, 20.13},$$

which is satisfied by $\pi = 0.954$.

Chapter 5

Poisson Counts

The distribution of event counts x observed over an area of opportunity of size n is often governed by the Poisson probability distribution:

$$Poisson\,(x;n,\lambda) = \frac{(n\lambda)^x e^{-n\lambda}}{x!} \text{ for } x = 0, 1, \ldots \tag{5.1}$$

where λ is the mean event rate per unit area of opportunity and the mean of x is $\mu_x = n\lambda$. The purpose of this chapter is to present sample size calculations for the confidence interval for the Poisson parameter λ and sample size and power calculations for one-, two-, and many-sample tests for λ for Poisson-distributed processes. Exact methods and normal approximation methods are presented.

5.1 One Poisson Count

Sample size calculations are presented for confidence intervals and hypothesis tests for the Poisson parameter λ using the exact Poisson method, the large sample normal approximation method, and the square root transformation method. Sample size calculations are presented for both the area of opportunity to be inspected and for the number of counts.

5.1.1 Confidence Intervals for the Poisson Mean

5.1.1.1 Confidence Interval for the Poisson Mean (Exact Method)

For a random sample of n units drawn from a Poisson process, the total number of counts x observed is Poisson-distributed with mean $\mu_x = n\lambda$ where λ is the mean count per unit. For a sample of n units, which gives x' total counts, the point estimate for the Poisson mean is

$$\widehat{\lambda} = \frac{x'}{n} \tag{5.2}$$

and the two-sided confidence interval for λ is

$$P\left(\widehat{\lambda}_l < \lambda < \widehat{\lambda}_u\right) = 1 - \alpha \tag{5.3}$$

where $\widehat{\lambda}_l$ and $\widehat{\lambda}_u$ are determined from

$$Poisson\left(x' \leq x < \infty; n\widehat{\lambda}_l\right) = \alpha/2 \tag{5.4}$$

and

$$Poisson\left(0 \leq x \leq x'; n\widehat{\lambda}_u\right) = \alpha/2. \tag{5.5}$$

These equations are tedious to solve, but the χ^2 formulation of the Poisson distribution provides an equivalent and more convenient form for determining the confidence limits:

$$P\left(\frac{1}{2n}\chi^2_{\alpha/2,2x'} < \lambda < \frac{1}{2n}\chi^2_{1-\alpha/2,2(x'+1)}\right) = 1 - \alpha \tag{5.6}$$

where the χ^2 distribution is indexed by its left tail area. Although this interval is useful for calculating the confidence interval after experimental data have been collected, it is not practical for sample size calculations because of the complex relationship between the upper and lower confidence limits but alternative approximate methods are available and are easy to use.

5.1.1.2 Confidence Interval for the Poisson Mean (Large-Sample Approximation)

For $n\lambda > 30$, the Poisson distribution of x, the total number of events observed in n units inspected, is approximately normal, so the confidence interval for λ is approximately

$$P\left(\frac{x - z_{\alpha/2}\sqrt{x}}{n} < \lambda < \frac{x + z_{\alpha/2}\sqrt{x}}{n}\right) = 1 - \alpha. \tag{5.7}$$

If these confidence limits for λ are expressed relative to x/n

$$P\left(\frac{x}{n}\left(1 - z_{\alpha/2}/\sqrt{x}\right) < \lambda < \frac{x}{n}\left(1 + z_{\alpha/2}/\sqrt{x}\right)\right) = 1 - \alpha, \tag{5.8}$$

then the confidence interval half-width relative to x/n is

$$\delta = z_{\alpha/2}/\sqrt{x}. \tag{5.9}$$

The number of events x required to obtain a specified value of δ is given by

$$x = \left(\frac{z_{\alpha/2}}{\delta}\right)^2 \tag{5.10}$$

5.1. One Poisson Count

and the corresponding sample size is

$$n = \frac{x}{\lambda}$$
$$= \frac{1}{\lambda}\left(\frac{z_{\alpha/2}}{\delta}\right)^2. \quad (5.11)$$

Because n depends on an estimate for λ but x does not, sampling should continue until the x condition is satisfied, just in case the λ estimate is wrong.

Example 5.1 How many Poisson events must be observed if the relative error of the estimate for λ must be no larger than $\pm 10\%$ with 95% confidence?
Solution: The desired confidence interval for λ has the form

$$P\left(\frac{x}{n}(1 - 0.10) < \lambda < \frac{x}{n}(1 + 0.10)\right) = 0.95,$$

so $\delta = 0.10$ and from Equation 5.10

$$x = \left(\frac{1.96}{0.10}\right)^2 = 385.$$

That is, if the Poisson process is sampled until $x = 385$ counts are obtained, then the 95% confidence limits for λ will be

$$UCL/LCL = \left(\frac{385}{n}\right)(1 \pm 0.10)$$

or

$$P\left(\frac{346}{n} < \lambda < \frac{424}{n}\right) = 0.95.$$

5.1.1.3 Confidence Interval for the Poisson Mean (Square-Root Transform)

If x is Poisson-distributed with mean $\mu_x = n\lambda$ and the sample size is large, then by the delta method (see Appendix G.3.6) the distribution of $x' = \sqrt{x}$ is approximately normal with mean $\mu' = \sqrt{n\lambda}$ and standard deviation $\sigma' \simeq \frac{1}{2}$. The corresponding $(1 - \alpha) 100\%$ confidence interval for $\mu' = \sqrt{n\lambda}$ based on the observation of x events in n units inspected is given by

$$\Phi\left(\sqrt{x} - z_{\alpha/2}\left(\frac{1}{2}\right) < \sqrt{n\lambda} < \sqrt{x} + z_{\alpha/2}\left(\frac{1}{2}\right)\right) = 1 - \alpha. \quad (5.12)$$

This interval can be solved for λ to obtain the desired confidence interval:

$$\Phi\left(\frac{x}{n}\left(1 - \frac{z_{\alpha/2}}{2\sqrt{x}}\right)^2 < \lambda < \frac{x}{n}\left(1 + \frac{z_{\alpha/2}}{2\sqrt{x}}\right)^2\right) = 1 - \alpha. \quad (5.13)$$

As long as x is sufficiently large, this interval is well approximated by Equation 5.8 and the sample size condition of Equation 5.10 is obtained.

5.1.2 Tests for the Poisson Mean

5.1.2.1 Test for the Poisson Mean (Exact Method)

The hypotheses to be tested are $H_0 : \lambda = \lambda_0$ versus $H_A : \lambda > \lambda_0$ where λ is the mean number of counts per unit area of opportunity. The decision to reject H_0 or not is made on the basis of a sample of size n units from which the observed number of events x is determined. Under H_0, the sampling distribution of x is Poisson with mean $\mu_x = n\lambda_0$. The critical value of the accept/reject boundary for the test is $x_{A/R}$, the largest value of x that meets the condition

$$1 - Poisson\left(0 \leq x \leq x_{A/R}; n\lambda_0\right) \leq \alpha \tag{5.14}$$

where α is the significance level. Then the acceptance interval for H_0 is $0 \leq x \leq x_{A/R}$ and, because of the discrete nature of the Poisson random variable, the exact value of α might be less than the prescribed value. The power to reject H_0 when $H_A : \lambda > \lambda_0$ is true is given by

$$\pi = 1 - Poisson\left(0 \leq x \leq x_{A/R}; n\lambda\right). \tag{5.15}$$

Example 5.2 For the hypothesis test of $H_0 : \lambda = 4$ versus $H_A : \lambda > 4$ based on a sample of size $n = 2$ units using $\alpha \leq 0.05$, determine the power to reject H_0 when $\lambda = 8$.

Solution: Under H_0, the distribution of the observed number of counts x will be Poisson with $\mu_x = n\lambda_0 = 2 \times 4 = 8$. The acceptance interval for H_0 will be $0 \leq x \leq 13$ because

$$(1 - Poisson\,(0 \leq x \leq 12; 8) = 0.064) > 0.05$$
$$(1 - Poisson\,(0 \leq x \leq 13; 8) = 0.034) < 0.05,$$

so the exact significance level for the test will be $\alpha = 0.034$. With $\lambda = 8$, the power to reject H_0 is

$$\pi = 1 - Poisson\,(0 \leq x \leq 13; n\lambda = 16) = 0.725.$$

The count distributions under H_0 and H_A are shown in Figure 5.1.

5.1.2.2 Test for the Poisson Mean (Square-Root Transform)

The hypotheses to be tested are $H_0 : \lambda = \lambda_0$ versus $H_A : \lambda > \lambda_0$ where the decision to reject H_0 or not is based on the total number of events x observed in n units inspected. The sampling distribution of x is Poisson with mean $\mu = n\lambda$,

5.1. One Poisson Count

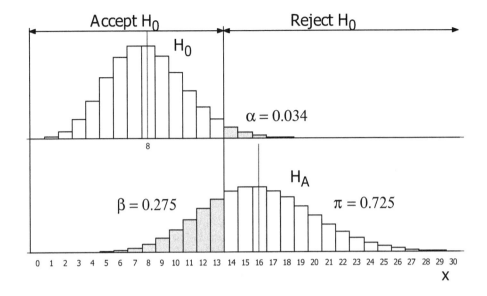

Figure 5.1: Count distributions under H_0 and H_A.

but the distribution of $x' = \sqrt{x}$ is approximately normal with mean $\mu' = \sqrt{n\lambda}$ and standard deviation $\sigma' \simeq \frac{1}{2}$. Under H_A, the power to reject H_0 is given by

$$\begin{aligned} \pi &= \Phi\left(-z_\beta < z < \infty\right) \\ &= \Phi\left(-\left(\frac{\mu' - \mu'_0}{\sigma'}\right) + z_\alpha < z < \infty\right) \\ &= \Phi\left(-2\sqrt{n}\left(\sqrt{\lambda} - \sqrt{\lambda_0}\right) + z_\alpha < z < \infty\right). \end{aligned} \quad (5.16)$$

From Equation 5.16, the number of units that must be inspected to reject H_0 with specified power for a specified shift in the Poisson mean is given by

$$n = \frac{1}{4}\left(\frac{z_\alpha + z_\beta}{\sqrt{\lambda} - \sqrt{\lambda_0}}\right)^2. \quad (5.17)$$

Example 5.3 For the hypothesis test of $H_0: \lambda = 4$ versus $H_A: \lambda > 4$ based on a sample of size $n = 5$ units, determine the power to reject H_0 when $\lambda = 9$. Use the square root transformation method with $\alpha = 0.05$.

Solution: By Equation 5.16, the power to reject $H_0: \lambda = 4$ when $\lambda = 9$ is

$$\begin{aligned} \pi &= \Phi\left(-2\sqrt{5}\left(\sqrt{9} - \sqrt{4}\right) + z_{0.05} < z < \infty\right) \\ &= \Phi\left(-2.83 < z < \infty\right) \\ &= 0.9977. \end{aligned}$$

Example 5.4 How many sampling units must be inspected to reject $H_0 : \lambda = 10$ with 90% power in favor of $H_A : \lambda > 10$ when in fact $\lambda = 15$?
Solution: By Equation 5.17 the necessary sample size is

$$n = \frac{1}{4}\left(\frac{1.645 + 1.282}{\sqrt{15} - \sqrt{10}}\right)^2 = 4.2,$$

which rounds up to $n = 5$ sampling units.

5.2 Two Poisson Counts

Exact and approximate methods are presented for calculating sample size for confidence intervals and for calculating sample size and power for hypothesis tests for two Poisson means including the difference between the means and the ratio of the means.

5.2.1 Confidence Intervals for Two Poisson Means

5.2.1.1 Confidence Interval for the Difference Between Two Poisson Means

The point estimate for the difference between two Poisson means is given by

$$\widehat{\Delta\lambda} = \widehat{\lambda}_1 - \widehat{\lambda}_2$$
$$= \frac{x_1}{n_1} - \frac{x_2}{n_2} \qquad (5.18)$$

where x_i is the number of events observed in area of opportunity n_i. When the x_i are large, the distribution of $\widehat{\Delta\lambda}$ is approximately normal, so the confidence interval for $\Delta\lambda$ has the form

$$P\left(\widehat{\Delta\lambda} - \delta < \Delta\lambda < \widehat{\Delta\lambda} + \delta\right) = 1 - \alpha \qquad (5.19)$$

where the confidence interval half-width is

$$\delta = z_{\alpha/2}\widehat{\sigma}_{\widehat{\Delta\lambda}} \qquad (5.20)$$

and the standard deviation of the $\widehat{\Delta\lambda}$ distribution is given by the delta method (see Appendix G.3.7):

$$\widehat{\sigma}_{\widehat{\Delta\lambda}} = \sqrt{\frac{\widehat{\lambda}_1}{n_1} + \frac{\widehat{\lambda}_2}{n_2}}. \qquad (5.21)$$

From the last two equations, the sample size n_1 required to obtain a specified value of the confidence interval half-width δ when the sample size ratio is n_1/n_2 is

$$n_1 = \left(\frac{z_{\alpha/2}}{\delta}\right)^2 \left(\lambda_1 + \frac{n_1}{n_2}\lambda_2\right), \qquad (5.22)$$

5.2. Two Poisson Counts

or, when the two sample sizes are equal,

$$n_1 = n_2 = \left(\frac{z_{\alpha/2}}{\delta}\right)^2 (\lambda_1 + \lambda_2). \tag{5.23}$$

The optimal allocation of observations to the two samples is determined by minimizing $\hat{\sigma}_{\Delta\hat{\lambda}}$ with respect to n_1 and n_2 when $n_1 + n_2$ is fixed resulting in the condition

$$\frac{n_1}{n_2} = \sqrt{\frac{\lambda_1}{\lambda_2}}. \tag{5.24}$$

In practice, when there is uncertainty about the true values of λ_1 and λ_2, sampling from the two groups should continue until the number of count events in both samples exceeds the predicted number of counts, that is, until $x_1 > n_1\lambda_1$ and $x_2 > n_2\lambda_2$.

Example 5.5 What optimal sample sizes are required to estimate the difference between two Poisson means with 30% precision if the means are expected to be $\lambda_1 = 25$ and $\lambda_2 = 16$?

Solution: The difference between the means is expected to be $\Delta\lambda = 9$, so the confidence interval half-width must be 30% of that, or

$$\delta = 0.3 \times 9 = 2.7.$$

From Equation 5.24, the optimal sample size ratio is

$$\frac{n_1}{n_2} = \sqrt{\frac{\lambda_1}{\lambda_2}} = \sqrt{\frac{25}{16}} = 1.25.$$

From Equation 5.22, with $\alpha = 0.05$, the sample size n_1 must be

$$n_1 = \left(\frac{1.96}{2.7}\right)^2 (25 + 1.25 \times 16) = 23.7$$

and the sample size n_2 must be

$$n_2 = \frac{n_1}{\left(\frac{n_1}{n_2}\right)} = \frac{23.7}{1.25} = 18.96,$$

which round up to $n_1 = 24$ and $n_2 = 19$.

5.2.1.2 Confidence Interval for the Ratio of Two Poisson Means

The two-sided confidence interval for the ratio of two independent Poisson means, each estimated by $\widehat{\lambda}_i = x_i/n_i$, is given by

$$P\left(\frac{\widehat{\lambda}_1}{\widehat{\lambda}_2} F_{\alpha/2} < \frac{\lambda_1}{\lambda_2} < \frac{\widehat{\lambda}_1}{\widehat{\lambda}_2} F_{1-\alpha/2}\right) = 1 - \alpha \tag{5.25}$$

where the F distribution has $2x_1$ numerator and $2x_2$ denominator degrees of freedom. This form of the confidence interval is inconvenient for the purpose of calculating sample sizes, so the following large-sample approximation method should be used instead.

If the sample counts from the two populations to be compared are both large, then an approximate $(1-\alpha)\,100\%$ confidence interval for $\ln(\lambda_1/\lambda_2)$ is given by

$$P\left(\ln\left(\widehat{\lambda}_1/\widehat{\lambda}_2\right) - \delta < \ln(\lambda_1/\lambda_2) < \ln\left(\widehat{\lambda}_1/\widehat{\lambda}_2\right) + \delta\right) = 1 - \alpha \tag{5.26}$$

where the confidence interval half-width is

$$\delta = z_{\alpha/2}\,\widehat{\sigma}_{\ln(\widehat{\lambda}_1/\widehat{\lambda}_2)} \tag{5.27}$$

and the standard deviation is given by the delta method (see Appendix G.3.8)

$$\widehat{\sigma}_{\ln(\widehat{\lambda}_1/\widehat{\lambda}_2)} = \sqrt{\frac{1}{n_1\lambda_1} + \frac{1}{n_2\lambda_2}}. \tag{5.28}$$

Then the sample size n_1 required to obtain a specified confidence interval half-width δ with sample size ratio n_1/n_2 is

$$n_1 = \left(\frac{1}{\lambda_1} + \frac{n_1}{n_2}\frac{1}{\lambda_2}\right)\left(\frac{z_{\alpha/2}}{\delta}\right)^2. \tag{5.29}$$

The number of counts x_1 required to obtain confidence interval half-width δ with sample size ratio n_1/n_2 is given by

$$\begin{aligned} x_1 &= n_1\lambda_1 \\ &= \left(1 + \frac{n_1}{n_2}\frac{\lambda_1}{\lambda_2}\right)\left(\frac{z_{\alpha/2}}{\delta}\right)^2. \end{aligned} \tag{5.30}$$

The confidence interval in Equation 5.26 is not in a useful form because it is expressed in terms of $\ln(\lambda_1/\lambda_2)$ instead of only λ_1/λ_2. However, if x_1 and x_2 are large enough so that δ is close to 0 and small compared to 1 (that is, $|\delta| \ll 1$), then Equation 5.26 may be approximated by

$$P\left(\frac{\widehat{\lambda}_1}{\widehat{\lambda}_2}(1-\delta) < \frac{\lambda_1}{\lambda_2} < \frac{\widehat{\lambda}_1}{\widehat{\lambda}_2}(1+\delta)\right) = 1 - \alpha \tag{5.31}$$

where δ has the same numerical value and is the relative half-width for the confidence interval for λ_1/λ_2.

5.2. Two Poisson Counts

Example 5.6 How many Poisson counts are required to estimate the ratio of the means of two independent Poisson distributions to within 20% of the true ratio with 95% confidence if the sample sizes will be the same and the ratio of the means is expected to be $\lambda_1/\lambda_2 \simeq 2$?

Solution: With $n_1/n_2 = 1$, $\lambda_1/\lambda_2 = 2$, $z_{0.025} = 1.96$, and $\delta = 0.02$ in Equation 5.30, the number of Poisson counts required in the first sample is

$$x_1 = (1 + 1 \times 2) \left(\frac{1.96}{0.2}\right)^2$$
$$= 289.$$

The corresponding required counts in the second sample are about half of the counts in the first: $x_2 = 289/2 = 145$.

5.2.2 Tests for Two Poisson Means

Power and sample size calculations are presented for tests for differences and ratios of two Poisson means using exact and approximate test methods.

5.2.2.1 Test for a Difference Between Two Poisson Means (Exact Binomial Method)

The hypotheses to be tested are $H_0 : \lambda_1 = \lambda_2$ versus $H_A : \lambda_1 < \lambda_2$ where λ_1 and λ_2 are Poisson rates, that is, the counts per unit area of opportunity. If the sample sizes are n_1 and n_2, respectively, and the total observed counts from the two samples are x_1 and x_2, respectively, then x_1 and x_2 are Poisson distributed with means $\mu_1 = n_1\lambda_1$ and $\mu_2 = n_2\lambda_2$, respectively. The p value for the two-sample test is given by the binomial probability

$$\begin{aligned} p &= b\left(0 \leq x \leq x_1; x_1 + x_2, \frac{n_1}{n_1 + n_2}\right) \\ &= \sum_{x=0}^{x_1} \binom{x_1 + x_2}{x} \left(\frac{n_1}{n_1 + n_2}\right)^x \left(1 - \frac{n_1}{n_1 + n_2}\right)^{x_1 + x_2 - x}. \end{aligned} \quad (5.32)$$

The power of the test by the binomial method is given by

$$\pi = \sum_{x_1=0}^{\infty} \sum_{x_2=0}^{\infty} \delta\left(p \leq \alpha\right) Poisson\left(x_1; \mu_1\right) Poisson\left(x_2; \mu_2\right) \quad (5.33)$$

where $\delta\left(p < \alpha\right)$ is a Boolean function that is equal to 1 if the expression in parentheses is true and 0 otherwise. In practice, the summation over x_1 needs to be considered only where the probability of x_1 is substantial, that is, in the neighborhood of μ_1. The summation over x_2 can be similarly limited. Even with these limitations, the only practical way to determine the exact power is with a suitable computer program or spreadsheet, however, the approximate methods that follow generally give similar results and are much easier to use.

5.2.2.2 Test for the Ratio of Two Poisson Means (F Test Method)

The hypotheses to be tested are $H_0 : \lambda_1/\lambda_2 = 1$ versus $H_A : \lambda_1/\lambda_2 > 1$ where λ_1 and λ_2 are Poisson mean event rates. If the sample sizes are n_1 and n_2 and the observed counts are x_1 and x_2, then under H_0 the distribution of

$$F = \frac{\widehat{\lambda}_1}{\widehat{\lambda}_2} = \frac{x_1/n_1}{x_2/n_2} \tag{5.34}$$

follows the F distribution with $2x_1$ numerator and $2x_2$ denominator degrees of freedom. Under H_A the distribution of

$$F = \frac{\lambda_2}{\lambda_1} \frac{\widehat{\lambda}_1}{\widehat{\lambda}_2} \tag{5.35}$$

also follows the F distribution, so the power to reject $H_0 : \lambda_1/\lambda_2 = 1$ in favor of $H_A : \lambda_1/\lambda_2 > 1$ is given by

$$\begin{aligned} \pi &= P\left(F_\beta < F < \infty\right) \\ &= P\left(\frac{\lambda_2}{\lambda_1} F_{1-\alpha} < F < \infty\right) \end{aligned} \tag{5.36}$$

where the F distribution has $2n_1\lambda_1$ numerator and $2n_2\lambda_2$ denominator degrees of freedom. This form of the hypothesis test is inconvenient for the purpose of calculating power and sample size so one of the following approximate methods should be used instead.

5.2.2.3 Test for the Ratio of Two Poisson Means (Log Transformation Method)

When the Poisson counts x_1 and x_2 are large, the hypotheses of Section 5.2.2.2 may be tested using the z statistic:

$$z = \frac{\ln\left(\widehat{\lambda}_1/\widehat{\lambda}_2\right)}{\widehat{\sigma}_{\ln(\widehat{\lambda}_1/\widehat{\lambda}_2)}} = \frac{\ln\left(\frac{x_1/n_1}{x_2/n_2}\right)}{\sqrt{\frac{1}{x_1} + \frac{1}{x_2}}} \tag{5.37}$$

where the point estimates for the λ_i are given by

$$\widehat{\lambda}_i = \frac{x_i}{n_i} \tag{5.38}$$

and $\widehat{\sigma}_{\ln(\widehat{\lambda}_1/\widehat{\lambda}_2)}$ is given by the delta method (see Appendix G.3.8). Then the power $\pi = 1 - \beta$ to reject $H_0 : \lambda_1/\lambda_2 = 1$ in favor of $H_A : \lambda_1/\lambda_2 > 1$ when in fact $\lambda_1 > \lambda_2$ is given by

$$\pi = \Phi\left(-\infty < z < z_\beta\right) \tag{5.39}$$

where
$$z_\beta = \frac{\ln(\lambda_1/\lambda_2)}{\sqrt{\frac{1}{n_1\lambda_1} + \frac{1}{n_2\lambda_2}}} - z_\alpha. \tag{5.40}$$

From Equation 5.40 the sample size n_1 required to obtain power $\pi = 1 - \beta$ for specified sample size ratio n_1/n_2 is

$$n_1 = \left(\frac{1}{\lambda_1} + \frac{n_1}{n_2}\frac{1}{\lambda_2}\right)\left(\frac{z_\alpha + z_\beta}{\ln(\lambda_1/\lambda_2)}\right)^2. \tag{5.41}$$

Use of this equation requires specification of λ_1 and λ_2 which is inconsistent with the λ_1/λ_2 form of the hypotheses being tested, but this problem can be resolved by expressing the sample size requirement directly in terms of the number of Poisson counts. From Equations 5.38 and 5.41 the number of count events x_1 required to obtain power $\pi = 1 - \beta$ with sample size ratio n_1/n_2 is

$$x_1 = \left(1 + \frac{n_1}{n_2}\frac{\lambda_1}{\lambda_2}\right)\left(\frac{z_\alpha + z_\beta}{\ln(\lambda_1/\lambda_2)}\right)^2 \tag{5.42}$$

and the corresponding number of x_2 counts is

$$x_2 = \frac{x_1}{\left(\frac{n_1}{n_2}\right)\left(\frac{\lambda_1}{\lambda_2}\right)}. \tag{5.43}$$

The optimal allocation of samples is obtained when $x_1 = x_2$ which is equivalent to

$$\frac{n_1}{n_2} = \frac{\lambda_2}{\lambda_1}. \tag{5.44}$$

When the sample sizes are equal, the number of x_1 and x_2 counts must be

$$x_1 = \left(1 + \frac{\lambda_1}{\lambda_2}\right)\left(\frac{z_\alpha + z_\beta}{\ln(\lambda_1/\lambda_2)}\right)^2 \tag{5.45}$$

and

$$x_2 = \frac{x_1}{\left(\frac{\lambda_1}{\lambda_2}\right)}. \tag{5.46}$$

5.2.2.4 Test for the Difference Between Two Poisson Means (Large Samples)

If the count responses x_1 and x_2 are Poisson distributed with means μ_1 and μ_2, respectively, where $(\mu_1 = n_1\lambda_1) > 30$ and $(\mu_2 = n_2\lambda_2) > 30$, then the distributions of x_1 and x_2 are approximately normal. By application of scale changes, the distributions of $x_1' = x_1/n_1$ and $x_2' = x_2/n_2$ are approximately normal with means $\mu_1' = \lambda_1$ and $\mu_2' = \lambda_2$ and standard deviations $\sigma_1' = \sqrt{\lambda_1/n_1}$

and $\sigma'_2 = \sqrt{\lambda_2/n_2}$, respectively. Then the hypotheses $H_0 : \lambda_1 = \lambda_2$ versus $H_A : \lambda_1 < \lambda_2$ may be tested using a z statistic:

$$z = \frac{\widehat{\mu'_2} - \widehat{\mu'_1}}{\sqrt{(\sigma'_1)^2 + (\sigma'_2)^2}} = \frac{\widehat{\lambda_2} - \widehat{\lambda_1}}{\sqrt{\frac{\widehat{\lambda_1}}{n_1} + \frac{\widehat{\lambda_2}}{n_2}}} \tag{5.47}$$

where $\widehat{\lambda_i} = x_i/n_i$. The power to reject H_0 when $\lambda_1 < \lambda_2$ may be approximated from

$$\pi = \Phi\left(-\infty < z < z_\beta\right) \tag{5.48}$$

where

$$z_\beta = \frac{\lambda_2 - \lambda_1}{\sqrt{\frac{\lambda_1}{n_1} + \frac{\lambda_2}{n_2}}} - z_\alpha. \tag{5.49}$$

From Equation 5.49 the sample size n_1 required to obtain power $\pi = 1 - \beta$ with sample size ratio n_1/n_2 is

$$n_1 = \left(\lambda_1 + \frac{n_1}{n_2}\lambda_2\right)\left(\frac{z_\alpha + z_\beta}{\lambda_2 - \lambda_1}\right)^2. \tag{5.50}$$

When the sample sizes are equal,

$$n_1 = n_2 = (\lambda_1 + \lambda_2)\left(\frac{z_\alpha + z_\beta}{\lambda_2 - \lambda_1}\right)^2. \tag{5.51}$$

The optimal sample size ratio is given by Equation 5.24.

5.2.2.5 Test for the Difference Between Two Poisson Means (Square-Root Transform)

If x_1 and x_2 are Poisson-distributed with means $\mu_1 = n_1\lambda_1$ and $\mu_2 = n_2\lambda_2$, respectively, then by application of square root transforms and scale changes, the distributions of $x'_1 = \sqrt{x_1/n_1}$ and $x'_2 = \sqrt{x_2/n_2}$ are approximately normal with means $\mu'_1 = \sqrt{\lambda_1}$ and $\mu'_2 = \sqrt{\lambda_2}$ and standard deviations $\sigma'_1 = \frac{1}{2\sqrt{n_1}}$ and $\sigma'_2 = \frac{1}{2\sqrt{n_2}}$, respectively. Then the hypotheses $H_0 : \lambda_1 = \lambda_2$ versus $H_A : \lambda_1 < \lambda_2$ may be tested using a z statistic:

$$z = \frac{\widehat{\mu'_2} - \widehat{\mu'_1}}{\sqrt{(\sigma'_1)^2 + (\sigma'_2)^2}} = \frac{\sqrt{\widehat{\lambda_2}} - \sqrt{\widehat{\lambda_1}}}{\frac{1}{2}\sqrt{\frac{1}{n_1} + \frac{1}{n_2}}}. \tag{5.52}$$

The power to reject H_0 when $\lambda_1 < \lambda_2$ may be approximated from Equation 5.48 with

$$z_\beta = \frac{\sqrt{\lambda_2} - \sqrt{\lambda_1}}{\frac{1}{2}\sqrt{\frac{1}{n_1} + \frac{1}{n_2}}} - z_\alpha. \tag{5.53}$$

5.2. Two Poisson Counts

From Equation 5.53, the sample size n_1 required to obtain power $\pi = 1 - \beta$ with sample size ratio n_1/n_2 is

$$n_1 = \frac{1}{4}\left(1 + \frac{n_1}{n_2}\right)\left(\frac{z_\alpha + z_\beta}{\sqrt{\lambda_2} - \sqrt{\lambda_1}}\right)^2. \tag{5.54}$$

When the sample sizes are equal,

$$n_1 = n_2 = \frac{1}{2}\left(\frac{z_\alpha + z_\beta}{\sqrt{\lambda_2} - \sqrt{\lambda_1}}\right)^2. \tag{5.55}$$

Example 5.7 Find the power to reject $H_0 : \lambda_1 = \lambda_2$ in favor of $H_A : \lambda_1 < \lambda_2$ when $\lambda_1 = 10$, $n_1 = 8$ and $\lambda_2 = 15$, $n_2 = 6$. Use the large-sample normal approximation, square root transform, and F test methods with $\alpha = 0.05$.
Solution: The expected number of counts from the first (x_1) and second (x_2) populations are both large enough to justify the large sample approximation method. By this method the power is

$$\pi = \Phi\left(-\infty < z < \frac{15 - 10}{\sqrt{\frac{15}{6} + \frac{10}{8}}} - 1.645\right)$$

$$= \Phi(-\infty < z < 0.937)$$

$$= 0.826.$$

By the log-transformation method the power is

$$\pi = \Phi\left(-\infty < z < \frac{\log(15/10)}{\sqrt{\frac{1}{6 \times 15} + \frac{1}{8 \times 10}}} - 1.645\right)$$

$$= \Phi(-\infty < z < 0.994)$$

$$= 0.840.$$

By the square-root transform method the power is

$$\pi = \Phi\left(-\infty < z < \frac{\sqrt{15} - \sqrt{10}}{\frac{1}{2}\sqrt{\frac{1}{8} + \frac{1}{6}}} - 1.645\right)$$

$$= \Phi(-\infty < z < 1.01)$$

$$= 0.838.$$

By the F test method the power is

$$\pi = P\left(\frac{10}{15}F_{0.95, 2(6)(15), 2(8)(10)} < F < \infty\right)$$

$$= P(0.86 < F < \infty)$$

$$= 0.837.$$

For reference, the power by the exact method of Equation 5.33 is $\pi = 0.816$.

Example 5.8 What minimum total counts are required for the two-sample counts test to detect a factor of two difference between the count rates with 90% power? Assume that the two sample sizes will be equal.
Solution: The hypotheses to be tested are $H_0 : \lambda_1/\lambda_2 = 1$ versus $H_A : \lambda_1/\lambda_2 > 1$. From Equation 5.45, which is expressed in terms of the ratio of the two means, the number of count events x_1 required to reject H_0 when $\lambda_1/\lambda_2 = 2$ is

$$\begin{aligned} x_1 &= \left(1 + \frac{\lambda_1}{\lambda_2}\right)\left(\frac{z_\alpha + z_\beta}{\ln(\lambda_1/\lambda_2)}\right)^2 \\ &= (1+2)\left(\frac{1.645 + 1.282}{\ln(2)}\right)^2 \\ &= 54. \end{aligned}$$

Because $\lambda_2 = \lambda_1/2$, the corresponding number of x_2 counts is $x_2 = 54/2 = 27$.

5.3 Tests for Many Poisson Counts

Two methods are presented to test for location differences among three or more Poisson populations: ANOVA applied to the square-root transformed observations and the χ^2 contingency table method.

5.3.1 Test for Differences Between Many Poisson Means (Square-Root Transform)

The hypotheses to be tested are $H_0 : \lambda_i = \lambda_j$ for all i, j pairs versus $H_A : \lambda_i \neq \lambda_j$ for at least one i, j pair. The similarity to ANOVA is obvious, however, ANOVA requires that the populations being studied are normal and homoscedastic which is not generally the case for Poisson populations, and certainly not under H_A. These issues can be simply resolved by application of the square root transform $x' = \sqrt{x}$, which recovers at least approximate normality and homoscedasticity because $\sigma' \simeq \frac{1}{2}$ for all treatments under both H_0 and H_A. Thus, the power and sample size calculations developed for one-way ANOVA can be used to analyze the three-or-more count responses problem by use of the square root transform.[1]

Example 5.9 In a test for differences between mean counts from five different processes, determine the power to reject $H_0 : \lambda_i = \lambda_j$ for all i, j pairs when $\lambda_1 = \lambda_2 = \lambda_3 = 16$, $\lambda_4 = 9$, $\lambda_5 = 25$ and $n = 3$ units from each process are inspected. The number of counts will be reported for each unit. Assume that the test will be performed using one-way ANOVA applied to the square root transformed counts.

---
[1]Section 8.1 reviews one-way ANOVA power and sample size calculations.

5.3. Tests for Many Poisson Counts

Solution: After the square root transform, the transformed treatment means are $\lambda_1' = \lambda_2' = \lambda_3' = 4$, $\lambda_4' = 3$, and $\lambda_5' = 5$. The grand transformed mean is $\overline{\lambda'} = 4$, so the treatment biases relative to the grand mean are $0, 0, 0, -1$, and 1, respectively. The ANOVA F test noncentrality parameter is then

$$\phi = \frac{E(SS_{Treatment})}{E(MS_\epsilon)} = \frac{3\left(0^2 + 0^2 + 0^2 + (-1)^2 + (1)^2\right)}{\left(\frac{1}{2}\right)^2} = 24$$

where $E(MS_\epsilon) = (\sigma')^2 = \left(\frac{1}{2}\right)^2$ is the error variance of the transformed counts. The ANOVA will have $df_{Treatment} = 4$ and $df_\epsilon = 15 - 1 - 4 = 10$, so the F test critical value will be $F_{0.95, 4, 24} = 3.48$. The power to reject H_0 is then given by Equation 8.1:

$$\begin{aligned} F_{1-\alpha} &= F_{1-\pi, \phi} \\ 3.48 &= F_{1-\pi, 24} \\ 3.48 &= F_{0.11, 24}, \end{aligned}$$

so the power is $\pi = 0.89$ to reject H_0 for the specified set of means.

5.3.2 Test for Differences Between Many Poisson Means (Chi-square Method)

If there are x_i Poisson events observed in a sample of size n_i, where the subscript $i = 1$ to k distinguishes k qualitative treatments, then the test of $H_0 : \lambda_i = \lambda_0$ for all i versus $H_A : \lambda_i \neq \lambda_0$ for at least one i, where λ_0 is a constant, may be performed using the χ^2 goodness-of-fit test from Section 4.5.3. The test statistic is

$$\chi^2 = \sum_{i=1}^{k} \frac{\left(x_i - n_i \widehat{\lambda_0}\right)^2}{n_i \widehat{\lambda_0}} \tag{5.56}$$

where

$$\widehat{\lambda_0} = \frac{\sum_{i=1}^{k} x_i}{\sum_{i=1}^{k} n_i}. \tag{5.57}$$

Under H_0, the χ^2 statistic follows the χ^2 distribution with $\nu = k - 1$ degrees of freedom so we would reject H_0 in favor of H_A if $\chi^2 > \chi^2_{1-\alpha}$. Under H_A the χ^2 statistic follows the noncentral χ^2 distribution with noncentrality parameter,

$$\phi = \sum_{i=1}^{k} \frac{\left(n_i \left(\lambda_{A,i} - \lambda_0\right)\right)^2}{n_i \lambda_0} \tag{5.58}$$

where the $\lambda_{A,i}$ are the Poisson means under the alternate hypothesis and must meet the constant-total-counts constraint:

$$\lambda_0 \sum_{i=1}^{k} n_i = \sum_{i=1}^{k} n_i \lambda_{A,i}. \tag{5.59}$$

When the n_i are all equal, as is frequently the case, this constraint simplifies to

$$\lambda_0 = \frac{1}{k} \sum_{i=1}^{k} \lambda_{A,i}. \tag{5.60}$$

Then the power to reject H_0 comes from the condition

$$\chi^2_{1-\alpha} = \chi^2_{1-\pi,\phi}. \tag{5.61}$$

Example 5.10 In a test for differences among the means of five Poisson populations, determine the probability of rejecting $H_0 : \lambda_i = \lambda_0$ for all i when $\lambda_i = \{16, 16, 16, 12, 20\}$. The number of units inspected is $n_i = 4$ for all i.
Solution: Given the Poisson means specified under H_A, the value of λ_0 under H_0 is given by

$$\lambda_0 = \frac{1}{5}(16 + 16 + 16 + 12 + 20) = 16.$$

With $n_i = n = 4$, the noncentrality parameter is given by

$$\begin{aligned}\phi &= n \sum_{i=1}^{k} \frac{(\lambda_{A,i} - \lambda_0)^2}{\lambda_0} \\ &= 4\left(\frac{(0)^2}{16} + \frac{(0)^2}{16} + \frac{(0)^2}{16} + \frac{(-4)^2}{16} + \frac{(4)^2}{16}\right) = 8.\end{aligned}$$

The power is determined from Equation 5.61:

$$\chi^2_{0.95} = 9.49 = \chi^2_{0.395,8}$$

where the central and noncentral χ^2 distributions both have $\nu = 5 - 1 = 4$ degrees of freedom, so the power is $\pi = 0.605$.

5.4 Correcting for Background Counts

In some cases, such as radioactivity measurements or incidences of disease, Poisson distributed count data are collected in the presence of a background source of counts. Because the sum of two Poisson random variables is also Poisson, all of the methods described above are still relevant with the correction that each Poisson distribution be biased by the appropriate background count rate. For example, if a test involves one or more Poisson distributions with means given by

5.4. Correcting for Background Counts

λ_i, then the corresponding background-biased Poisson distributions will have means $\lambda'_i = \lambda_i + \lambda_0$ where λ_0 is the common background count rate.

Example 5.11 In a two-sample test for counts, what common sample size $n = n_1 = n_2$ is required to distinguish $\lambda_1 = \lambda_2 = 6$ from $\lambda_1 = 6, \lambda_2 = 15$ with 90% power in the presence of a background count rate of $\lambda_0 = 10$?

Solution: From Equation 5.55, modified to account for the background count rate, the necessary sample size to reject $H_0 : \lambda_1 = \lambda_2$ in favor of $H_A : \lambda_1 < \lambda_2$ with 90% power and $\alpha = 0.05$ is given by

$$n = \frac{1}{2}\left(\frac{z_\alpha + z_\beta}{(\sqrt{\lambda_2 + \lambda_0} - \sqrt{\lambda_1 + \lambda_0})}\right)^2 \qquad (5.62)$$

$$= \frac{1}{2}\left(\frac{1.645 + 1.282}{\sqrt{25} - \sqrt{16}}\right)^2 = 5.$$

Chapter 6

Regression

The purpose of this chapter is to present power and sample size calculation methods for regression analysis including least squares linear regression and binary logistic regression.

6.1 Linear Regression

For paired (x, y) observations where a linear model is anticipated of the form

$$y = \beta_0 + \beta_1 x, \tag{6.1}$$

the standard method of estimating the β_i parameters from experimental data is by least squares linear regression. The two statistical analysis goals for the β_i parameters are

- To form confidence intervals of the form

$$P(b_i - \delta < \beta_i < b_i + \delta) = 1 - \alpha \tag{6.2}$$

  where b_i is the regression coefficient that estimates β_i.

- To perform tests of the hypotheses $H_0 : \beta_i = 0$ versus $H_A : \beta_i \neq 0$.

The confidence interval problem is relevant when the goal of the experiment is to quantify a β_i value with a specified degree of precision. The hypothesis testing problem is relevant when the goal of the experiment is to determine if x has any influence at all on y.

Sample size calculations are presented for determining a confidence interval for the slope of the linear regression equation and for tests of the value of the slope.

6.1.1 Confidence Interval for the Slope

Sample size calculations for the confidence interval for the linear regression slope parameter depend on the pattern of x observations planned for the experiment. Sample size calculations are presented for the general case, for uniformly distributed x values over a specified range, and for equally spaced, evenly weighted levels of x.

6.1.1.1 General Formula

The confidence interval for the linear regression slope parameter β_1 has the form

$$P(b_1 - \delta < \beta_1 < b_1 + \delta) = 1 - \alpha \tag{6.3}$$

where b_1 is the estimate for β_1 obtained from the experimental (x, y) data and

$$\begin{aligned} \delta &= t_{\alpha/2} \hat{\sigma}_{b_1} \\ &= t_{\alpha/2} \left(\frac{\hat{\sigma}_\epsilon}{\sqrt{SS_x}} \right) \end{aligned} \tag{6.4}$$

where $\hat{\sigma}_\epsilon$ is the standard error of the regression, the t distribution has df_ϵ degrees of freedom, and SS_x is the x sum of squares

$$\begin{aligned} SS_x &= \sum_{i=1}^{N} (x_i - \bar{x})^2 \\ &= N\sigma_x^2 \end{aligned} \tag{6.5}$$

where N is the total number of (x, y) observations in the data set. Obviously, SS_x depends on the method used to select the x levels and their weights, which could be quite different from experiment to experiment; however, for some common experiment designs, SS_x can be simplified and then Equation 6.4 can be solved to determine the sample size.

6.1.1.2 x Is Normally Distributed

If x is normally distributed with standard deviation σ_x, then

$$SS_x = N\sigma_x^2. \tag{6.6}$$

By substituting this expression into Equation 6.4, the number of (x, y) observations required to obtain a confidence interval for β_1 with desired half-width δ is

$$N \geq \left(\frac{t_{\alpha/2} \hat{\sigma}_\epsilon}{\delta \hat{\sigma}_x} \right)^2. \tag{6.7}$$

6.1. Linear Regression

6.1.1.3 x Is Uniformly Distributed

If x is uniformly distributed with lower and upper limits x_{min} and x_{max}, respectively, then $\sigma_x = (x_{max} - x_{min})/\sqrt{12}$ and

$$SS_x = N\sigma_x^2 = \frac{N}{12}(x_{max} - x_{min})^2, \tag{6.8}$$

so the number of observations required to obtain β_1 confidence interval half-width δ is

$$N \geq 12 \left(\frac{t_{\alpha/2}\hat{\sigma}_\epsilon}{\delta(x_{max} - x_{min})} \right)^2. \tag{6.9}$$

6.1.1.4 Equally Weighted, Evenly Spaced Levels of x

In many designed experiments the levels of x are fixed, evenly spaced, and an equal number of observations is assigned to each x level. If there are k levels of x over the interval from x_{min} to x_{max} with equal spacing Δx between adjacent levels and n observations at each level, then

$$\Delta x = \frac{x_{max} - x_{min}}{k-1} \tag{6.10}$$

and

$$SS_x = \frac{n}{12}(k-1)k(k+1)\Delta x^2 \tag{6.11}$$

from which the following sample size condition is obtained:

$$n \geq \frac{12}{(k-1)k(k+1)} \left(\frac{t_{\alpha/2}\hat{\sigma}_\epsilon}{\delta \Delta x} \right)^2. \tag{6.12}$$

The total number of (x,y) observations required will be $N = kn$ or

$$N \geq \frac{12}{(k-1)(k+1)} \left(\frac{t_{\alpha/2}\hat{\sigma}_\epsilon}{\delta \Delta x} \right)^2. \tag{6.13}$$

Example 6.1 Designed experiments frequently involve two or three equally weighted levels of x. Compare the sample sizes required for these two important special cases if they must both deliver a β_1 confidence interval half-width δ and the observations are taken over the same x range from x_{min} to x_{max}. For the three-level case, assume that the middle level will be midway between x_{min} and x_{max}.

Solution: The subscripts 2 and 3 will be used to indicate parameters from the two-level and three-level cases, respectively. For the two-level case, from Equation 6.12 with $k_2 = 2$ and $\Delta x_2 = x_{max} - x_{min}$, the sample size per x level will be

$$n_2 \geq 2\left(\frac{t_{\alpha/2}\widehat{\sigma}_\epsilon}{\delta \Delta x_1}\right)^2$$

$$\geq 2\left(\frac{t_{\alpha/2}\widehat{\sigma}_\epsilon}{\delta(x_{max}-x_{min})}\right)^2. \quad (6.14)$$

For the three-level case with $k_3 = 3$ and $\Delta x_3 = \frac{1}{2}(x_{max} - x_{min})$ the sample size per x level will be

$$n_3 \geq \frac{1}{2}\left(\frac{t_{\alpha/2}\widehat{\sigma}_\epsilon}{\delta \Delta x_2}\right)^2$$

$$\geq 2\left(\frac{t_{\alpha/2}\widehat{\sigma}_\epsilon}{\delta(x_{max}-x_{min})}\right)^2. \quad (6.15)$$

Because $n_2 = n_3$, $N_2 = 2n_2$, and $N_3 = 3n_2$, the two experiments appear to have the same ability to resolve β_1 even though the three-level experiment requires 50% more observations! This means that the middle observations in the three-level experiment are effectively wasted for the purpose of estimating β_1. This statement is not entirely true because the middle observations in the three-level experiment do add error degrees of freedom, which potentially decrease n_3 compared to n_2 for the same δ. In general, the purpose of using three levels of x in an experiment is not to improve the precision of the β_1 estimate; rather, three levels are used to allow a linear lack of fit test, which is not possible using just two levels of x.

Example 6.2 Compare the sample sizes required to estimate the slope parameter with equal precision for two experiments if x is uniformly distributed over the interval from x_{min} to x_{max} in the first experiment and if x has two levels, x_{min} and x_{max}, in the second experiment.

Solution: From Equations 6.9 and 6.13 the ratio of the total number of observations required by the two experiments is

$$\frac{N_{\text{uniform } x}}{N_{\text{two levels of } x}} \simeq \frac{12\left(\frac{t_{\alpha/2}\widehat{\sigma}_\epsilon}{\delta(x_{max}-x_{min})}\right)^2}{4\left(\frac{t_{\alpha/2}\widehat{\sigma}_\epsilon}{\delta \Delta x}\right)^2}$$

where the $t_{\alpha/2}$ values may differ a bit because of the difference in error degrees of freedom. Both experiments cover the same x range, so $\Delta x = x_{max} - x_{min}$ and the sample size ratio reduces to:

$$\frac{N_{\text{uniform } x}}{N_{\text{two levels of } x}} \simeq 3.$$

6.1. Linear Regression

That is, three times as many observations are required in an experiment that uses uniformly distributed x values than if the x values are concentrated at the ends of the x range. Because we saw in Example 6.1 that the experiment with three evenly spaced, equally weighted levels of x requires 1.5 times as many observations as the two-level experiment, other methods of taking evenly spaced, equally weighted observations of x must give experiment sample size ratios between 1.5 and 3. Obviously, the two-level equally weighted method is the most efficient method for specifying x values for an experiment.

6.1.2 Test for the Slope

The hypotheses to be tested are $H_0 : \beta_1 = 0$ versus $H_A : \beta_1 \neq 0$, where β_1 is the regression slope parameter to be tested using the b_1 statistic. The sampling distribution of b_1 follows Student's t distribution with $df_\epsilon = N - 2$ degrees of freedom and has standard error given by

$$\sigma_{b_1} = \frac{\sigma_\epsilon}{\sqrt{SS_x}}. \tag{6.16}$$

The power to reject H_0 when H_A is true is given by[1]

$$\pi = P(-\infty < t < t_\beta) \tag{6.17}$$

where

$$\begin{aligned} t_\beta &= \frac{|\beta_1|}{\sigma_{b_1}} - t_{\alpha/2} \\ &= \frac{|\beta_1|\sqrt{SS_x}}{\sigma_\epsilon} - t_{\alpha/2}. \end{aligned} \tag{6.18}$$

Because $SS_x = N\sigma_x^2$, the number of observations required to obtain a specified power value for a given β_1 value is the smallest value of N that satisfies

$$N \geq \left(t_{\alpha/2} + t_\beta\right)^2 \left(\frac{\sigma_\epsilon}{\beta_1 \sigma_x}\right)^2. \tag{6.19}$$

This expression is transcendental and must be solved by iteration. $t \simeq z$ provides a good starting point for iterations.

Example 6.3 For an experiment to be analyzed by linear regression with a single predictor, how many observations are required to reject $H_0 : \beta_1 = 0$ in favor of $H_A : \beta_1 \neq 0$ with 90% power for $\beta_1 = 10$ when a) the distribution of x values will be normal with $\mu_x \simeq 15$ and $\sigma_x \simeq 2$; b) an equal number of observations

---

[1] The notation is confusing here. Be careful to distinguish between the slope parameter β_1 and the type II error rate β.

will be taken at $x = 10$ and $x = 20$; c) uniformly distributed values of x will be used over the interval $10 \leq x \leq 20$; and d) an equal number of observations will be taken at $x = 10$, 15, and 20. Experience with the process tells us the standard error of the model is expected to be $\sigma_\epsilon = 30$.

Solution:

a) With $t \simeq z$ in Equation 6.19, the first iteration to find N gives

$$N = (z_{0.025} + z_{0.10})^2 \left(\frac{30}{10 \times 2}\right)^2 = 24.$$

Further iterations indicate that the required sample size is $N = 26$.

b) The standard deviation of the x values will be

$$\sigma_x = \sqrt{\frac{SS_x}{N}} = \sqrt{\frac{1}{N}\frac{N}{2}\left((-5)^2 + (5)^2\right)} = 5.$$

The first iteration to find N, with $t \simeq z$, gives

$$N = (z_{0.025} + z_{0.10})^2 \left(\frac{30}{10 \times 5}\right)^2 = 4.$$

Further iterations indicate that $N = 7$ observations are required.

c) For uniformly distributed x, the standard deviation of the x values is

$$\sigma_x = \frac{x_{\max} - x_{\min}}{\sqrt{12}} = \frac{10}{\sqrt{12}} = 2.89.$$

With $t \simeq z$, the first iteration to find N gives

$$N = (z_{0.025} + z_{0.10})^2 \left(\frac{30}{10 \times 2.89}\right)^2 = 12.$$

Further iterations indicate that $N = 14$ observations are required.

d) The standard deviation of the x values will be

$$\sigma_x = \sqrt{\frac{SS_x}{N}} = \sqrt{\frac{1}{N}\frac{N}{3}\left((-5)^2 + (0)^2 + (5)^2\right)} = 4.0825.$$

The first iteration to find N, with $t \simeq z$, gives

$$N = (z_{0.025} + z_{0.10})^2 \left(\frac{30}{10 \times 4.0825}\right)^2 = 6.$$

Further iterations indicate that $N = 9$ observations are required.

6.2. Logistic Regression

Example 6.4 What is the power to reject H_0 for the situation described in Example 6.3a if the sample size is $N = 20$?

Solution: From Equation 6.18 with $SS_x = N\sigma_x^2$ and $df_\epsilon = 20 - 2 = 18$,

$$t_\beta = \frac{|\beta_1|\sqrt{N}\sigma_x}{\sigma_\epsilon} - t_{0.025,18}$$

$$= \frac{10\sqrt{202}}{30} - 2.10$$

$$= 0.881.$$

The power, as given by Equation 6.17, is

$$\pi = P(-\infty < t < 0.881)$$

$$= 0.805.$$

6.2 Logistic Regression

If a dichotomous (i.e., binary) response y depends on a continuous variable x where the probability of success on any trial, $p = P(y = 1)$, is a function of x, then $p(x)$ is generally S-shaped or *sigmoidal*. However, the logistic-transformed response given by

$$p'(x) = \ln\left(\frac{p}{1-p}\right) \tag{6.20}$$

is approximately linear and may be fitted with a binary logistic regression (BLR) model of the form

$$\ln\left(\frac{p}{1-p}\right) = \beta_0 + \beta_1 x. \tag{6.21}$$

The hypotheses to be tested are $H_0 : \beta_1 = 0$ versus $H_A : \beta_1 \neq 0$.

The purpose of this section is to present sample size calculation methods for experiments to be analyzed by binary logistic regression.

6.2.1 Dichotomous Independent Variable

When x has only two states, the test of $H_0 : \beta_1 = 0$ versus $H_A : \beta_1 \neq 0$ coincides with the two-sample test for proportions $H_0 : p_1 = p_2$ versus $H_A : p_1 \neq p_2$ where the subscripts distinguish the two levels of x, then the sample size and power calculation methods from the two-sample test for proportions are appropriate for the hypothesis test of the logistic regression model's slope. The most common sample-size calculation method used in this case is the log odds ratio method of Section 4.3.2.5.

Example 6.5 What sample size is required for an experiment to be analyzed by logistic regression if $H_0 : \beta_1 = 0$ should be rejected in favor of $H_A : \beta_1 \neq 0$ with 90% power when x is dichotomous with associated proportions $p_1 = 0.04$ and $p_2 = 0.08$?

Solution: The odds ratio for the given proportions is

$$OR = \frac{p_1/(1-p_1)}{p_2/(1-p_2)} = \frac{0.04/0.96}{0.08/0.92} = 0.479.$$

The required sample size is given by Equation 4.97:

$$n = \left(\frac{z_{0.025} + z_{0.10}}{\ln(0.479)}\right)^2 \left(\frac{1}{0.04(0.96)} + \frac{1}{0.08(0.92)}\right) = 770.$$

6.2.2 Normally Distributed Independent Variable

When the distribution of x is normal with mean μ and standard deviation σ, then Hsieh [29] has shown that the approximate sample size required to reject $H_0 : \beta_1 = 0$ in favor of $H_A : \beta_1 \neq 0$ is given by

$$n = \frac{(z_{\alpha/2} + z_\beta)^2}{p_\mu(1-p_\mu)\left(\ln\left(\frac{p_\mu/(1-p_\mu)}{p_{\mu+\sigma}/(1-p_{\mu+\sigma})}\right)\right)^2} \tag{6.22}$$

where p_μ is the expected success probability evaluated at $x = \mu$ and $p_{\mu+\sigma}$ is the expected success probability evaluated at $x = \mu+\sigma$. Hsieh also gives an exact but more complicated sample size calculation that gives similar large-sample results.

Example 6.6 What sample size is required for an experiment to be analyzed by logistic regression if $H_0 : \beta_1 = 0$ should be rejected in favor of $H_A : \beta_1 \neq 0$ with 90% power when x is normally distributed with expected success proportions $p(x = \mu) = 0.14$ and $p(x = \mu+\sigma) = 0.22$.

Solution: From Equation 6.22 the required sample size is

$$n = \frac{(1.96 + 1.282)^2}{0.14(0.86)\left(\ln\left(\frac{0.14/0.86}{0.22/0.78}\right)\right)^2} = 289.$$

Chapter 7

Correlation and Agreement

Correlation and agreement analyses are used to quantify the agreement between paired observations. The methods of analysis that are presented are

- Pearson's correlation.
- intraclass correlation.
- Cohen's kappa (κ).
- receiver operating characteristic (ROC) curves.
- Bland-Altman plots.

Pearson's correlation, intraclass correlation, and Bland-Altman plots use quantitative data, Cohen's κ is used to measure the agreement between paired attribute observations, and ROC curves are used to evaluate the accuracy of a binary classification method.

7.1 Pearson's Correlation

Pearson's correlation is used to quantify the association between two continuous paired random variables. The sample correlation coefficient r provides an estimate of the population correlation parameter ρ, where r and ρ are confined to the interval $[-1, 1]$. $\rho = 0$ indicates no correlation, $\rho = 1$ indicates perfect positive correlation, and $\rho = -1$ indicates perfect negative correlation.

The sampling distribution of ρ is skewed, however, the distribution of Fisher's Z transform given by

$$Z = \tanh^{-1}(r) \tag{7.1}$$
$$= \frac{1}{2} \ln\left(\frac{1+r}{1-r}\right) \tag{7.2}$$

is approximately normal with mean

$$\mu_Z = \frac{1}{2}\ln\left(\frac{1+\rho}{1-\rho}\right) \quad (7.3)$$

and standard deviation

$$\sigma_Z = \frac{1}{\sqrt{n-3}} \quad (7.4)$$

where n is the number of paired observations used to determine r. Because Z is better behaved than r, sample size and power calculations are performed in terms of Z.

The equation for r in terms of Z is

$$r = \frac{e^{2Z}-1}{e^{2Z}+1}. \quad (7.5)$$

However, transformations between Z and r are usually performed by using an appropriate table or a scientific calculator that has hypergeometric trigonometry functions.

7.1.1 Confidence Interval for Pearson's Correlation

A calculation for the number of paired observations is desired to obtain a confidence interval for the correlation ρ with specified confidence interval half-width of the form

$$P(r-\delta < \rho < r+\delta) = 1-\alpha. \quad (7.6)$$

As simple as this confidence interval appears, the sampling distribution of r is too skewed, so this approach is not realistic and other methods must be considered. If numerical values are chosen for the upper and lower confidence limits on ρ,

$$P(LCL_\rho < \rho < UCL_\rho) = 1-\alpha. \quad (7.7)$$

Then the limits may be Z-transformed to obtain

$$P\left(Z_{LCL_\rho} < Z_\rho < Z_{UCL_\rho}\right) = 1-\alpha. \quad (7.8)$$

Because Z is approximately normal with standard deviation given by Equation 7.4, the confidence interval width must satisfy the condition

$$\begin{aligned} Z_{UCL_\rho} - Z_{LCL_\rho} &= 2z_{\alpha/2}\sigma_Z \\ &= \frac{2z_{\alpha/2}}{\sqrt{n-3}}, \end{aligned} \quad (7.9)$$

so the sample size to obtain the desired confidence interval is

$$n = 4\left(\frac{z_{\alpha/2}}{Z_{UCL_\rho} - Z_{LCL_\rho}}\right)^2 + 3. \quad (7.10)$$

7.1. Pearson's Correlation

Example 7.1 Determine the number of paired observations required to obtain the following confidence interval for the population correlation:

$$P(0.9 < \rho < 0.99) = 0.95.$$

Solution: The Fisher's Z-transformed confidence interval is

$$P(Z_{0.9} < Z_\rho < Z_{0.99}) = 0.95$$
$$P(1.472 < Z_\rho < 2.647) = 0.95.$$

Then with $\alpha = 0.05$ in Equation 7.10, the required sample size is

$$n = 4\left(\frac{1.96}{2.647 - 1.472}\right)^2 + 3$$
$$= 15.$$

7.1.2 One-Sample Test for Pearson's Correlation

Power and sample size calculations for the one-sample test for correlation use the usual one-sample z test methods after the stated correlations are transformed using Fisher's Z transform. The following example demonstrates a typical power calculation for a one-sample test for correlation.

Example 7.2 An experiment is planned to test $H_0 : \rho = 0.9$ versus $H_A : \rho < 0.9$ on the basis of $n = 28$ paired observations. Determine the power of the test to reject H_0 when $\rho = 0.7$.
Solution: Under H_0 following Fisher's transform we have

$$(\mu_Z)_0 = \frac{1}{2}\ln\left(\frac{1 + 0.9}{1 - 0.9}\right) = 1.472$$

and by Equation 7.4

$$\sigma_Z = \frac{1}{\sqrt{28 - 3}} = 0.2.$$

For the one-sided left-tailed test, the critical value of Z that distinguishes the accept/reject regions is given by

$$Z_{A/R} = (\mu_Z)_0 - z_\alpha \sigma_z$$
$$= 1.472 - z_{0.05}(0.2)$$
$$= 1.472 - 1.645(0.2)$$
$$= 1.143.$$

The corresponding Z value under H_A when $\rho = 0.7$ is

$$(\mu_Z)_A = \frac{1}{2}\ln\left(\frac{1+0.7}{1-0.7}\right) = 0.867.$$

Then the power to reject H_0 when $\rho = 0.7$ is

$$\begin{aligned}\pi &= \Phi\left(-\infty < Z < Z_{A/R}; (\mu_Z)_A, \sigma_Z\right) \\ &= \Phi(-\infty < Z < 1.143; 0.867, 0.2) \\ &= \Phi(-\infty < z < 1.38) \\ &= 0.916.\end{aligned}$$

7.1.3 Two-Sample Test for Pearson's Correlation

Power and sample size calculations for two-sample tests for correlation may be performed by applying Fisher's Z transform to the correlations and using the usual two-sample normal distribution methods modified to account for the dependence of σ_Z on the sample size.

For a specified difference ΔZ between two correlations and specified sample sizes, the power to reject $H_0 : \rho_1 = \rho_2$ in favor of $H_A : \rho_1 \neq \rho_2$ is given by

$$\pi = \Phi\left(-\infty < z < z_\beta\right) \tag{7.11}$$

where

$$z_\beta = \frac{\Delta Z}{\sigma_{\Delta Z}} - z_{\alpha/2} \tag{7.12}$$

and

$$\begin{aligned}\sigma_{\Delta Z} &= \sqrt{\sigma_{Z_1}^2 + \sigma_{Z_2}^2} \\ &= \sqrt{\frac{1}{n_1 - 3} + \frac{1}{n_2 - 3}}.\end{aligned} \tag{7.13}$$

When $n_1 = n_2 = n$ we have $\sigma_{\Delta Z} = \sqrt{\frac{2}{n-3}}$ and z_β simplifies to

$$z_\beta = \frac{\Delta Z}{\sqrt{\frac{2}{n-3}}} - z_{\alpha/2}. \tag{7.14}$$

If the common sample size n is to be determined for specified values of ΔZ and the power, then the required sample size is

$$n = 2\left(\frac{z_{\alpha/2} + z_\beta}{\Delta Z}\right)^2 + 3. \tag{7.15}$$

7.1. Pearson's Correlation

Example 7.3 Find the power to reject $H_0 : \rho_1 = \rho_2$ in favor of $H_A : \rho_1 \neq \rho_2$ when $\rho_1 = 0.99$, $\rho_2 = 0.95$, and $n_1 = n_2 = 30$.

Solution: The Fisher-transformed difference between the two correlations under H_A is

$$\begin{aligned} \Delta Z &= Z_1 - Z_2 \\ &= \frac{1}{2} \ln\left(\frac{1+0.99}{1-0.99}\right) - \frac{1}{2} \ln\left(\frac{1+0.95}{1-0.95}\right) \\ &= 0.815. \end{aligned}$$

From Equations 7.11 and 7.14 with $\alpha = 0.05$ the power is

$$\begin{aligned} \pi &= \Phi\left(-\infty < z < \left(\frac{0.815}{\sqrt{\frac{2}{30-3}}} - 1.96\right)\right) \\ &= \Phi(-\infty < z < 1.03) \\ &= 0.85. \end{aligned}$$

Example 7.4 What sample size should be drawn from two populations to perform the two-sample test for correlation ($H_0 : \rho_1 = \rho_2$ versus $H_A : \rho_1 \neq \rho_2$) with 90% power to reject H_0 when $\rho_1 = 0.9$ and $\rho_2 = 0.8$?

Solution: The Fisher-transformed difference between the two correlations is

$$\begin{aligned} \Delta Z &= Z_1 - Z_2 \\ &= \frac{1}{2} \ln\left(\frac{1+0.9}{1-0.9}\right) - \frac{1}{2} \ln\left(\frac{1+0.8}{1-0.8}\right) \\ &= 0.374. \end{aligned}$$

From Equation 7.15 with $\alpha = 0.05$ and $\beta = 0.10$ the required common sample size is

$$\begin{aligned} n &= 2 \left(\frac{1.96 + 1.28}{0.374}\right)^2 + 3 \\ &= 154. \end{aligned}$$

7.1.4 Multiple Correlation

The coefficient of multiple determination for a response y, which is a function of k continuous predictors $x_1, x_2, \ldots, x_k$, is given in terms of the ANOVA sums of squares (SS) by

$$r^2 = \frac{SS_{model}}{SS_{total}} = 1 - \frac{SS_\epsilon}{SS_{total}} \tag{7.16}$$

where the x_i are random, as opposed to fixed, variables. In this application, the purpose of calculating r^2 is to determine whether y is indeed a function of one or more of the x_i, that is, we desire to test the hypotheses $H_0 : \rho^2 = 0$ versus $H_A : \rho^2 > 0$. This test may be performed using an omnibus F statistic that simultaneously tests for the significance of all of the predictors in the model. If there are n observations in the data set, then this omnibus F statistic is given by

$$\begin{align} F &= \frac{MS_{model}}{MS_\epsilon} \\ &= \frac{SS_{model}/df_{model}}{SS_\epsilon/df_\epsilon} \\ &= \frac{df_\epsilon}{df_{model}} \frac{SS_{model}/SS_{total}}{SS_\epsilon/SS_{total}} \\ &= \frac{n-k-1}{k} \frac{r^2}{1-r^2}. \end{align} \qquad (7.17)$$

The power π to reject H_0 for a specified value of ρ^2 is given by the condition

$$F_{1-\alpha} = F_{1-\pi,\phi} \qquad (7.18)$$

where the central and noncentral F distributions have $df_{model} = k$ numerator and $df_\epsilon = n - k - 1$ denominator degrees of freedom and the F distribution noncentrality parameter ϕ is

$$\phi = n\frac{\rho^2}{1-\rho^2}. \qquad (7.19)$$

Example 7.5 Determine the power to reject $H_0 : \rho^2 = 0$ when in fact $\rho^2 = 0.6$ based on a sample of $n = 20$ observations taken with four random covariates.
Solution: The regression model for y as a function of the four predictors will have $df_{model} = k = 4$ model degrees of freedom and $df_\epsilon = n-k-1 = 20-4-1 = 15$ error degrees of freedom. The F distribution noncentrality parameter from Equation 7.19 with $\rho^2 = 0.6$ is

$$\phi = 20\frac{0.6}{1-0.6} = 30.$$

From Equation 7.18 we have

$$\begin{align} F_{0.95} &= F_{1-\pi,30} \\ 3.056 &= F_{0.024,30}, \end{align}$$

so the power is $\pi = 1 - 0.024 = 0.976$.

7.2 Intraclass Correlation

Pearson's correlation (Section 7.1) is used to determine the correlation between paired observations when there is a natural distinction between the two observations that make up each pair. When there is no natural distinction between the paired observations, Pearson's correlation is not appropriate and the *intraclass correlation coefficient* (ICC) is used instead. The intraclass correlation may also be used when there are more than two observations in each group. For example, Pearson's correlation would be used to determine the correlation between the heights of brothers and sisters from the same family (by randomly sampling one brother and one sister from each of many families), but the intraclass correlation would be necessary to determine the correlation between two or more brothers from the same family.

There are several different types of intraclass correlation coefficients. All of them are based on one-way or two-way classification designs analyzed by mixed or random effects ANOVA. Shrout and Fleiss [62] identified six intraclass correlations determined by the combination of three ANOVA models and two statistics. The language adopted here is taken from measurement reliability analysis, where observers or raters are rating subjects and the ICC quantifies the agreement between raters. The three experiment designs and their ANOVA models are as follows:

- Ratings are assigned by different raters chosen at random who potentially do not inspect all of the subjects. Subjects are random. Analysis is by one-way random effects ANOVA for subjects, so rater biases are pooled with the error.

- Every subject is rated by every rater in a balanced complete two-way crossed design analyzed by two-way ANOVA where subjects and raters are both random.

- Every subject is rated by every rater in a balanced complete two-way crossed design analyzed by two-way mixed effects ANOVA where subjects are random and raters are fixed.

The two types of ICC statistics that can be calculated for each of these three cases are

- ICC for a single rating.

- ICC for the average rating taken over all raters.

The only case considered here is the single-rater ICC for the one-way ANOVA random effects model. See the literature for details about other intraclass correlation coefficients.

Source	df	SS	MS	E(MS)	F
Subject	$n-1$	SS_τ	MS_τ	$n\sigma_\tau^2 + \sigma_\epsilon^2$	F_τ
Error	$n(r-1)$	SS_ϵ	MS_ϵ	σ_ϵ^2	
Total	$nr-1$	SS_{total}			

Table 7.1: ANOVA table for calculating the intraclass correlation coefficient.

If each of n subjects is rated once by each of r raters who are randomly selected from a large population of raters, then the model for the rating response is

$$y_{ij} = \mu + \tau_i + \epsilon_{ij} \tag{7.20}$$

where μ is the population mean rating, τ_i is the bias of the ith subject relative to μ, and ϵ_{ij} is the rating error attributed to a combination of rater bias and random error.[1] The τ_i are assumed to be independent and normally distributed with mean equal to zero and standard deviation σ_τ. The ϵ_{ij} values are assumed to be independent and normally distributed with mean equal to zero and standard deviation σ_ϵ.

Table 7.1 shows the one-way random effects ANOVA table layout and the expected mean squares ($E(MS)$). The intraclass correlation coefficient is defined as

$$ICC = \frac{\sigma_\tau^2}{\sigma_\tau^2 + \sigma_\epsilon^2} \tag{7.21}$$

where σ_τ and σ_ϵ are estimated by variance components analysis, that is, by solving the system of equations given by equating the mean squares with their expected mean squares. Equation 7.21 shows that the intraclass correlation may be interpreted as the proportion of the total variance in a measurement that is explained by differences between subjects. ICC will be near 1 when there are large differences between subjects and small differences within subjects, that is, when $\sigma_\tau^2 \gg \sigma_\epsilon^2$. ICC will be near 0 when there are small differences between subjects and large differences within subjects, that is, when $\sigma_\tau^2 \ll \sigma_\epsilon^2$.

From variance components analysis, ICC can be estimated from the experimental mean squares by

$$\widehat{ICC} = \frac{MS_\tau - MS_\epsilon}{MS_\tau + (n-1)MS_\epsilon} \tag{7.22}$$

or from the ANOVA F statistic for the random subject effect by

$$\widehat{ICC} = \frac{F_\tau - 1}{F_\tau + n - 1} \tag{7.23}$$

where

$$F_\tau = \frac{MS_\tau}{MS_\epsilon}. \tag{7.24}$$

---

[1] For those familiar with the language of gage error studies, ϵ_{ij} has two components, one from operator reproducibility and another from measurement repeatability.

7.2. Intraclass Correlation

Equation 7.23 is especially important. It shows that ICC is simply a mathematical transformation of F_τ, so that the results of confidence interval and hypothesis test analyses for F_τ may be transformed to give the corresponding analyses for ICC. Intraclass correlations also have similar distribution behavior to Pearson's correlation, so the approximate Fisher's transformation method from Section 7.1 is applicable with some minor changes.

7.2.1 Confidence Interval for the Intraclass Correlation

The exact confidence interval for ICC is determined from the confidence interval for $(\sigma_\tau/\sigma_\epsilon)^2$. The quantity

$$\frac{MS_\tau/\left(n\sigma_\tau^2 + \sigma_\epsilon^2\right)}{MS_\epsilon/\sigma_\epsilon^2}$$

follows the F distribution with $n-1$ numerator and $n(r-1)$ denominator degrees of freedom, so

$$P\left(F_{\alpha/2} < \frac{MS_\tau/\left(n\sigma_\tau^2 + \sigma_\epsilon^2\right)}{MS_\epsilon/\sigma_\epsilon^2} < F_{1-\alpha/2}\right) = 1-\alpha. \tag{7.25}$$

This equation can be manipulated to obtain the following confidence interval for $(\sigma_\tau/\sigma_\epsilon)^2$:

$$P\left(LCL_{(\sigma_\tau/\sigma_\epsilon)^2} < (\sigma_\tau/\sigma_\epsilon)^2 < UCL_{(\sigma_\tau/\sigma_\epsilon)^2}\right) = 1-\alpha \tag{7.26}$$

where the lower and upper confidence limits are

$$LCL_{(\sigma_\tau/\sigma_\epsilon)^2} = \frac{1}{r}\left(\frac{F_\tau}{F_{1-\alpha/2}} - 1\right) \tag{7.27}$$

and

$$UCL_{(\sigma_\tau/\sigma_\epsilon)^2} = \frac{1}{r}\left(\frac{F_\tau}{F_{\alpha/2}} - 1\right). \tag{7.28}$$

From Equation 7.21, ICC may be expressed as

$$ICC = \frac{(\sigma_\tau/\sigma_\epsilon)^2}{(\sigma_\tau/\sigma_\epsilon)^2 + 1}, \tag{7.29}$$

so the confidence limits for ICC can be calculated from the upper and lower confidence limits for $(\sigma_\alpha/\sigma_\epsilon)^2$ as

$$P\left(\frac{LCL_{(\sigma_\tau/\sigma_\epsilon)^2}}{LCL_{(\sigma_\tau/\sigma_\epsilon)^2} + 1} < ICC < \frac{UCL_{(\sigma_\tau/\sigma_\epsilon)^2}}{UCL_{(\sigma_\tau/\sigma_\epsilon)^2} + 1}\right) = 1-\alpha. \tag{7.30}$$

Equation 7.30 is useful for calculating exact confidence limits for ICC from experimental data, but it is not practical for calculating sample sizes for experiments.

An approximate confidence interval for ICC can be calculated using a slightly modified form of the Fisher's Z transformation method of Section 7.1. The Fisher's transform for ICC is

$$Z_{ICC} = \frac{1}{2} \ln \left(\frac{1 + (r-1) ICC}{1 - ICC} \right) \tag{7.31}$$

and the standard deviation of the sampling distribution of Z_{ICC} is given by

$$\sigma_{\widehat{Z}_{ICC}} = \begin{cases} \frac{1}{\sqrt{n - \frac{3}{2}}} & \text{if } r = 2 \\ \sqrt{\frac{r}{2(r-1)(n-2)}} & \text{if } r > 2 \end{cases} \tag{7.32}$$

Then the $(1 - \alpha)\,100\%$ confidence interval for ICC may be found from the inverse transform of the corresponding confidence interval for Z_{ICC} given by

$$\Phi \left(\widehat{Z}_{LCL} < Z_{ICC} < \widehat{Z}_{UCL} \right) = 1 - \alpha \tag{7.33}$$

where

$$\widehat{Z}_{LCL} = \widehat{Z}_{ICC} - z_{\alpha/2} \sigma_{\widehat{Z}_{ICC}} \tag{7.34}$$
$$\widehat{Z}_{UCL} = \widehat{Z}_{ICC} + z_{\alpha/2} \sigma_{\widehat{Z}_{ICC}}. \tag{7.35}$$

The confidence interval width will be

$$\widehat{Z}_{UCL} - \widehat{Z}_{LCL} = 2 z_{\alpha/2} \sigma_{\widehat{Z}_{ICC}},$$

so from Equation 7.32, the number of subjects n required to obtain a confidence interval for ICC with specified upper and lower confidence limits is approximately

$$n = \begin{cases} 4 \left(\frac{z_{\alpha/2}}{\widehat{Z}_{UCL} - \widehat{Z}_{LCL}} \right)^2 + \frac{3}{2} & \text{if } r = 2 \\ \frac{2r}{r-1} \left(\frac{z_{\alpha/2}}{\widehat{Z}_{UCL} - \widehat{Z}_{LCL}} \right)^2 + 2 & \text{if } r > 2 \end{cases} \tag{7.36}$$

Because the number of subjects n will depend on the number of raters r who rate each subject, a unique value of r will minimize the total number of ratings $n \times r$. This might be an important consideration if the cost of obtaining individual ratings is relatively high.

Donner and Koval [18] give the following large-sample formula for $\sigma_{\widehat{ICC}}$:

$$\sigma_{\widehat{ICC}} = \sqrt{\frac{2}{r(r-1)n}} \left(1 - \widehat{ICC} \right) \left(1 + (r-1) \widehat{ICC} \right), \tag{7.37}$$

7.2. Intraclass Correlation

which leads to the following sample size requirement:

$$n = \frac{8}{r(r-1)} \left(\frac{z_{\alpha/2}(1-ICC)(1+(r-1)ICC)}{UCL_{ICC} - LCL_{ICC}} \right)^2. \qquad (7.38)$$

This solution gives sample sizes similar to those determined by the Fisher's transformation method.

Example 7.6 An experiment will be performed to determine the single-rater intraclass correlation in a one-way design with $r = 2$ observations per subject. How many subjects must be sampled if the desired confidence interval for ICC is $P(0.7 < ICC < 0.9) = 0.95$?

Solution: By Equation 7.31, the desired confidence interval for ICC transforms into the following confidence interval for Z_{ICC}:

$$P(0.867 < Z_{ICC} < 1.472) = 0.95.$$

Then, from Equation 7.36 with $r = 2$ observations per subject and $\alpha = 0.05$, the number of subjects required is

$$\begin{aligned} n &= 4 \left(\frac{1.96}{1.472 - 0.867} \right)^2 + \frac{3}{2} \\ &= 44. \end{aligned}$$

Example 7.7 Confirm the answer to Example 7.6 using the method of Donner and Koval.

Solution: Assuming that $\widehat{ICC} = 0.8$, the sample size according to Donner and Koval is given by Equation 7.38:

$$\begin{aligned} n &= \frac{8}{2(2-1)} \left(\frac{1.96(1-0.8)(1+(2-1)0.8)}{0.9 - 0.7} \right)^2 \\ &= 50, \end{aligned}$$

which is in reasonable agreement with the sample size determined by the Fisher's transformation method.

7.2.2 Test for the Intraclass Correlation

The hypotheses to be tested are $H_0 : ICC = ICC_0$ versus $H_A : ICC > ICC_0$. When the single-rater ICC is determined from n subjects that are each evaluated by r randomly selected raters, the experimental data are analyzed by one-way random effects ANOVA for subjects, and the ICC is estimated from the variance components estimates. The power associated with the ANOVA F test for the random subjects effect is given by the methods of Section 8.4.2, but F_τ and $\widehat{ICC}$ are related by Equation 7.23, so the power for the test of ICC is determined by the power for the F test for the subjects effect. This analysis shows that the exact power π to reject $H_0 : ICC = ICC_0$ when $ICC = ICC_1$ is given by

$$\pi = P\left(\frac{1+r\left(\frac{ICC_0}{1-ICC_0}\right)}{1+r\left(\frac{ICC_1}{1-ICC_1}\right)} F_{1-\alpha} < F < \infty\right) \tag{7.39}$$

where the F distribution has $df_1 = n - 1$ and $df_2 = n(r - 1)$, r is the number of raters, and n is the number of subjects being rated.

Equation 7.39 is not practical for the purpose of calculating experiment sample sizes; however, the power may be approximated using the Fisher's Z transformation method which is more practical for calculating sample size. By Fisher's method, the approximate power is

$$\pi = \Phi\left(-z_\beta < z < \infty\right) \tag{7.40}$$

where

$$z_\beta = \frac{(Z_1 - Z_0)}{\sigma_{\widehat{Z}_{ICC}}} - z_\alpha \tag{7.41}$$

where the Z_i are given by Equation 7.31 and $\sigma_{\widehat{Z}_{ICC}}$ is given by Equation 7.32. Then the number of subjects required to reject $H_0 : ICC = ICC_0$ with power $\pi = 1 - \beta$ when $ICC = ICC_1$ is approximately

$$n = \begin{cases} \left(\frac{z_\alpha + z_\beta}{Z_1 - Z_0}\right)^2 + \frac{3}{2} & \text{if } r = 2 \\ \frac{r}{2(r-1)}\left(\frac{z_\alpha + z_\beta}{Z_1 - Z_0}\right)^2 + 2 & \text{if } r > 2 \end{cases}. \tag{7.42}$$

Example 7.8 How many subjects are required in an experiment to reject $H_0 : ICC = 0.6$ with 80% power when $ICC = 0.8$ and two raters will rate each subject? Confirm the sample size by calculating the exact power.
Solution: From Equation 7.31 the Z_{ICC} values corresponding to $ICC = 0.6$ and $ICC = 0.8$ are $Z_0 = 0.693$ and $Z_1 = 1.099$, respectively. From Equation 7.42 with $r = 2$ and $\alpha = 0.05$, the approximate sample size is

$$\begin{aligned} n &= \left(\frac{z_{0.05} + z_{0.20}}{Z_1 - Z_0}\right)^2 + \frac{3}{2} \\ &= \left(\frac{1.645 + 0.84}{1.099 - 0.693}\right)^2 + \frac{3}{2} \\ &= 39. \end{aligned}$$

The exact power is given by Equation 7.39 where the F distribution has $df_1 = 39 - 1 = 38$ numerator degrees of freedom and $df_2 = 39(2-1) = 39$ denomi-

7.3. Cohen's Kappa

		\multicolumn{3}{c}{j}		
		1	2	3
i	1	f_{11}	f_{12}	f_{13}
	2	f_{21}	f_{22}	f_{23}
	3	f_{31}	f_{32}	f_{33}

Table 7.2: Frequency table for $k = 3$ categories for κ calculation.

nator degrees of freedom. The power is given by

$$\begin{aligned}\pi &= P\left(\frac{1+2\left(\frac{0.6}{1-0.6}\right)}{1+2\left(\frac{0.8}{1-0.8}\right)}F_{0.95} < F < \infty\right) \\ &= P\left(0.760 < F < \infty\right) \\ &= 0.80,\end{aligned}$$

which is in excellent agreement with the target power.

7.3 Cohen's Kappa

Cohen's kappa (κ) is used to test for or quantify the agreement between paired observations of a categorical response by two raters. If the ratings are summarized in a square category-versus-category table of frequencies f_{ij}, such as in Table 7.2 for $k = 3$ categories, where i and j represent the categories of the first and second rater, respectively, then the proportion of the total number of observations in the ijth cell of the table is

$$p_{ij} = \frac{f_{ij}}{n} \tag{7.43}$$

where $n = f_{\bullet\bullet}$ is the total number of paired observations.[2] The experimental κ is defined as

$$\widehat{\kappa} = \frac{p_o - p_e}{1 - p_e} \tag{7.44}$$

where the observed proportion of the cases in which the raters are in agreement is

$$p_o = \sum_{i=1}^{k} p_{ii} \tag{7.45}$$

and the expected proportional agreement by chance is

$$p_e = \sum_{i=1}^{k} p_{i\bullet} p_{\bullet i}. \tag{7.46}$$

---

[2] The dot notation indicates summation over the dotted subscript, that is, $f_{\bullet\bullet} = \sum_{i}\sum_{j} f_{ij}$.

The numerator in Equation 7.44, $p_o - p_e$, gives the proportion of excess agreement and the denominator, $1 - p_e$, gives the potential excess agreement. When the raters are in perfect agreement with each other, then $p_o = 1$ and $\hat{\kappa} = 1$. If the agreement between raters is no better than chance, then $p_o \simeq p_e$ and $\kappa \simeq 0$. If the agreement is worse than chance, then $p_o < p_e$ and $\hat{\kappa} < 0$, but such cases are rare.

Fleiss [22] gives the equation for calculating the standard error for $\hat{\kappa}$ from the experimental p_{ij}, but that equation is not very useful for estimating the sample size when planning an experiment. A large-sample approximation for $\sigma_{\hat{\kappa}}$ is given by Sheskin:

$$\sigma_{\hat{\kappa}} \simeq \frac{1}{1 - p_e} \sqrt{\frac{p_o(1 - p_o)}{n}} \tag{7.47}$$

where it might be necessary to express p_o in terms of p_e and κ:

$$p_o = \kappa(1 - p_e) + p_e. \tag{7.48}$$

In many practical situations, κ takes on an extreme value and an approximation for $\sigma_{\hat{\kappa}}$ is available. When κ is expected to be near 0, then $p_o \simeq p_e$ and Equation 7.47 simplifies to

$$\sigma_{\hat{\kappa}} \simeq \sqrt{\frac{p_e}{n(1 - p_e)}}. \tag{7.49}$$

Hanley [25] observes that when κ is large, say $\kappa > 0.8$, then $\sigma_{\hat{\kappa}}$ may be approximated by

$$\sigma_{\hat{\kappa}} \simeq \sqrt{\frac{\hat{\kappa}(1 - \hat{\kappa})}{n(1 - p_e)}}. \tag{7.50}$$

The value of p_e used in sample size calculations may be estimated from a preliminary experiment, but estimates for p_e may also be determined from the expected allocation of units to the k categories. For example, if the units to be inspected by the two raters are selected in a balanced manner, so that there are about n/k units in each category, then, from Equation 7.46,

$$p_e \simeq k\left(\frac{1}{k}\right)^2$$
$$\simeq \frac{1}{k}. \tag{7.51}$$

7.3.1 Confidence Interval for Cohen's Kappa

The sampling distribution of $\hat{\kappa}$ for large samples is approximately normal, so an approximate $(1 - \alpha)\,100\%$ confidence interval for κ is given by

$$\Phi(\hat{\kappa} - \delta < \kappa < \hat{\kappa} + \delta) = 1 - \alpha \tag{7.52}$$

7.3. Cohen's Kappa

with confidence interval half-width

$$\delta = z_{\alpha/2}\sigma_{\hat{\kappa}}. \tag{7.53}$$

When κ is expected to be near 0, $\sigma_{\hat{\kappa}}$ may be estimated using Equation 7.49, so the approximate sample size required to estimate κ with specified confidence interval half-width δ is

$$n = \frac{p_e}{1 - p_e}\left(\frac{z_{\alpha/2}}{\delta}\right)^2 \tag{7.54}$$

where an estimate for p_e is required to complete the calculation.

When κ is expected to be near 1, $\sigma_{\hat{\kappa}}$ may be estimated using Equation 7.50, so the approximate sample size required to estimate κ with specified confidence interval half-width δ is

$$n = \frac{\hat{\kappa}(1 - \hat{\kappa})}{1 - p_e}\left(\frac{z_{\alpha/2}}{\delta}\right)^2 \tag{7.55}$$

where estimates for p_e and κ are required.

Example 7.9 How many units should two operators evaluate in an attribute inspection agreement experiment to be analyzed using Cohen's κ if the true value of the unknown κ must be determined to within ± 0.10 with 95% confidence? A preliminary experiment indicated that $\kappa \simeq 0.85$ and $p_e \simeq 0.5$.
Solution: With $\alpha = 0.05$ and $\delta = 0.10$ in Equation 7.55, the required sample size is

$$n = \frac{0.85(1 - 0.85)}{1 - 0.5}\left(\frac{1.96}{0.10}\right)^2 = 98.$$

7.3.2 Test for Cohen's Kappa

The hypotheses to be tested are $H_0 : \kappa = \kappa_0$ versus $H_A : \kappa > \kappa_0$ where κ_0 is a specified value. If the sample size is large, the distribution of $\hat{\kappa}$ is approximately normal, so the decision to reject H_0 or not is based on the standard normal test statistic

$$z = \frac{\hat{\kappa} - \kappa_0}{\sigma_{\hat{\kappa}}}. \tag{7.56}$$

Figure 7.1 shows the distributions of $\hat{\kappa}$ under H_0 and H_A, where $\kappa = \kappa_1$ under H_A. The effect size δ is

$$\delta = \kappa_1 - \kappa_0 = z_\alpha \sigma_{\hat{\kappa}_0} + z_\beta \sigma_{\hat{\kappa}_1},$$

so the approximate power to reject H_0 when $\kappa = \kappa_1$ is given by

$$\pi = \Phi(-\infty < z < z_\beta) \tag{7.57}$$

where

$$z_\beta = \frac{\delta - z_\alpha \sigma_{\hat{\kappa}_0}}{\sigma_{\hat{\kappa}_1}}. \tag{7.58}$$

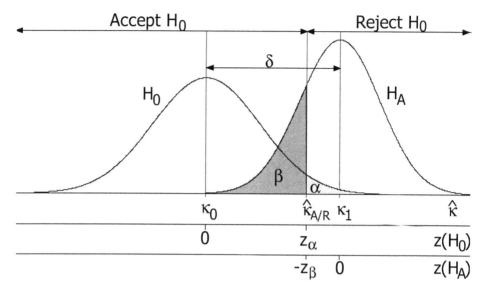

Figure 7.1: Distributions of $\hat{\kappa}$ under H_0 and H_A.

Example 7.10 Calculate the power to reject $H_0: \kappa = 0.4$ in favor of $H_A: \kappa > 0.4$ when $\kappa = 0.7$ if a sample of size $n = 70$ is allocated to $k = 3$ categories in the ratio $0.4 : 0.5 : 0.1$.

Solution: The expected chance agreement by Equation 7.46 is $p_e = 0.4^2 + 0.5^2 + 0.1^2 = 0.42$. The two κ values of interest have intermediate values not covered by the large- or small-κ approximations, so it is necessary to estimate $\sigma_{\hat{\kappa}}$ using Equation 7.47. Under H_0 with $\kappa = 0.4$ and $p_e = 0.42$ in Equation 7.44, we have

$$\begin{aligned} p_o &= 0.4(1 - 0.42) + 0.42 \\ &= 0.652, \end{aligned}$$

so

$$\begin{aligned} \sigma_{\hat{\kappa}_0} &\simeq \frac{1}{1 - 0.42}\sqrt{\frac{0.652(1 - 0.652)}{70}} \\ &\simeq 0.0982. \end{aligned}$$

Under H_A with $\kappa = 0.7$ we have

$$\begin{aligned} p_o &= 0.7(1 - 0.42) + 0.42 \\ &= 0.826, \end{aligned}$$

7.3. Cohen's Kappa

so

$$\sigma_{\hat{\kappa}_1} \simeq \frac{1}{1-0.42}\sqrt{\frac{0.826\,(1-0.826)}{70}}$$
$$\simeq 0.0781.$$

Then with $\alpha = 0.05$, z_β is given by Equation 7.58:

$$z_\beta = \frac{(0.7-0.4) - 1.645\,(0.0982)}{0.0781}$$
$$= 1.77$$

and the power is given by Equation 7.57:

$$\pi = \Phi(-\infty < z < 1.77)$$
$$= 0.962.$$

When κ_0 and κ_1 are sufficiently small so that their $\sigma_{\hat{\kappa}}$ estimates may be obtained from Equation 7.49, the sample size required to obtain power $\pi = 1 - \beta$ is

$$n \simeq \frac{p_e}{1-p_e}\left(\frac{z_\alpha + z_\beta}{\delta}\right)^2. \tag{7.59}$$

When κ_0 and κ_1 are large enough that their $\sigma_{\hat{\kappa}}$ estimates can be obtained from Equation 7.50, the sample size required to obtain power $\pi = 1 - \beta$ is

$$n \simeq \frac{1}{1-p_e}\left(\frac{z_\alpha\sqrt{\kappa_0\,(1-\kappa_0)} + z_\beta\sqrt{\kappa_1\,(1-\kappa_1)}}{\kappa_1 - \kappa_0}\right)^2. \tag{7.60}$$

Example 7.11 An experiment is to be performed to test for agreement between two methods of categorizing a dichotomous response. The hypotheses to be tested are $H_0 : \kappa = 0$ versus $H_A : \kappa > 0$ where κ is Cohen's kappa. How many units must be inspected if the test should have 90% power to reject H_0 when $\kappa = 0.40$ and the total number of units to be inspected is evenly split between the two categories?

Solution: Because the units will be balanced between the two categories, the agreement expected by chance from Equation 7.51 is $p_e \simeq 0.5$. With $\alpha = 0.05$, $\pi = 0.90$, $\beta = 1 - \pi = 0.10$, and $\delta = 0.4 - 0 = 0.4$ in Equation 7.59, the required sample size is

$$n \simeq \frac{0.5}{1-0.5}\left(\frac{z_{0.05} + z_{0.10}}{\delta}\right)^2 = \left(\frac{1.645 + 1.282}{0.40}\right)^2 = 54.$$

Example 7.12 An experiment is to be performed to test for agreement between two raters using a categorical four-state response. The hypotheses to be tested are $H_0 : \kappa = 0.8$ versus $H_A : \kappa > 0.8$. How many units must be inspected if the test should have 90% power to reject H_0 when $\kappa = 0.9$? The units to be inspected are evenly balanced across the four categories.

Solution: From Equation 7.51 with $k = 4$ categories, $p_e \simeq 0.25$. With $\alpha = 0.05$, $\beta = 1 - \pi = 0.10$, and $\delta = 0.9 - 0.8 = 0.1$ in Equation 7.60, the required sample size is

$$n \simeq \frac{1}{1 - 0.25} \left(\frac{1.645\sqrt{0.8 \times 0.2} + 1.282\sqrt{0.9 \times 0.1}}{0.9 - 0.8} \right)^2$$
$$\sim 145.$$

7.4 Receiver Operating Characteristic (ROC) Curves

An inspection method that uses a quantitative measurement to sort objects into two complementary categories, referred to as positive and negative states, is called a *diagnostic test*. Examples of diagnostic tests (and their positive and negative states) are an e-mail filter to block spam (spam and not spam), a blood test to detect a disease (diseased and healthy), and a measurement to identify defective units (defective and not defective). A model for the two-state response as a function of the quantitative predictor may be obtained by binary logistic regression (BLR) or discriminant analysis. However, the related method of receiver operating characteristic curves provides a better way to study the effects of different predictor cutoff values.

Most diagnostic tests are not perfectly accurate. The performance of a diagnostic test is determined by comparing its predictions to the known true states of experimental units. Figure 7.2a shows distributions of a diagnostic measurement for units with known positive and negative states. The proportion of the positives that are classified correctly relative to the indicated cutoff value is called the *true positive rate* (TPR). The proportion of the negatives that are classified incorrectly with respect to the same cutoff value is called the *false positive rate* (FPR).[3] An ROC curve is a plot of TPR versus FPR for all possible cutoff values. Figure 7.2b shows the ROC curve corresponding to Figure 7.2a.

The accuracy of a diagnostic test is often expressed in terms of the ROC curve's area under the curve (AUC). This area is equal to the probability that a randomly chosen positive unit will be given a higher diagnostic test score than a randomly chosen negative unit. The area under the curve of a perfectly accurate diagnostic test is $AUC = 1$. The area under the curve of a diagnostic test that

---
[3]TPR is also called the test's sensitivity and $1 - FPR$ is called its specificity. TPR is analogous to the power of a hypothesis test and FPR is analogous to the significance level.

7.4. Receiver Operating Characteristic (ROC) Curves

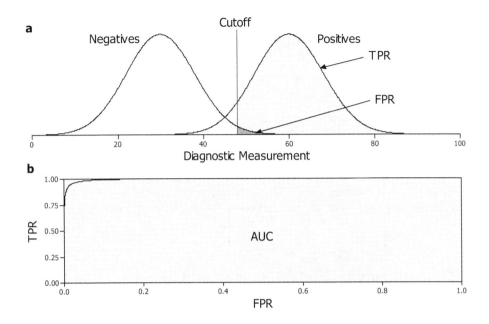

Figure 7.2: a) Diagnostic test distributions and b) their ROC curve.

has no predictive value is $AUC = 0.5$. The shaded area under the ROC curve in Figure 7.2b is $AUC = 0.996$.

The purpose of this section is to present sample size calculation methods for one-sample confidence intervals and hypothesis tests for the ROC curve's area under the curve.

7.4.1 Confidence Interval for the ROC Curve's AUC

The sampling distribution of the ROC curve's AUC is approximately normal for large samples, so the $(1-\alpha)\,100\%$ confidence interval for the ROC curve's AUC has the form

$$P\left(\widehat{AUC} - \delta < AUC < \widehat{AUC} + \delta\right) = 1 - \alpha \quad (7.61)$$

where the confidence interval half-width is

$$\delta = z_{\alpha/2}\widehat{\sigma}_\epsilon. \quad (7.62)$$

Hanley and McNeil [26] give the standard error for the ROC curve's AUC as

$$\sigma_\epsilon = \sqrt{\frac{AUC\,(1-AUC) + (n_P - 1)\left(\frac{AUC}{2-AUC} - AUC^2\right) + (n_N - 1)\left(\frac{2AUC^2}{1+AUC}\right)}{n_P n_N}} \quad (7.63)$$

where n_P and n_N are the number of positives and negatives, respectively. When the sample sizes are equal ($n_P = n_N = n$) and large (say $n \geq 30$), then the standard error of the AUC is approximately

$$\sigma_\epsilon \simeq \sqrt{\frac{AUC(1 - AUC)}{n(1 + AUC)}} \qquad (7.64)$$

$$\simeq \begin{cases} \sqrt{\frac{1}{6n}} & \text{for } AUC \simeq 0.5 \\ \sqrt{\frac{1-AUC}{2n}} & \text{for } AUC > 0.7 \end{cases} \qquad (7.65)$$

If the AUC is expected to be large, then, from Equations 7.65 and 7.62, the sample size required to obtain confidence interval half-width δ must be

$$n \simeq \frac{1 - AUC}{2} \left(\frac{z_{\alpha/2}}{\delta}\right)^2. \qquad (7.66)$$

Example 7.13 What sample size is required to estimate the value of an ROC curve's AUC to within ± 0.05 with 95% confidence if the AUC value is expected to be about 90%?

Solution: The desired confidence interval has the form

$$P\left(\widehat{AUC} - 0.05 < AUC < \widehat{AUC} + 0.05\right) = 0.95.$$

With $\alpha = 0.05$, $AUC = 0.90$, and $\delta = 0.05$ in Equation 7.66, the required sample size is

$$n \simeq \frac{1 - AUC}{2} \left(\frac{z_{0.025}}{\delta}\right)^2$$

$$\simeq \frac{1 - 0.90}{2} \left(\frac{1.96}{0.05}\right)^2$$

$$\simeq 77.$$

That is, about 77 positives and 77 negatives are required. The large-sample and large AUC assumptions are reasonably satisfied, so this approximate sample size should be accurate.

7.4.2 Test for the ROC Curve's AUC

The hypotheses to be tested are $H_0 : AUC = AUC_0$ versus $H_A : AUC \neq AUC_0$ and we wish to determine the sample size required to reject H_0 with power π for a specified value of AUC. The sample size calculation is complicated slightly because the standard error of AUC is different under H_0 and H_A. However, the sample size calculation for the AUC test is analogous to the sample size calculation for the test for one proportion given in Equation 4.22.

7.4. Receiver Operating Characteristic (ROC) Curves

When the sample sizes of positive and negative units are large and equal, by analogy to the one-sample test for a proportion where the standard errors are different under H_0 and H_A (Equation 4.22), the sample size required to reject $H_0 : AUC = AUC_0$ with power π when $AUC = AUC_1$ is given by

$$n = \begin{cases} \left(\dfrac{\sqrt{\frac{1}{6}} z_{\alpha/2} + \sqrt{\frac{1-AUC_1}{2}} z_\beta}{AUC_1 - \frac{1}{2}} \right)^2 & \text{for } AUC_0 = 0.5 \\ \left(\dfrac{\sqrt{\frac{1-AUC_0}{2}} z_{\alpha/2} + \sqrt{\frac{1-AUC_1}{2}} z_\beta}{AUC_1 - AUC_0} \right)^2 & \text{for } AUC_0 > 0.7. \end{cases} \quad (7.67)$$

Example 7.14 What sample size is required to reject $H_0 : AUC = 0.9$ in favor of $H_A : AUC \neq 0.9$ with 90% power when $AUC = 0.95$?
Solution: With $\alpha = 0.05$ in Equation 7.67, the required sample size is approximately

$$n = \left(\dfrac{\sqrt{\frac{1-0.90}{2}} z_{0.025} + \sqrt{\frac{1-0.95}{2}} z_{0.10}}{0.95 - 0.90} \right)^2$$

$$= \left(\dfrac{\sqrt{\frac{1-0.90}{2}} 1.96 + \sqrt{\frac{1-0.95}{2}} 1.282}{0.95 - 0.90} \right)^2$$

$$= 165.$$

Example 7.15 What sample size is required to reject $H_0 : AUC = 0.5$ versus $H_A : AUC > 0.5$ with 90% power when $AUC = 0.75$?
Solution: With $\beta = 0.10$ when $AUC = 0.75$ in Equation 7.67, the required sample size is approximately

$$n = \left(\dfrac{\sqrt{\frac{1}{6}} z_{0.05} + \sqrt{\frac{1-0.75}{2}} z_{0.10}}{0.75 - 0.50} \right)^2$$

$$= \left(\dfrac{\sqrt{\frac{1}{6}} 1.645 + \sqrt{\frac{1-0.75}{2}} 1.282}{0.75 - 0.50} \right)^2$$

$$= 21.$$

The large-sample assumption is only marginally satisfied, so this sample size may be somewhat inaccurate.

7.5 Bland-Altman Plots

Bland-Altman plots, also known as *Tukey mean-difference plots*, are used to assess the agreement between two measurement methods that have comparable precision error. The method is performed by obtaining paired observations from n samples using the two measurement methods. Then the difference d_i between the paired observations given by

$$d_i = y_{1i} - y_{2i}, \qquad (7.68)$$

where y_{1i} and y_{2i} are the observations by methods 1 and 2 on the ith sample, is plotted versus the mean $\bar{y}_i$ of the paired observations given by

$$\bar{y}_i = \frac{y_{1i} + y_{2i}}{2}. \qquad (7.69)$$

If the distribution of d_i is normal, then upper and lower limits of agreement (LOA) that contain $(1-p)\,100\%$ of the d_i are given by

$$LOA_{U/L} = \mu_d \pm z_{p/2} \sigma_d. \qquad (7.70)$$

Because μ_d and σ_d are usually unknown, they must be estimated with $\bar{d}$ and s_d, respectively, and $(1-\alpha)\,100\%$ confidence limits for the limits of agreement may be obtained by the two-sided normal distribution tolerance interval method (see Section 10.3.2.1). If the limits of agreement are chosen in advance, the sample size required to demonstrate the agreement between the two methods may be estimated (see Table E.7 of Appendix E).

Example 7.16 What minimum sample size is required to demonstrate the agreement between two measurement methods by the Bland-Altman method if the limits of agreement are $LOA_{U/L} = 0 \pm 3$ and the standard deviation of the differences had been estimated to be $\hat{\sigma}_d = 0.65$ from historical data? Assume that the limits of agreement must cover 99% of the samples with 95% confidence and that there is no bias between the two methods, i.e., $\mu_d = 0$.
Solution: The two-sided normal distribution tolerance interval factor k_2 is given by

$$\begin{aligned} k_2 &= \frac{LOA}{\hat{\sigma}_d} \\ &= \frac{3cm}{0.65cm} \\ &= 4.615. \end{aligned}$$

With $\alpha = 0.05$ and $Y = 0.99$ in Appendix E, Table E.7, the smallest sample size that gives $k_2 \le 4.615$ is $n = 10$.

Chapter 8

Designed Experiments

Sample size and power calculations are presented for the completely-randomized one-way classification design, the randomized block design, balanced full factorial designs with fixed and random effects, nested designs, two-level factorial designs, and response surface designs.

8.1 One-Way Fixed Effects ANOVA

For the completely randomized one-way classification design, one-way ANOVA provides a test of the hypotheses $H_0 : \mu_i = \mu_j$ *for all i, j pairs of treatments* versus $H_A : \mu_i \neq \mu_j$ *for at least one i, j pair of treatments*. The decision to reject H_0 or not is based on the ANOVA F statistic. If the results from a post-ANOVA multiple comparisons test are more important than the ANOVA's omnibus F test, the power and sample size should be evaluated using the methods of Section 2.6.

The ANOVA method assumes that the treatment populations are homoscedastic, in which case the optimal allocation of observations to treatments is to use a balanced design. When the treatments are heteroscedastic, a variable transformation that recovers approximate homoscedasticity should be considered. If such a transform can be found, then a balanced design is still appropriate; otherwise, the treatment sample sizes should be set in proportion to the treatment standard deviations, analogous to the case of the two-sample t test for means with unequal population standard deviations (Section 2.3.2.1).

8.1.1 Balanced One-Way Design

Power calculations for balanced one-way classification designs to be analyzed by one-way ANOVA are presented for three common treatment bias patterns: the general case with specified treatment biases, two treatments biased symmetrically about the grand mean, and one treatment biased with respect to all of the others. Sample sizes must be determined by iteration to obtain the target power.

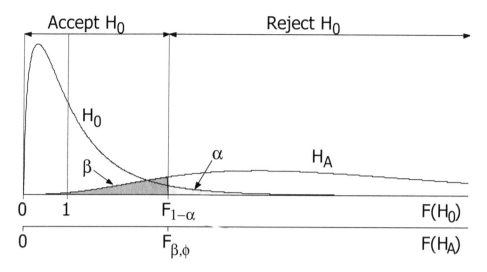

Figure 8.1: Distributions of F under H_0 and H_A.

8.1.1.1 General Case

For a specified set of k treatments with n observations per treatment and biases τ_i relative to the grand mean given by $\tau_i = \mu_i - \bar{\mu}$, the power $\pi = 1 - \beta$ to reject H_0 is given by the condition

$$F_{1-\alpha} = F_{\beta,\phi} \tag{8.1}$$

where the F distribution noncentrality parameter is

$$\phi = \frac{E\left(SS_{treatments}\right)}{E\left(MS_\epsilon\right)} = \frac{n \sum_{i=1}^{k} \tau_i^2}{\sigma_\epsilon^2} \tag{8.2}$$

and the F distributions have $df_{treatments} = k - 1$ numerator and $df_\epsilon = df_{total} - df_{treatment} = k(n-1)$ denominator degrees of freedom. By the way the τ_i are defined, they must satisfy the constraint $\sum \tau_i = 0$. The distributions of F under H_0 and H_A are shown in Figure 8.1. Under H_0, $F_{A/R} = F_{1-\alpha}$ and under H_A, $F_{A/R} = F_{\beta,\phi}$.

Some software packages require the treatment biases to be expressed as the standard deviation of the biases which may be obtained from

$$s_\tau = \sqrt{\frac{\sum_{i=1}^{k} \tau_i^2}{k-1}}. \tag{8.3}$$

And some software packages divide the sum of squares of treatment biases by k instead of by $k - 1$.

8.1. One-Way Fixed Effects ANOVA

Example 8.1 In a one-way classification design with four treatments and five observations per treatment, determine the power of the ANOVA to reject H_0 if the treatment means are $\mu_i = \{40, 55, 55, 50\}$. The four populations are expected to be normal and homoscedastic with $\sigma_\epsilon = 8$.

Solution: The grand mean is $\bar{\mu} = 50$ so the treatment biases relative to the grand mean are $\tau_i = \{-10, 5, 5, 0\}$. From Equation 8.2 the F distribution noncentrality parameter is,

$$\phi = \frac{n \sum_{i=1}^{k} \tau_i^2}{\sigma_\epsilon^2} = \frac{5\left(-10^2 + (5)^2 + (5)^2 + (0)^2\right)}{8^2} = 11.72.$$

The F statistic will have $df_{treatments} = 4 - 1 = 3$ and $df_\epsilon = 4(5-1) = 16$ degrees of freedom. The power is 72% as determined from Equation 8.1:

$$F_{0.95} = 3.239 = F_{0.280, 11.72}.$$

8.1.1.2 Two Treatments Biased Symmetrically About the Grand Mean

In a balanced design with two of the treatments biased symmetrically about the grand mean and the other treatment means equal to the grand mean, the treatment biases relative to the grand mean are

$$\tau_i = \left\{-\frac{\delta}{2}, \frac{\delta}{2}, 0, 0, \cdots\right\} \quad (8.4)$$

where δ is the distance between the two biased treatment means. Then, from Equation 8.2, the noncentrality parameter is

$$\phi = \frac{n}{2}\left(\frac{\delta}{\sigma_\epsilon}\right)^2 \quad (8.5)$$

and the power to reject H_0 is given by Equation 8.1.

Example 8.2 Determine the power of the ANOVA to reject H_0 in a one-way classification design with four treatments and five observations per treatment if the treatment biases from the grand mean are $\tau_i = \{-12, 12, 0, 0\}$. The four populations are expected to be normal and homoscedastic with $\sigma_\epsilon = 8$.

Solution: From Equation 8.5 with $\delta = 24$, the F distribution noncentrality parameter is

$$\phi = \frac{n}{2}\left(\frac{\delta}{\sigma_\epsilon}\right)^2 = \frac{5}{2}\left(\frac{24}{8}\right)^2 = 22.5.$$

The F statistic will have $df_{treatments} = 4 - 1 = 3$ and $df_\epsilon = 4(5-1) = 16$ degrees of freedom. The power is 95.4% as determined from Equation 8.1:

$$F_{0.95} = 3.239 = F_{0.046, 22.5}.$$

8.1.1.3 One Treatment Biased Relative to the Others

When one of k treatments is biased low or high with respect to the $k - 1$ others, the treatment biases relative to the grand mean are

$$\tau_i = \left\{ -\frac{(k-1)\delta}{k}, \frac{\delta}{k}, \frac{\delta}{k}, \cdots \right\} \tag{8.6}$$

where δ is the difference between the one and the $k - 1$ treatment means. Then, from Equation 8.2, the noncentrality parameter is

$$\phi = \frac{n(k-1)}{k} \left(\frac{\delta}{\sigma_\epsilon} \right)^2 \tag{8.7}$$

and the power to reject H_0 is given by Equation 8.1.

Example 8.3 In a one-way classification design with four treatments and five observations per treatment, determine the power of the ANOVA to reject H_0 if the treatment biases from the grand mean are $\tau_i = \{18, -6, -6, -6\}$. The four populations are expected to be normal and homoscedastic with $\sigma_\epsilon = 8$.
Solution: From Equation 8.6 with $\delta = 24$, the F distribution noncentrality parameter is

$$\phi = \frac{n(k-1)}{k} \left(\frac{\delta}{\sigma_\epsilon} \right)^2 = \frac{5 \times 3}{4} \left(\frac{24}{8} \right)^2 = 33.75.$$

The F statistic will have $df_{treatments} = 4 - 1 = 3$ and $df_\epsilon = 4(5-1) = 16$ degrees of freedom. The power is 99.5% as determined from Equation 8.1:

$$F_{0.95} = 3.239 = F_{0.005, 33.75}.$$

8.1.2 Unbalanced One-Way Design

When the one-way classification design is unbalanced, the ANOVA power is still given by Equation 8.1 but with F distribution noncentrality parameter

$$\phi = \frac{\sum_{i=1}^{k} n_i \tau_i^2}{\sigma_\epsilon^2} \tag{8.8}$$

8.2. Randomized Block Design

subject to the constraint $\sum n_i \tau_i = 0$. To guarantee that this constraint is satisfied, it is easiest to specify an unbalanced design problem in terms of the μ_i and n_i and then determine the τ_i from

$$\tau_i = \mu_i - \frac{\sum n_i \mu_i}{\sum n_i}. \tag{8.9}$$

Example 8.4 Determine the power to reject H_0 by one-way ANOVA when the treatment means are $\mu_i = \{50, 30, 40, 40, 40\}$ and the sample sizes are $n_i = \{12, 12, 20, 20, 15\}$. The five populations are expected to be normal and homoscedastic with $\sigma_\epsilon = 13$.

Solution: The grand mean of the experimental data is expected to be

$$\frac{\sum n_i \mu_i}{\sum n_i} = \frac{(50 \times 12) + \cdots + (40 \times 15)}{12 + \cdots + 15} = 40.$$

The treatment biases relative to the grand mean are $\tau_i = \{10, -10, 0, 0, 0\}$ so the noncentrality parameter is

$$\phi = \frac{12(10)^2 + 12(-10)^2 + 20(0)^2 + 20(0)^2 + 15(0)^2}{13^2} = 14.2.$$

The ANOVA will have $df_{treatments} = 5 - 1 = 4$ and $df_\epsilon = \sum n_i - 1 - 4 = 74$ degrees of freedom. The power is 84.7% as determined from Equation 8.1:

$$F_{0.95} = 2.495 = F_{0.153, 14.2}.$$

8.2 Randomized Block Design

The one-way classification design of Section 8.1.1 is called the *completely randomized design* (CRD) because all of the observations from all replicates are run in completely random order. This design is rarely used because the power of the experiment can usually be improved by running it as a *randomized block design* (RBD) where each block contains one replicate and blocks are run consecutively. The benefit of the RBD is that it provides more homogeneous conditions within blocks and controls for possible differences between blocks, which increases the experiment's sensitivity to differences between treatments.

The RBD is analyzed using two-way ANOVA with terms in the ANOVA model for treatments and blocks. Table 8.1 shows the form of this ANOVA where there are k treatments and r replicates with one replicate in each block. The F statistic for block effects is not usually of interest and is often not reported. Because the ANOVA F test for treatment effects is still based on the ANOVA error mean square, MS_ϵ, the ANOVA power for fixed treatment effects is calculated the same way as was the power for the one-way CRD in Section 8.1.1.

Source	df	SS	MS	F
Treatment	$k-1$	$SS_{treatments}$	$MS_{treatments}$	$\frac{MS_{treatments}}{MS_\epsilon}$
Block	$r-1$	SS_{blocks}	MS_{blocks}	
Error	$(k-1)(r-1)$	SS_ϵ	MS_ϵ	
Total	$kr-1$	SS_{total}		

Table 8.1: ANOVA table for the randomized block design.

Example 8.5 Recalculate the power for Example 8.1 if the experiment is built as a randomized block design and the standard deviation of the population of block biases is $\sigma_{blocks} = 4$.

Solution: The F distribution noncentrality parameter ($\phi = 11.72$) and the treatment degrees of freedom ($df_{treatments} = 3$) will be unchanged from the original solution, but if the experiment is built in five blocks with one replicate in each block, the new error degrees of freedom for the RBD will be

$$\begin{aligned} df_\epsilon &= df_{total} - df_{treatments} - df_{blocks} \\ &= 19 - 3 - 4 \\ &= 12. \end{aligned}$$

The power of the RBD is 68% as determined from

$$F_{0.95} = 3.490 = F_{0.32, 11.72}.$$

This is slightly lower than the original power (72%) because the RBD has fewer error degrees of freedom than the CRD. The RBD's power is not affected by the block variation because it separates that variation from the error variation that is used to determine the power.

After an RBD has been performed, the *blocking efficiency* may be calculated. Blocking efficiency measures the size of the CRD relative to the size of the RBD required to obtain the same precision for estimating treatment effects if the CRD were run without controlling for the blocking variable. If the blocking efficiency is high, then the RBD is more efficient than the CRD and future experiments should definitely be built in blocks. If the blocking efficiency is low, then the RBD and CRD have comparable performance, so future experiments could use the simpler CRD. In practice, we almost always build the RBD anyway, even when the blocking efficiency is low.

If the CRD has k treatments and n replicates with standard error $\sigma_{\epsilon(CRD)}$ and the RBD has k treatments and r replicates in blocks with standard error $\sigma_{\epsilon(RBD)}$, then demanding equal precision for estimating treatment effects in the two experiments gives

$$\frac{\sigma_{\epsilon(CRD)}}{\sqrt{n}} = \frac{\sigma_{\epsilon(RBD)}}{\sqrt{r}}, \tag{8.10}$$

so the blocking efficiency E is

$$E = \frac{n}{r} = \left(\frac{\sigma_{\epsilon(CRD)}}{\sigma_{\epsilon(RBD)}}\right)^2. \tag{8.11}$$

The error variance from the CRD, which gets contributions from the blocking variable and the pure error from the RBD, may be estimated from

$$\sigma^2_{\epsilon(CRD)} = \frac{(r-1)\,MS_{blocks} + r\,(k-1)\,MS_{\epsilon(RBD)}}{kr-1} \tag{8.12}$$

which gives the blocking efficiency

$$\begin{aligned} E &= \frac{MS_{\epsilon(CRD)}}{MS_{\epsilon(RBD)}} \\ &= \frac{(r-1)\,MS_{blocks} + r\,(k-1)\,MS_{\epsilon(RBD)}}{(kr-1)\,MS_{\epsilon(RBD)}}. \end{aligned} \tag{8.13}$$

A correction can be made to the blocking efficiency to account for the difference in the error degrees of freedom between the RBD and CRD designs, but this correction is negligible unless the RBD's error degrees of freedom are very small.

Example 8.6 A 40-run experiment was performed using an RBD with $k = 5$ treatments and $r = 8$ blocks. The ANOVA table from the experiment is shown in Table 8.2. Calculate the blocking efficiency and the increase in the number of runs required to obtain the same estimation precision for treatment means using a CRD.

Solution: The blocking efficiency as determined from Equation 8.13 is

$$\begin{aligned} E &= \frac{(7 \times 14) + (8 \times 4 \times 4)}{(5 \times 8 - 1)\,4} \\ &= 1.45. \end{aligned}$$

That is, the CRD will require about 45% more runs than the RBD because it ignores the variation associated with block effects. Because the number of runs in the RBD was $kr = 40$, the number of runs required for the CRD to obtain the same estimation precision for the treatment means would be $Ekr = 1.45 \times 40 = 58$. Apparently the blocking was beneficial and should be used in future studies.

8.3 Balanced Full Factorial Design with Fixed Effects

In balanced full factorial designs with all factors fixed, the ANOVA power calculations are performed the same way as they are for one-way classification designs, so the methods of Section 8.1.1 are applicable. The only modification required is that the model to be fitted to the data must be anticipated so that the

Source	df	SS	MS	F
Treatment	4	80	20	5.0
Block	7	98	14	
Error	28	112	4	
Total	39	290		

Table 8.2: ANOVA table for the randomized block design.

degrees of freedom for the factor being considered in the power calculation and the error degrees of freedom for the ANOVA can be determined.

In general, for balanced full factorial designs with different numbers of levels for the different variables, the power to detect a difference between a pair of treatment means will be greater for a variable with fewer levels than for a variable with many levels. Consequently, if an experiment is required to detect a specified difference between a pair of treatment means with a minimum specified power for all study variables, then the minimum number of replicates will be determined by the variable with the most levels.

Example 8.7 A $2 \times 3 \times 5$ full factorial experiment with four replicates is planned. The experiment will be blocked on replicates and the ANOVA model will include main effects and two-factor interactions. Determine the power to detect a difference of $\delta = 300$ between two levels of each study variable if the standard error of the model is expected to be $\sigma_\epsilon = 500$.

Solution: If the three study variables are given the names A, B, and C and have $a = 2$, $b = 3$, and $c = 5$ levels, respectively, then the degrees of freedom associated with the terms in the model will be $df_{blocks} = 3$, $df_A = 1$, $df_B = 2$, $df_C = 4$, $df_{AB} = 2$, $df_{AC} = 4$, $df_{BC} = 8$, and

$$\begin{aligned} df_\epsilon &= df_{total} - df_{model} \\ &= (2 \times 3 \times 5 \times 4 - 1) - (3 + 1 + 2 + 4 + 2 + 4 + 8) \\ &= 119 - 24 \\ &= 95. \end{aligned}$$

From Equation 8.2, the F distribution noncentrality parameter for variable A with treatment biases $\alpha_1 = -150$ and $\alpha_2 = 150$ is

$$\begin{aligned} \phi_A &= \frac{bcn \sum_{i=1}^{2} \alpha_i^2}{\sigma_\epsilon^2} \\ &= \frac{3 \times 5 \times 4 \times \left((-150)^2 + 150^2 \right)}{500^2} \\ &= 10.8. \end{aligned}$$

8.3. Balanced Full Factorial Design with Fixed Effects

The distribution of F_A will have $df_A = 1$ numerator and $df_\epsilon = 95$ denominator degrees of freedom, so the power associated with A is given by Equation 8.1:

$$F_{0.95} = 3.942 = F_{1-\pi_A, 10.8},$$

which is satisfied by $\pi_A = 0.908$ or 90.8%.

Similarly, the F distribution noncentrality parameter for B with biases $\beta_1 = -150$, $\beta_2 = 150$, and $\beta_3 = 0$ is

$$\phi_B = \frac{acn \sum_{i=1}^{3} \beta_i^2}{\sigma_\epsilon^2}$$

$$= \frac{2 \times 5 \times 4 \times \left((-150)^2 + 150^2 + 0^2\right)}{500^2}$$

$$= 7.2.$$

The distribution of F_B will have $df_B = 2$ numerator and $df_\epsilon = 95$ denominator degrees of freedom, so the power associated with B is given by

$$F_{0.95} = 3.093 = F_{1-\pi_B, 7.2},$$

which is satisfied by $\pi_B = 0.654$.

Finally, the F distribution noncentrality parameter for C with biases $\gamma_1 = -150$, $\gamma_2 = 150$, and $\gamma_3 = \gamma_4 = \gamma_5 = 0$ is

$$\phi_C = \frac{abn \sum_{i=1}^{5} \gamma_i^2}{\sigma_\epsilon^2}$$

$$= \frac{2 \times 3 \times 4 \times \left((-150)^2 + 150^2 + 0^2 + 0^2 + 0^2\right)}{500^2}$$

$$= 4.32.$$

The distribution of F_C will have $df_C = 4$ numerator and $df_\epsilon = 95$ denominator degrees of freedom, so the power associated with C is given by

$$F_{0.95} = 2.469 = F_{1-\pi_C, 4.32},$$

which is satisfied by $\pi_C = 0.328$. These three power calculations confirm by example that the power to detect a variable effect decreases as the number of variable levels increases.

8.4 Random and Mixed Models

An experimental variable is said to be *fixed* when all of its possible levels are included in an experiment so that future observations may be corrected for biases between levels. In contrast, a variable is said to be *random* when it has so many levels that only a sample of them may be included in an experiment. With respect to a random variable, the goals of an experiment are limited to: 1) testing hypotheses about the standard deviation of the population of level biases and/or 2) estimating that standard deviation using a method called *variance components analysis*.

ANOVA models that have all fixed, all random, or at least one fixed and one random variable are called fixed, random, and mixed models, respectively. The purpose of this section is to provide an introduction to power and sample size calculations for fixed and random variable effects in mixed and random models. This topic is too large to be covered comprehensively in this book, but the methods presented here are general and may be extended to more complicated designs.

8.4.1 Fixed Effects in Mixed Models

Whether an ANOVA model is fixed or mixed, the ANOVA F statistic for a fixed variable A is calculated from

$$F_A = \frac{MS_A}{MS_{\epsilon(A)}} \tag{8.14}$$

where $MS_{\epsilon(A)}$ is the mean square associated with the error for estimating the A effect. In a model with all variables fixed, $MS_{\epsilon(A)}$ is always equal to the mean square error, MS_ϵ. However, in a mixed or random model that includes at least one interaction between A and a random variable, $MS_{\epsilon(A)}$ will be different from MS_ϵ. The appropriate $MS_{\epsilon(A)}$ for calculating F_A is determined from the equations for the ANOVA expected mean squares.

For a specified set of a A-level biases $\alpha_i = \{\alpha_1, \alpha_2, \cdots, \alpha_a\}$ subject to the constraint $\sum \alpha_i = 0$, the power to reject the hypothesis $H_0 : \alpha_i = 0$ *for all* $i = 1$ to a is given by Equation 8.1 with df_A numerator and $df_{\epsilon(A)}$ denominator degrees of freedom where the F distribution noncentrality parameter ϕ_A is given by

$$\phi_A = \frac{N}{a} \frac{\sum_{i=1}^{a} \alpha_i^2}{MS_{\epsilon(A)}} \tag{8.15}$$

and N is the total number of observations. The expected value of the ANOVA F_A statistic is related to the noncentrality parameter by

$$E(F_A) = 1 + \frac{\phi_A}{df_A}. \tag{8.16}$$

8.4. Random and Mixed Models

Source	df	$E(MS)$	F
A	$a-1$	$\sigma_\epsilon^2 + n\sigma_{AB}^2 + \frac{bn}{a-1}\sum_{i=1}^{a}\alpha_i^2$	$\frac{MS_A}{MS_{AB}}$
B	$b-1$	$\sigma_\epsilon^2 + n\sigma_{AB}^2 + an\sigma_B^2$	$\frac{MS_B}{MS_{AB}}$
AB	$(a-1)(b-1)$	$\sigma_\epsilon^2 + n\sigma_{AB}^2$	$\frac{MS_{AB}}{MS_\epsilon}$
$Error(\epsilon)$	$ab(n-1)$	σ_ϵ^2	
$Total$	$abn-1$		

Table 8.3: ANOVA table for one fixed and one random variable.

Of course, under the null hypothesis $H_0 : \alpha_i = 0$ for all i, the noncentrality parameter is $\phi_A = 0$, so the expected value of F_A is $E(F_A) = 1$.

Example 8.8 A balanced full factorial experiment is to be performed using $a = 3$ levels of a fixed variable A, $b = 5$ randomly selected levels of a random variable B, and $n = 4$ replicates. Determine the power to reject $H_0 : \alpha_i = 0$ for all i when the A-level biases are $\alpha_i = \{-20, 20, 0\}$ with $\sigma_B = 25$, $\sigma_{AB} = 0$, and $\sigma_\epsilon = 40$. Assume that the AB interaction term will be included in the ANOVA even though its expected variance component is 0.

Solution: The ANOVA table with the equations for the expected mean squares is shown in Table 8.3.[1] From the ANOVA table, the error mean square used for testing the A effect (that is, the denominator of F_A) is

$$MS_{\epsilon(A)} = MS_{AB}$$
$$= \hat{\sigma}_\epsilon^2 + n\hat{\sigma}_{AB}^2.$$

The noncentrality parameter for the test of the fixed effect A is given by Equation 8.15:

$$\phi_A = \frac{N\sum_{i=1}^{a}\alpha_i^2}{a\,MS_{\epsilon(A)}}$$
$$= \frac{3 \times 5 \times 4}{3} \frac{(-20)^2 + (20)^2 + (0)^2}{(40)^2 + 4(0)^2}$$
$$= 10.$$

With $df_A = 2$, $df_{AB} = 8$, and $\alpha = 0.05$ in Equation 8.1

$$F_{0.95} = 4.459 = F_{1-\pi, 10.0},$$

which is satisfied by $\pi = 0.64$.

---

[1] This solution uses the unrestricted model. The mean squares equations for the restricted model are different, but the unrestricted model is more conservative with respect to power. See Neter et al. [51] for details.

8.4.2 Random Effects in Mixed and Random Models

For a random variable R with r randomly selected levels in an experiment, the hypotheses $H_0 : \sigma_R^2 = 0$ versus $H_A : \sigma_R^2 > 0$ are tested by ANOVA. The ANOVA calculations for degrees of freedom, sums of squares, and mean squares are done the same way for random effects as they are for fixed effects; however, the ANOVA F statistic for a random variable is calculated using a different ratio of mean squares from that for a fixed variable. Furthermore, the distribution of the ANOVA F statistic under H_A for a fixed variable follows the noncentral F distribution, but the distribution of the ANOVA F statistic under H_A for a random variable follows a transformed central F distribution. In general, the random variable's F statistic is given by

$$F_R = \frac{MS_R}{MS_{\epsilon(R)}} \tag{8.17}$$

where MS_R is the ANOVA mean square associated with R and $MS_{\epsilon(R)}$ is the mean square associated with the error term for testing the R effect. Under $H_A : \sigma_R^2 > 0$, the distribution of

$$F'_R = \frac{F_{1-\alpha}}{F_R} \tag{8.18}$$

follows the central F distribution with df_R numerator and $df_{\epsilon(R)}$ denominator degrees of freedom. Then for specified values of the variances required to estimate MS_R and $MS_{\epsilon(R)}$ under H_A, the expected value of F_R is[2]

$$\begin{aligned} E(F_R) &= E\left(\frac{MS_R}{MS_{\epsilon(R)}}\right) \\ &\simeq \frac{E(MS_R)}{E(MS_{\epsilon(R)})} \end{aligned} \tag{8.19}$$

and the corresponding power to reject H_0 is approximately

$$\pi \simeq P\left(\frac{F_{1-\alpha}}{E(F_R)} < F < \infty\right). \tag{8.20}$$

Example 8.9 Determine the power to reject $H_0 : \sigma_B^2 = 0$ when $\sigma_B = 25$, $\sigma_{AB} = 0$, and $\sigma_\epsilon = 40$ for Example 8.8. Retain the AB interaction term in the model even though its variance component is 0.
Solution: The ANOVA table with the equations for the expected mean squares is shown in Table 8.3. From Equation 8.19 under the specified conditions, the

---
[2] The expected value of the ratio of mean squares is never equal to the ratio of the expected values; however, under frequently satisfied conditions they are approximately equal to each other.

expected F_B value is approximately

$$E(F_B) \simeq \frac{E(MS_B)}{E(MS_{AB})}$$

$$\simeq \frac{\sigma_\epsilon^2 + n\sigma_{AB}^2 + an\sigma_B^2}{\sigma_\epsilon^2 + n\sigma_{AB}^2}$$

$$\simeq \frac{(40)^2 + 4(0)^2 + 3 \times 4 \times (25)^2}{(40)^2 + 4(0)^2}$$

$$\simeq 5.69.$$

With $df_B = 4$, $df_{AB} = 8$, and $\alpha = 0.05$, the critical F value for the test for the B effect is $F_{0.95,4,8} = 3.838$, so from Equation 8.20 the power is approximately

$$\pi \simeq P\left(\frac{3.838}{5.69} < F < \infty\right)$$

$$\simeq P(0.675 < F < \infty)$$

$$\simeq 0.618.$$

8.4.3 Confidence Intervals for Variance Components

A *variance component* is the variance associated with the level biases of a random variable. A point estimate for a variance component may be determined by solving the system of equations obtained by equating the ANOVA's expected mean squares with their observed mean squares. Unfortunately, there are no exact formulas for the variance component confidence intervals when a variance component must be estimated from the sum or difference of two or more mean squares. Approximate confidence intervals may be obtained using the Satterthwaite approximation or the modified least squares (MLS) method. Satterthwaite's method is easier to use, but it is inaccurate under some common conditions, so the MLS method is preferred. Neither method is well suited for performing sample size calculations. See Burdick [10] or Neter et al. [51] for details on use of the Satterthwaite and MLS methods.

8.5 Nested Designs

One variable is nested inside of another when the levels of the first variable are different within each of the levels of the second variable. For example, if variable A has a levels, each with b unique levels of another variable B within it, then variable B is nested within variable A. The usual notation used to indicate nesting is $B(A)$. A practical example of this situation is an experiment to compare three types of material using four different lots of each material type. The

ANOVA model for this experiment would include terms for material and for lot, so the model is: $Material + Lot(Material)$.

Designs may have multiply nested variables. For example, if a large batch of material is divided into subbatches and then those are subdivided into packages, then the model is: $Batch + SubBatch(Batch) + Package(SubBatch(Batch))$.

A nested design may be crossed with other variables. For example, if B is nested in A and both are crossed with another variable C, then the following model may be fitted to the data: $A + B(A) + C + AC + BC$.

The variables in a nested design may be fixed or random. After the appropriate ANOVA table with expected means squares equations is determined, then the methods of Section 8.4 may be used to determine the power.

Example 8.10 A is a fixed variable with three levels and B is a random variable with four levels nested within each level of A. The nested design is crossed with a five-level fixed variable C and one replicate of the experiment will be built. The model to be fitted is: $A + B(A) + C + AC + BC$. Find the power to detect a difference $\delta = 40$ between two levels of A assuming that the standard deviations for the random effects are $\sigma_B = 12$, $\sigma_{BC} = 4$, and $\sigma_\epsilon = 10$.

Solution: The ANOVA table with the equations for the expected mean squares is shown in Table 8.4 where the α_i are the A-level biases, the τ_i are the C-level biases, and the γ_i are the AC interaction biases. The error mean square used for testing the A effect is $MS_{\epsilon(A)} = MS_{B(A)}$. The noncentrality parameter for the test of the fixed effect A is given by Equation 8.15:

$$\phi_A = \frac{a \times b \times c \times n}{a} \frac{\sum_{i=1}^{a} \alpha_i^2}{\sigma_\epsilon^2 + n\sigma_{BC}^2 + cn\sigma_{B(A)}^2}$$

$$= \frac{3 \times 4 \times 5 \times 1}{3} \frac{(-20)^2 + (20)^2 + (0)^2}{(10)^2 + (4)^2 + 5(12)^2}$$

$$= 19.14.$$

With $df_A = 2$, $df_{B(A)} = 3(4-1) = 9$, and $\alpha = 0.05$ in Equation 8.1:

$$F_{0.95} = 4.256 = F_{1-\pi, 19.14}$$

which is satisfied by $\pi = 0.915$.

8.6 Two-Level Factorial Designs

There are two sample size calculation methods available for two-level factorial experiment designs that are distinguished by the goals of the experiments. The first method corresponds to calculating the power associated with the hypothesis test of $H_0 : \Delta\mu_i = 0$ (there is no difference between the response means of the low and

8.6. Two-Level Factorial Designs

Source	df	E(MS)	F
A	$a-1$	$\sigma_\epsilon^2 + n\sigma_{BC}^2 + cn\sigma_{B(A)}^2 + \frac{bcn}{a-1}\sum\alpha_i^2$	$\frac{MS_A}{MS_{B(A)}}$
B(A)	$a(b-1)$	$\sigma_\epsilon^2 + n\sigma_{BC}^2 + cn\sigma_{B(A)}^2$	$\frac{MS_B}{MS_{BC}}$
C	$c-1$	$\sigma_\epsilon^2 + n\sigma_{BC}^2 + \frac{abn}{c-1}\sum\tau_i^2$	$\frac{MS_C}{MS_{BC}}$
AC	$(a-1)(c-1)$	$\sigma_\epsilon^2 + n\sigma_{BC}^2 + \frac{bn}{(a-1)(c-1)}\sum\gamma_i^2$	$\frac{MS_{AC}}{MS_{BC}}$
BC	$a(b-1)(c-1)$	$\sigma_\epsilon^2 + n\sigma_{BC}^2$	$\frac{MS_{BC}}{MS_\epsilon}$
Error (ϵ)	$abc(n-1)$	σ_ϵ^2	
Total	$abcn-1$		

Table 8.4: ANOVA table and expected mean squares for a crossed nested design.

high levels of the ith variable) versus $H_A : \Delta\mu_i \neq 0$ (there is a difference between the response means of the low and high levels of the ith variable). That is, the goal of the experiment is simply to detect the presence of variable effects, such as in a screening experiment. This power calculation comes directly from the analysis for the balanced full factorial designs in Section 8.3.

The second sample size calculation method corresponds to calculating the sample size required to deliver a confidence interval for the regression coefficient of the ith variable with specified confidence interval half-width. This calculation is appropriate when a variable is already known or expected to have a significant effect on the response and the goal of the experiment is to quantify the size of that effect with a specified degree of precision. This sample size calculation comes directly from the analysis of the confidence interval for a regression coefficient in Section 6.1.1.

The sample size calculated to meet confidence interval requirements is approximately the same as the sample size calculated by the ANOVA power method if the power is set to 50%. This is a tricky way to get power calculation software to perform sample size calculations for confidence interval problems.

8.6.1 Test for a Main Effect

When the purpose of a two-level full factorial experiment is to detect significant effects, that is, to perform a test of the hypotheses $H_0 : \Delta\mu_i = 0$ *(there is no difference between the response means of the low and high levels of the ith variable)* versus $H_A : \Delta\mu_i \neq 0$ *(there is a difference between the response means of the low and high levels of the ith variable)*, the appropriate sample size calculation comes from the power of the ANOVA F test of the ith variable. For a 2^k balanced full factorial experiment with n replicates, the power to detect a difference $\Delta\mu_i = \delta$ between the two levels of ith design variable is given by the condition

$$F_{1-\alpha} = F_{1-\pi,\phi} \tag{8.21}$$

where the F distributions have $df_i = 1$ numerator and df_ϵ denominator degrees of freedom and the noncentral F distribution noncentrality parameter is

$$\begin{aligned}\phi &= E\left(\frac{SS_i}{MS_\epsilon}\right) \\ &\simeq \frac{n2^{k-1}\left(\left(-\frac{\delta}{2}\right)^2 + \left(\frac{\delta}{2}\right)^2\right)}{\sigma_\epsilon^2} \\ &\simeq n2^{k-2}\left(\frac{\delta}{\sigma_\epsilon}\right)^2.\end{aligned} \qquad (8.22)$$

The error degrees of freedom used in Equation 8.21 are calculated from

$$df_\epsilon = df_{total} - df_{model} \qquad (8.23)$$

where df_{total} is the total number of observations minus 1 and the value used for df_{model} forces us to specify the expected form of the model. If the experiment is to be built in blocks, then degrees of freedom for the block effects should be included in df_{model}. Because some insignificant model terms are usually expected in experiments with many variables, it might be useful to anticipate dropping some terms from the model to increase the error degrees of freedom, thereby increasing the power associated with the remaining terms.

The power condition given by Equation 8.21 may also be approximated in terms of the t distribution, giving a form that is more convenient for determining the number of replicates. This approach leads to

$$n \geq \frac{1}{2^{k-2}}\left(t_{\alpha/2} + t_\beta\right)^2 \left(\frac{\sigma_\epsilon}{\delta}\right)^2 \qquad (8.24)$$

where the t distribution has df_ϵ degrees of freedom. This expression is transcendental, but the approximation $t \simeq z$, valid when the error degrees of freedom is large, provides a good starting point that often delivers the correct value of n on the first iteration.

When the error degrees of freedom in Equation 8.24 is very large, the total number of observations required in a 2^k experiment to obtain a specified power value for a specified effect size is approximately

$$n2^k \simeq 4\left(z_{\alpha/2} + z_\beta\right)^2 \left(\frac{\sigma_\epsilon}{\delta}\right)^2. \qquad (8.25)$$

Because the right-hand side of this equation is independent of n and k, the power associated with any 2^k experiment depends only on the total number of observations and not on n or k individually! For example, 2^k experiments with one to five variables and 32 runs all have about the same power. This means that new variables may be added to 2^k experiments without adding runs or significantly reducing the power as long as the error degrees of freedom stays sufficiently large!

8.6. Two-Level Factorial Designs

Example 8.11 Use the method of Equation 8.24 to determine the number of replicates required to detect an effect of size $\delta = 6$ with 90% power in a 2^4 experiment when $\sigma_\epsilon = 10$. Assume that the ANOVA model will include main effects and two-factor interactions.

Solution: With $t \simeq z$ in the first iteration of Equation 8.24, the number of replicates required to deliver 90% power to detect the difference $\delta = 6$ between two levels of a design variable is

$$n \geq \frac{1}{2^{4-2}} (1.96 + 1.282)^2 \left(\frac{10}{6}\right)^2$$
$$\geq 8.$$

Another iteration (not shown) confirms that $n = 8$ is the correct number of replicates.

Example 8.12 Use the method of Equation 8.21 to confirm the solution to Example 8.11.

Solution: By Equation 8.22 the F distribution noncentrality parameter is

$$\phi = 8 \times 2^{4-2} \left(\frac{6}{10}\right)^2 = 11.5.$$

The central and noncentral F distributions will have $df_i = 1$ numerator and $df_\epsilon = df_{total} - df_{model} = (8 \times 2^4 - 1) - (4 + 6) = 117$ denominator degrees of freedom. The power, determined from the condition

$$F_{0.95} = F_{1-\pi, 11.5}$$
$$3.922 = F_{0.080, 11.5}$$

is $\pi = 0.920$ or 92.0%. This value is slightly larger than the 90% goal because the calculated value of n was fractional and was rounded up to the nearest integer. With $n = 7$ the power is slightly less than 90%.

Example 8.13 Suppose that two more two-level variables were added to the 2^4 experiment with $n = 8$ replicates from Example 8.11 without any increase in the total number of runs. Calculate the power for the resulting 2^6 experiment.

Solution: The 2^6 experiment must have $n = 2$ replicates to maintain the same number of runs as the original experiment. Because $8 \times 2^4 = 2 \times 2^6$, the F distribution noncentrality parameter will be unchanged. The new error degrees of freedom for the F distributions will be $df_\epsilon = (2 \times 2^6 - 1) - (6 + 15) = 106$.

The power, determined from

$$F_{0.95} = F_{1-\pi, 11.5}$$
$$3.931 = F_{0.081, 11.5}$$

is $\pi = 0.919$ or 91.9%. This example confirms that adding variables to a 2^k design without increasing the total number of observations has little effect on the power provided that the error degrees of freedom remains large.

Example 8.14 Derive a simplified expression for the total number of observations required for a 2^k experiment to detect a difference δ between two levels of a design variable assuming $\alpha = 0.05$ and $\beta = 0.10$. Under what conditions should this expression be valid?

Solution: From Equation 8.25 the total number of observations required for a 2^k design to have 90% power to detect a difference δ between two levels of a design variable is approximately

$$n2^k \geq 4(z_{0.025} + z_{0.10})^2 \left(\frac{\sigma_\epsilon}{\delta}\right)^2$$
$$\geq 42 \left(\frac{\sigma_\epsilon}{\delta}\right)^2. \qquad (8.26)$$

This condition will be strictly valid when df_ϵ is large so that the $t \simeq z$ approximation is well satisfied.

8.6.2 Confidence Interval for a Regression Coefficient

Suppose that a 2^k design with n replicates is to be performed for the purpose of obtaining a confidence interval for the regression coefficient associated with a main effect. If the analysis is performed by multiple linear regression, then the model will have the form

$$y = b_0 + b_1 x_1 + b_2 x_2 + \cdots + b_{12} x_{12} + \cdots \qquad (8.27)$$

and the regression coefficient confidence interval will have the form

$$P(b_i - \delta < \beta_i < b_i + \delta) = 1 - \alpha. \qquad (8.28)$$

If the design is specified in coded units, then this confidence interval formula applies to every one of the design variables. From Section 6.1.1, the confidence interval half-width is given by

$$\delta = t_{\alpha/2} \hat{\sigma}_{b_i}$$
$$= t_{\alpha/2} \frac{\hat{\sigma}_\epsilon}{\sqrt{nSS_x}} \qquad (8.29)$$

8.6. Two-Level Factorial Designs

where SS_x is the sum of squares associated with the design variable for one replicate of the experiment design, $\hat{\sigma}_\epsilon$ is the standard error of the regression, and the t distribution has df_ϵ degrees of freedom. Then the number of replicates is the smallest value of n which meets the condition

$$n \geq \frac{1}{SS_x} \left(\frac{t_{\alpha/2} \hat{\sigma}_\epsilon}{\delta} \right)^2. \tag{8.30}$$

When the design variables are expressed in terms of their coded ± 1 variable levels, then for one replicate

$$SS_x = \frac{2^k}{2} \left((-1)^2 + (1)^2 \right) = 2^k, \tag{8.31}$$

and the required number of replicates for the 2^k design is the smallest value of n that meets the condition

$$n \geq \frac{1}{2^k} \left(\frac{t_{\alpha/2} \hat{\sigma}_\epsilon}{\delta} \right)^2. \tag{8.32}$$

Example 8.15 How many replicates of a 2^3 design are required to determine the regression coefficient for a main effect with precision $\delta = 300$ with 95% confidence when the standard error of the model is expected to be $\sigma_\epsilon = 600$?
Solution: If the error degrees of freedom are sufficiently large that $t_{0.025} \simeq z_{0.025}$ then

$$n \geq \frac{1}{2^3} \left(\frac{1.96 \times 600}{300} \right)^2$$
$$\geq 2.$$

With only $2 \times 2^3 = 16$ total runs, the $t_{0.025} \simeq z_{0.025}$ assumption is not satisfied. Another iteration shows that the transcendental sample size condition is satisfied for $n = 3$ replicates of the 2^3 design.

8.6.3 Two-Level Fractional Factorial Designs

The sample size and power calculations for two-level fractional factorial designs are performed using the same methods that are used for full factorial designs. However, a little more thought might be required to determine what model will be fitted to the data, especially if the design is of low resolution or if there are few error degrees of freedom associated with the proposed model. Once the model to be fitted has been determined, the error degrees of freedom must be calculated assuming some value for the number of replicates. When the error degrees of freedom is very small, it might be helpful to assume that some of the model terms will be dropped to increase the error degrees of freedom.

Example 8.16 What is the power for the 2_{IV}^{4-1} design with two replicates to detect a difference of $\delta = 10$ between two levels of a design variable if $\sigma_\epsilon = 5$?

Solution: With two replicates the total number of experimental runs will be $2(2^{4-1}) = 16$. Because the experiment design is resolution IV, the model can include main effects and only three of the six possible two-factor interactions, so $df_{model} = 4 + 3 = 7$. Then, the error degrees of freedom will be $df_\epsilon = (16 - 1) - 7 = 8$. The F distribution noncentrality parameter associated with a difference of $\delta = 10$ between two levels of a design variable is given by a slightly modified form of Equation 8.22:

$$\phi = n2^{(k-p)-2}\left(\frac{\delta}{\sigma_\epsilon}\right)^2 \qquad (8.33)$$

$$= 2 \times 2^{(4-1)-2}\left(\frac{10}{5}\right)^2$$

$$= 16.0$$

where $p = 1$ accounts for the half-fractionation of the full factorial design. Then, by Equation 8.21

$$F_{0.95} = F_{1-\pi, 16}$$
$$5.318 = F_{0.063, 16}.$$

The power is $\pi = 1 - 0.063 = 0.937$.

Example 8.17 How many replicates of a 2_V^{5-1} design are required to have 90% power to detect a difference $\delta = 0.4$ between two levels of a design variable? Assume that ten of the fifteen possible terms will drop out of the model and that the standard error will be $\sigma_\epsilon = 0.18$.

Solution: If a model with main effects and two factor interactions is fitted to one replicate of the 2_V^{5-1} design, there will not be any degrees of freedom left to estimate the error, so either the experiment must be replicated or some terms must be dropped from the model. Under the assumption that the number of replicates is large, so that we can take $t \simeq z$ in the first iteration of Equation 8.24, we have

$$n \geq \frac{1}{2^{(5-1)-2}}\left(z_{0.025} + z_{0.10}\right)^2 \left(\frac{0.18}{0.4}\right)^2$$

$$\geq 0.532.$$

Obviously, the $t \simeq z$ approximation is not satisfied, so at least one more iteration is required. If only one replicate of the half-fractional factorial design is built and ten of the fifteen possible terms are dropped from the model, the error degrees of freedom will be $df_\epsilon = 15 - 10 = 5$. Then, for the second iteration of

8.6. Two-Level Factorial Designs

Equation 8.24, we have

$$n \geq \frac{1}{2^{(5-1)-2}} (t_{0.025} + t_{0.10})^2 \left(\frac{0.18}{0.4}\right)^2$$

$$\geq \frac{1}{2^{(5-1)-2}} (2.228 + 1.372)^2 \left(\frac{0.18}{0.4}\right)^2$$

$$\geq 0.656,$$

which rounds up to $n = 1$. Calculation of the power (not shown) confirms $\pi = 0.98$ for one replicate.

8.6.4 Plackett-Burman Designs

Power and sample size calculations for the Plackett-Burman designs are similar to those for the two-level fractional factorial designs. The Plackett-Burman designs are resolution III designs, so the models might include only main effects. If R is the number of runs in one replicate of a Plackett-Burman design, then the number of replicates n required to detect a difference δ between two levels of a design variable when the standard error is expected to be σ_ϵ is given by a variation on Equation 8.24:

$$n \geq \frac{4}{R} \left(t_{\alpha/2} + t_\beta\right)^2 \left(\frac{\sigma_\epsilon}{\delta}\right)^2. \tag{8.34}$$

Of course, if the number of replicates is sufficiently large, then the first iteration of this equation can use $t \simeq z$.

Example 8.18 How many replicates of a 9-variable 12-run Plackett-Burman design are required to detect a difference $\delta = 7000$ between two levels of a variable with 90% power if the standard error is expected to be $\sigma_\epsilon = 4000$?

Solution: Plackett-Burman designs are resolution III, so their models may contain only main effects. If the experiment is run with only one replicate, then $df_\epsilon = (12-1) - 9 = 2$ and the large-sample approximation is obviously not satisfied. If enough replicates are run so that the large-sample approximation is satisfied, then with $\alpha = 0.05$, $\beta = 0.10$ and $t \simeq z$, the approximate number of replicates required is

$$n \geq \frac{4}{12} (1.96 + 1.282)^2 \left(\frac{4000}{7000}\right)^2$$

$$\geq 2.$$

Another iteration confirms that two replicates are sufficient to achieve 90% power.

8.7 Two-Level Factorial Designs with Centers

Two-level factorial designs with centers are two-level factorial designs with added center cells, where all of the design variables must be quantitative and the center cell runs are made with all of the variables at their 0 levels midway between their ±1 levels. This means that the name *two-level factorial design with centers* is a misnomer because these are really three-level designs. For example, in a 2^3 design with centers, the experimental runs from the 2^3 part of the design requires $2^3 = 8$ runs with the variables at their coded ±1 levels: $(x_1, x_2, x_3) = (-,-,-)$, $(-,-,+), \ldots, (+,+,+)$, plus the center cells run with $(x_1, x_2, x_3) = (0, 0, 0)$. Any number of center cells can be added to the 2^k design, because, for the purpose of estimating main effects and interactions, they do not unbalance the design.

The inclusion of a third level for each variable in these designs provides the opportunity to perform a linear lack of fit (LOF) test, that is, a test of the assumption that the response is linear with respect to each quantitative design variable. This assumption is important because when it is true, the model can be used to interpolate the response within that part of the design space covered by the experiment. When the assumption is false, the model cannot be used for interpolation and another experiment, a response surface experiment, must be performed to resolve the source and magnitude of the curvature.

The model obtained from a two-level factorial plus centers design has the form

$$y = b_0 + b_1 x_1 + b_2 x_2 + \cdots + b_{12} x_{12} + \cdots + b_{**} x_*^2 \qquad (8.35)$$

where x_*^2 is a generic curvature or lack-of-linear-fit term. This term is the sum of the curvature effects from the individual design variables, hence the use of the wild card asterisk subscript. Testing for lack of linear fit is simply a matter of testing the statistical significance of the b_{**} regression coefficient; if the hypothesis $H_0 : \beta_{**} = 0$ cannot be rejected, then there is insufficient evidence to conclude that there is lack of linear fit.

8.7.1 Main Effects

Sample size and power calculations for main effects in two-level factorial designs with centers are performed the same way as for the designs without centers except for small modifications to the model and error degrees of freedom. These modifications are 1) the model for a two-level factorial design with centers consumes one extra degree of freedom for estimating the β_{**} coefficient, and 2) in the two-level factorial design with centers, the center points add degrees of freedom for the error estimate. It is somewhat counterintuitive that, except for their contribution to the error degrees of freedom, the center points do not improve the estimates for the main effects.

Example 8.19 Calculate the power to detect a difference $\delta = 1400$ between two levels of a study variable in a 2^3 design with three replicates built in blocks

8.7. Two-Level Factorial Designs with Centers

with two center points per block. Include terms for main effects, two-factor interactions, lack of fit, and blocks in the model. The standard error is expected to be $\sigma_\epsilon = 1000$.

Solution: The experiment will have $3(2^3 + 2) = 30$ total observations, so the total degrees of freedom will be $df_{total} = 29$. The degrees of freedom for the model will be

$$\begin{aligned} df_{model} &= df_{blocks} + df_{main\ effects} + df_{interactions} + df_{LOF} \\ &= 2 + 3 + 3 + 1 \\ &= 9, \end{aligned}$$

so the error degrees of freedom will be

$$df_\epsilon = df_{total} - df_{model} = 29 - 9 = 20.$$

The power π to reject $H_0 : \delta = 0$ for the main effect of any one of the study variables is given by Equation 8.21 with one numerator and twenty denominator degrees of freedom where the F distribution noncentrality parameter, as given in Equation 8.22, is

$$\begin{aligned} \phi &= 3 \times 2^{3-2} \left(\frac{1400}{1000} \right)^2 \\ &= 11.76. \end{aligned}$$

With $\alpha = 0.05$ and

$$\begin{aligned} F_{1-\alpha} &= F_{1-\pi,\phi} \\ F_{0.95} &= 4.351 = F_{1-\pi, 8.17}, \end{aligned}$$

we find the power to be $\pi = 0.903$.

8.7.2 Lack of Fit Test

In the analysis of power and confidence interval precision for the two-level factorial designs in Section 8.6, the sample size was determined to be inversely proportional to SS_x (one replicate), which for the main effects is $SS_x = 2^k$. The sum of squares for the quadratic term in the regression model $SS_{x_{**}}$ (one replicate) is dependent on the number of corner points in the 2^k part of the design (for which $x_{**} = 1$) and the number of center points (for which $x_{**} = 0$) according to

$$\begin{aligned} SS_{x_{**}} &= \sum (x_{**} - \overline{x_{**}})^2 \\ &= \frac{1}{\frac{1}{2^k} + \frac{1}{n_0}}. \end{aligned} \quad (8.36)$$

Then, from Equation 8.30, the number of replicates n required to resolve the quadratic term with specified precision δ_{**} must meet the condition

$$n \geq \frac{1}{SS_{x_{**}}} \left(\frac{t_{\alpha/2}\sigma_\epsilon}{\delta_{**}}\right)^2$$

$$\geq \left(\frac{1}{2^k} + \frac{1}{n_0}\right) \left(\frac{t_{\alpha/2}\sigma_\epsilon}{\delta_{**}}\right)^2. \tag{8.37}$$

When the number of corner points 2^k is much larger than the number of center points n_0, this condition approximates to the more manageable and conservative form

$$n \geq \frac{1}{n_0} \left(\frac{t_{\alpha/2}\sigma_\epsilon}{\delta_{**}}\right)^2. \tag{8.38}$$

Equations 8.32 and 8.37 indicate that the ratio of the estimates of the precisions of the quadratic and main effect terms is

$$\frac{\delta_{**}}{\delta} = \sqrt{\frac{\frac{1}{SS_{x_{**}}}}{\frac{1}{SS_x}}}$$

$$= \sqrt{\frac{SS_x}{SS_{x_{**}}}}$$

$$= \sqrt{1 + \frac{2^k}{n_0}}. \tag{8.39}$$

This result indicates that the precision of the estimate for the quadratic effect will always be worse than the precision of the estimate for main effects.

Example 8.20 Determine the ratio of the precisions of the estimates for the lack of fit and main effects in a 2^4 plus centers design when two center points are used per replicate.
Solution: From Equation 8.39 with $k = 4$ and $n_0 = 2$ the ratio of the lack of fit and main effect precision estimates will be

$$\frac{\delta_{**}}{\delta} = \sqrt{1 + \frac{2^4}{2}}$$
$$= 3.$$

That is, the confidence interval for the lack of fit estimate will be three times wider than the confidence interval for the main effects.

8.8 Response Surface Designs

Response surface designs are used to determine a model for a quantitative response as a function of two or more quantitative predictor variables when the response is expected to be a curved function of the predictors. The model fitted to the response in such an experiment includes main effects, two-factor interactions, and quadratic terms:

$$y = b_0 + b_1 x_1 + b_2 x_2 + \cdots + b_{12} x_{12} + \cdots + b_{11} x_1^2 + b_{22} x_2^2 + \cdots. \tag{8.40}$$

Response surface designs are usually used at the end of a series of designed experiments, after the few most important design variables have been identified.

The most common response surface designs are Box-Behnken and central composite designs. Box-Behnken designs use three evenly-spaced levels of each design variable. In coded units, the three levels are $-1, 0$, and 1. Central composite designs are based on the two-level factorial designs and have additional center points and star points. This gives the central composite designs five levels of each design variable, indicated in coded units as $-\alpha, -1, 0, 1$, and α, where α depends on the number of runs in the 2^k part of the experiment. See Montgomery [45] or Mathews [42] for details on the design and analysis of response surface designs.

In most circumstances, the purpose of a response surface design is to quantify the regression coefficients associated with model terms that are already known or suspected to be significant, so the relevant sample size calculation is for a regression coefficient confidence interval which has the form

$$P(b_i - \delta < \beta_i < b_i + \delta) = 1 - \alpha \tag{8.41}$$

where the confidence interval half-width is given by

$$\begin{aligned}\delta &= t_{\alpha/2} \hat{\sigma}_{b_i} \\ &= t_{\alpha/2} \left(\frac{\hat{\sigma}_\epsilon}{\sqrt{n SS_{x_i}}} \right), \end{aligned} \tag{8.42}$$

$\hat{\sigma}_\epsilon$ is the standard error of the regression, the t distribution has df_ϵ degrees of freedom, SS_{x_i} is the sum of squares associated with variable or model term x_i for one replicate of the experiment design, and n is the number of replicates. The number of replicates required to obtain a specified confidence interval half-width δ is the smallest value of n that satisfies

$$n \geq \frac{1}{SS_{x_i}} \left(\frac{t_{\alpha/2} \hat{\sigma}_\epsilon}{\delta} \right)^2. \tag{8.43}$$

Table 8.5 gives values for SS_{x_i} for main effects, two-factor interactions, and quadratic terms for one replicate of some common response surface designs assuming that all of the variable levels are expressed in coded units as indicated in column *Levels*. Box-Behnken designs are indicated with the notation $BB\,(k)$ and central composite designs are indicated with the notation $CC\,(2^k)$ where k is the number of design variables. Values for SS_{x_i} for some other designs are shown for reference.

Example 8.21 How many replicates of a three-variable Box-Behnken design are required to estimate the regression coefficients associated with main effects, two-factor interactions, and quadratic terms to within $\delta = \pm 2$ with 95% confidence if the standard error is expected to be $\sigma_\epsilon = 5$?

Solution: A first estimate for the number of replicates required to estimate the regression coefficients associated with main effects is given by Equation 8.43 with $t_{0.025} \simeq 2$ and, from Table 8.5 for the $BB\,(3)$ design, $SS_{Main\ Effects} = 8$ is

$$n \geq \frac{1}{8}\left(\frac{2 \times 5}{2}\right)^2$$
$$\geq 4.$$

With $n = 4$ replicates, the error degrees of freedom will be

$$\begin{aligned} df_\epsilon &= df_{total} - df_{model} \\ &= (4 \times 15 - 1) - 9 \\ &= 50, \end{aligned}$$

so the approximation for $t_{0.025}$ is justified. Another iteration with $n = 3$ replicates indicates that the precision of the regression coefficient estimates would be slightly greater than $\delta = 2$, so $n = 4$ replicates are required.

From Table 8.5 for two-factor interactions, $SS_{Interaction} = 4$, so the number of replicates required to estimate the regression coefficients associated with two-factor interactions with confidence interval half-width $\delta = 2$ with 95% confidence is

$$n \geq \frac{1}{4}\left(\frac{t_{0.025} \times 5}{2}\right)^2$$
$$\geq 6.$$

From Table 8.5 for quadratic terms, $SS_{Quadratic} = 3.694$, so the number of replicates required to estimate the regression coefficients associated with quadratic terms is

$$n \geq \frac{1}{3.694}\left(\frac{t_{0.025} \times 5}{2}\right)^2$$
$$\geq 7.$$

8.8. Response Surface Designs

Design	Runs	Levels	$SS_{Main\,Effects}$	$SS_{Interaction}$	$SS_{Quadratic}$
2^k	2^k	± 1	2^k	2^k	NA
2^{k-p}	2^{k-p}	± 1	2^{k-p}	2^{k-p}	NA
2^k plus n_0 centers	$2^k + n_0$	$0, \pm 1$	2^k	2^k	$\left(\frac{1}{2^k} + \frac{1}{n_0}\right)^{-1}$
3^k	3^k	$0, \pm 1$	$2 \times 3^{k-1}$	$4 \times 3^{k-2}$	$2 \times 3^{k-2}$
$BB(3)$	15	$0, \pm 1$	8	4	3.694
$BB(4)$	27	$0, \pm 1$	12	4	5.335
$BB(5)$	46	$0, \pm 1$	16	4	8.725
$CC(2^2)$	13	$0, \pm 1, \pm 1.414$	8	4	6.955
$CC(2^3)$	20	$0, \pm 1, \pm 1.682$	13.67	8	14.41
$CC(2^4)$	31	$0, \pm 1, \pm 2$	24	16	28.58
$CC(2_V^{5-1})$	32	$0, \pm 1, \pm 2$	24	16	29.35
$CC(2_V^{8-2})$	90	$0, \pm 1, \pm 2.828$	80	64	116.3

Table 8.5: Sums of squares factors for estimating effects in designed experiments.

Chapter 9

Reliability and Survival

Sample size calculations are presented for the following reliability engineering methods: reliability parameter and percentile estimation, reliability demonstration tests, two-sample reliability tests, and strength/load interference. These analyses are presented for the exponential, Weibull, and normal reliability distributions.

Although this chapter is presented using time to failure as the measured response, these methods are also applicable to other measures of the duration of a test, such as the number of cycles to failure, the number of miles until a car tire wears out, or the strength of a material in a tensile test.

9.1 Reliability Parameter Estimation

Sample size calculations are presented for experiments to determine confidence intervals for distribution parameters, percentiles, and reliability for the exponential, Weibull, and normal distributions. For the exponential and Weibull distributions the critical test variable that determines the confidence interval width is the number of failures and not the number of units tested. This approach permits some flexibility in reliability study design: experiments with complete failure data and suspensions (i.e., right censored observations) will have the same estimation precision as long as they have the same number of failures.

9.1.1 Exponential Reliability

9.1.1.1 Confidence Interval for the Exponential Mean

Under the exponential reliability model the reliability at time t, that is, the probability of survival to time t, is given by

$$R(t; \mu) = e^{-t/\mu} \qquad (9.1)$$

where μ is the mean life. A point estimate for μ based on observing r failures among n units tested is

$$\bar{t} = \frac{1}{r} \sum_{i=1}^{n} t_i. \tag{9.2}$$

The exact $(1-\alpha)\,100\%$ confidence interval for μ is given by

$$P\left(\frac{2r\bar{t}}{\chi^2_{1-\alpha/2}} < \mu < \frac{2r\bar{t}}{\chi^2_{\alpha/2}}\right) = 1 - \alpha \tag{9.3}$$

where the χ^2 distribution has $\nu = 2r$ degrees of freedom if the test is failure terminated or $\nu = 2(r+1)$ degrees of freedom if the test is time terminated.

Equation 9.3 is impractical for the purpose of calculating the number of failures r required to estimate μ with specified precision, however, the exponential distribution's standard deviation is equal to its mean, so when r is large, Equation 9.3 may be approximated by

$$P\left(\bar{t} - z_{\alpha/2}\frac{\bar{t}}{\sqrt{r}} < \mu < \bar{t} + z_{\alpha/2}\frac{\bar{t}}{\sqrt{r}}\right) = 1 - \alpha$$

$$P\left(\bar{t}(1-\delta) < \mu < \bar{t}(1+\delta)\right) = 1 - \alpha \tag{9.4}$$

where the confidence interval relative half-width is

$$\delta = \frac{z_{\alpha/2}}{\sqrt{r}}. \tag{9.5}$$

Then the approximate number of failures required to obtain a specified confidence interval half-width is

$$r = \left(\frac{z_{\alpha/2}}{\delta}\right)^2. \tag{9.6}$$

Example 9.1 How many units must be tested to failure to determine, with 20% precision and 95% confidence, the exponential mean life μ?

Solution: From Equation 9.6 with $\alpha = 0.05$ and $\delta = 0.2$, the required number of failures is

$$r = \left(\frac{1.96}{0.20}\right)^2 = 97.$$

9.1.1.2 Confidence Interval for an Exponential Percentile

The confidence interval for an exponential failure percentile is the general case from which the confidence interval for the exponential mean is derived. When μ is unknown, the point estimate for the $100f^{th}$ percentile t_f, that is, the time at which $100f\%$ of the units are expected to fail, is given by

$$\hat{t}_f = -\bar{t}\ln(1-f) \tag{9.7}$$

9.1. Reliability Parameter Estimation

where $\bar{t}$ is the mean time to failure based on observing r failures among n units tested as given by Equation 9.2.

The exact $(1-\alpha)\,100\%$ confidence interval for the $100 f^{th}$ percentile t_f is

$$P\left(\frac{2r\hat{t}_f}{\chi^2_{1-\alpha/2}} < t_f < \frac{2r\hat{t}_f}{\chi^2_{\alpha/2}}\right) = 1 - \alpha \tag{9.8}$$

where the χ^2 distribution has $\nu = 2r$ degrees of freedom if the test is failure terminated or $\nu = 2(r+1)$ degrees of freedom if the test is time terminated. This confidence interval has the same form as the confidence interval for μ in Equation 9.3, so the sample size calculation in Equation 9.6 applies. This means that the confidence intervals for all failure percentiles under the exponential distribution have the same relative width.

Example 9.2 How many units must be tested to failure to determine, with 20% precision and 95% confidence, any failure percentile under the assumption that the reliability distribution is exponential?

Solution: The conditions required to estimate the failure percentiles are the same as those in Example 9.1, so the same number of failures required is $r = 97$.

9.1.1.3 Confidence Interval for the Exponential Reliability

When the number of failures r is large and μ is unknown, by the delta method the sampling distribution of the reliability $R(t;\mu)$ is approximately normal with standard deviation:

$$\hat{\sigma}_{\widehat{R}} = \hat{\sigma}_{\hat{\mu}} \frac{dR}{d\mu} = -\frac{\widehat{R}\ln\left(\widehat{R}\right)}{\sqrt{r}}, \tag{9.9}$$

so the approximate large-sample $(1-\alpha)\,100\%$ confidence interval for $R(t;\mu)$ is

$$P\left(\widehat{R}(1-\delta) < R < \widehat{R}(1+\delta)\right) = 1 - \alpha \tag{9.10}$$

where the confidence interval's relative half-width is

$$\delta = -\frac{z_{\alpha/2}\ln\left(\widehat{R}\right)}{\sqrt{r}}. \tag{9.11}$$

Then the number of failures r required to obtain a specified confidence interval half-width is

$$r = \left(\frac{z_{\alpha/2}\ln\left(\widehat{R}\right)}{\delta}\right)^2. \tag{9.12}$$

Example 9.3 How many units must be tested to failure in an experiment to determine, with 95% confidence, the exponential reliability to within 10% of its true value if the expected reliability is 80%?

Solution: From Equation 9.12 with $\alpha = 0.05$, $\delta = 0.10$, and $\widehat{R} = 0.80$, the required number of failures is

$$r = \left(\frac{1.96 \ln(0.80)}{0.10}\right)^2 = 20. \tag{9.13}$$

9.1.2 Weibull Reliability

9.1.2.1 Confidence Interval for the Weibull Scale Parameter

The Weibull reliability at time t (i.e., the probability of survival to time t) is given by

$$R(t; \eta, \beta) = e^{-(t/\eta)^\beta} \tag{9.14}$$

where η is the scale factor and β is the shape factor. When β is known but η is not, which is often the case, then after the variable transformations $t' = t^\beta$ and $\eta' = \eta^\beta$ the Weibull distribution is transformed into the exponential distribution and the results from Section 9.1.1.1 apply. The resulting confidence interval for the scale parameter takes the form

$$P\left(\widehat{\eta}'(1-\delta') < \eta' < \widehat{\eta}'(1+\delta')\right) = 1 - \alpha \tag{9.15}$$

where $\delta' = z_{\alpha/2}/\sqrt{r}$ as before in the exponential case. By applying the inverse transform $(\eta')^{1/\beta} = \eta$ we obtain the desired confidence interval for η:

$$P\left(\widehat{\eta}(1-\delta')^{1/\beta} < \eta < \widehat{\eta}(1+\delta')^{1/\beta}\right) = 1 - \alpha, \tag{9.16}$$

which, when $|\delta'| \ll 1$, may be approximated by

$$P(\widehat{\eta}(1-\delta) < \eta < \widehat{\eta}(1+\delta)) = 1 - \alpha \tag{9.17}$$

where the confidence interval's relative half-width is

$$\delta = \frac{\delta'}{\beta}$$
$$= \frac{z_{\alpha/2}}{\beta\sqrt{r}}. \tag{9.18}$$

Then the number of failures required to obtain specified relative precision δ is

$$r = \left(\frac{z_{\alpha/2}}{\beta\delta}\right)^2. \tag{9.19}$$

9.1. Reliability Parameter Estimation

Example 9.4 How many units must be tested to failure to estimate, with 20% precision and 95% confidence, the Weibull scale factor if the shape factor is known to be $\beta = 2$?

Solution: The goal of the experiment is to obtain a confidence interval for the Weibull scale factor of the form given by Equation 9.17 with $\delta = 0.20$ and $\alpha = 0.05$. From Equation 9.19 the required number of failures is

$$r = \left(\frac{1.96}{2 \times 0.20}\right)^2 = 25.$$

9.1.2.2 Confidence Interval for the Weibull Shape Parameter

A $(1 - \alpha)\,100\%$ confidence interval for the Weibull shape parameter β is required of the form

$$P\left(\widehat{\beta}(1 - \delta) < \beta < \widehat{\beta}(1 + \delta)\right) = 1 - \alpha \tag{9.20}$$

where δ is the relative precision of the β estimate. If the sample size is sufficiently large, then the $\widehat{\beta}$ distribution is approximately normal, so we may take the confidence interval's relative half-width to be

$$\delta = z_{\alpha/2}\frac{\widehat{\sigma}_{\widehat{\beta}}}{\widehat{\beta}} \tag{9.21}$$

where the standard error of the β estimate is

$$\widehat{\sigma}_{\widehat{\beta}} = \frac{\widehat{\beta}}{\pi}\sqrt{\frac{6}{r}} \tag{9.22}$$

and r is the number of failures. Then the confidence interval's relative half-width is

$$\delta = \frac{z_{\alpha/2}}{\pi}\sqrt{\frac{6}{r}} \tag{9.23}$$

and the number of failures required to obtain a specified relative precision for the β estimate is

$$r = 6\left(\frac{z_{\alpha/2}}{\pi\delta}\right)^2. \tag{9.24}$$

Example 9.5 How many units must be tested to failure to estimate, with 95% confidence, the Weibull shape parameter to within 20% of its true value?

Solution: The goal of the experiment is to produce a 95% confidence interval for β of the form given by Equation 9.20 with $\delta = 0.20$. From Equation 9.24 with $\alpha = 0.05$, the required number of failures is

$$r = 6\left(\frac{1.96}{\pi \times 0.20}\right)^2 = 59.$$

9.1.2.3 Confidence Interval for a Weibull Percentile

Just as the confidence interval for the exponential reliability mean μ is a special case of the confidence interval for the failure percentile, the confidence interval for the Weibull scale parameter η is a special case of the confidence interval for the Weibull failure percentile.

When the Weibull shape parameter β is known but the scale parameter η is unknown, which is often the case, the point estimate for the $100f^{th}$ percentile t_f, that is, the time at which $100f\%$ of the units are expected to fail, is given by

$$\hat{t}_f = \hat{\eta} \left(-\ln(1-f) \right)^{1/\beta}. \tag{9.25}$$

From Equation 9.8 the exact $(1-\alpha)\,100\%$ confidence interval for t_f^β is

$$P\left(\frac{2r\hat{t}_f^\beta}{\chi^2_{1-\alpha/2}} < t_f^\beta < \frac{2r\hat{t}_f^\beta}{\chi^2_{\alpha/2}} \right) = 1-\alpha \tag{9.26}$$

where the χ^2 distribution has $\nu = 2r$ degrees of freedom if the test is failure terminated or $\nu = 2(r+1)$ degrees of freedom if the test is time terminated. The corresponding large-sample normal approximation is

$$P\left(\hat{t}_f^\beta (1-\delta') < t_f^\beta < \hat{t}_f^\beta (1+\delta') \right) = 1-\alpha \tag{9.27}$$

where $\delta' = z_{\alpha/2}/\sqrt{r}$. Then the approximate confidence interval for t_f is

$$P\left(\hat{t}_f (1-\delta')^{\frac{1}{\beta}} < t_f < \hat{t}_f (1+\delta')^{\frac{1}{\beta}} \right) = 1-\alpha, \tag{9.28}$$

which, when $|\delta'| \ll 1$, can be approximated by

$$P\left(\hat{t}_f \left(1 - \frac{\delta'}{\beta}\right) < t_f < \hat{t}_f \left(1 + \frac{\delta'}{\beta}\right) \right) = 1-\alpha$$
$$P\left(\hat{t}_f (1-\delta) < t_f < \hat{t}_f (1+\delta) \right) = 1-\alpha \tag{9.29}$$

where

$$\begin{aligned}\delta &= \frac{\delta'}{\beta} \\ &= \frac{z_{\alpha/2}}{\beta\sqrt{r}}.\end{aligned} \tag{9.30}$$

This is the same confidence interval half-width as was obtained for the Weibull scale parameter, so the number of failures required to estimate any Weibull failure percentile is given by Equation 9.19.

9.1. Reliability Parameter Estimation

9.1.2.4 Confidence Interval for the Weibull Reliability

A $(1 - \alpha)\,100\%$ confidence interval for the Weibull reliability $R(t; \eta, \beta)$ at time t is required of the form

$$P\left(\widehat{R}(1 - \delta) < R < \widehat{R}(1 + \delta)\right) = 1 - \alpha \quad (9.31)$$

where δ is the relative precision of the estimate. The exact confidence interval for the Weibull reliability when there are r failures and β is known is given by

$$P\left(\widehat{R}^{\frac{\chi^2_{1-\alpha/2, 2r}}{2r}} < R < \widehat{R}^{\frac{\chi^2_{\alpha/2, 2r}}{2r}}\right) = 1 - \alpha. \quad (9.32)$$

When the number of failures is large, say $r \geq 15$, this confidence interval may be approximated with

$$P\left(\widehat{R}^{1+\frac{z_{\alpha/2}}{\sqrt{r}}} < R < \widehat{R}^{1-\frac{z_{\alpha/2}}{\sqrt{r}}}\right) \simeq 1 - \alpha. \quad (9.33)$$

These confidence limits do not simplify over the entire range of R, but if R is expected to be large, say $0.8 \leq R < 1$, which is the usual range of interest, then this confidence interval approximates to the form given in Equation 9.31 with relative confidence interval half-width

$$\delta = \frac{z_{\alpha/2}}{\sqrt{r}}\left(\frac{1 - \widehat{R}}{\widehat{R}}\right). \quad (9.34)$$

Then the number of failures required to determine the Weibull reliability with relative precision δ is

$$r = \left(\frac{z_{\alpha/2}}{\delta}\left(\frac{1 - \widehat{R}}{\widehat{R}}\right)\right)^2. \quad (9.35)$$

Example 9.6 How many units must be tested to failure to estimate the Weibull reliability with 5% precision and 95% confidence when the expected reliability is 90%? Assume that the Weibull shape factor is known.
Solution: The desired confidence interval will have the form of Equation 9.31. From Equation 9.35 with $\delta = 0.05$, $\alpha = 0.05$, and $\widehat{R} = 0.90$, the required number of failures is

$$r = \left(\frac{1.645}{0.05}\left(\frac{1 - 0.9}{0.9}\right)\right)^2 = 19.$$

9.1.3 Normal Reliability

Methods are presented here for calculating sample sizes for experiments to estimate failure probabilities for complete normally distributed failure data. See Nelson [50] for analysis methods for censored normal data.

9.1.3.1 Confidence Interval for a Normal Percentile

An approximate $(1 - \alpha)\,100\%$ confidence interval for the $100f^{th}$ failure percentile t_f, that is, the time at which $100f\%$ of the units are expected to fail, is given by

$$P\left(\widehat{t}_f - \delta < t_f < \widehat{t}_f + \delta\right) = 1 - \alpha \tag{9.36}$$

where the confidence interval half-width is

$$\delta = z_{\alpha/2}\widehat{\sigma}_{\widehat{t}_f} \tag{9.37}$$

$$= z_{\alpha/2}\widehat{\sigma}_t \sqrt{\frac{1}{n}\left(1 + \frac{1}{2}z_f^2\right)} \tag{9.38}$$

and the standard deviation of the $\widehat{t}_f$ estimate is given by the delta method (see Appendix G.3.9) where z_f is the z-transformed value associated with t_f:

$$z_f = \frac{t_f - \mu_t}{\sigma_t}. \tag{9.39}$$

Then, from Equation 9.38, the sample size is

$$n = \left(\frac{z_{\alpha/2}\widehat{\sigma}_t}{\delta}\right)^2 \left(1 + \frac{1}{2}z_f^2\right). \tag{9.40}$$

Example 9.7 An experiment is planned to estimate, with 95% confidence, the time at which 10% of units will fail to within 1000 hours. The life distribution is expected to be normal with $\widehat{\sigma}_t = 2000$ and all units will be tested to failure.
Solution: With $z_{\alpha/2} = z_{0.025} = 1.96$ and $z_f = z_{0.10} = 1.282$ in Equation 9.40, the sample size is

$$n = \left(\frac{1.96 \times 2000}{1000}\right)^2 \left(1 + \frac{(1.282)^2}{2}\right)$$

$$= 28.$$

9.1.3.2 Confidence Interval for Normal Probability

An approximate $(1 - \alpha)\,100\%$ confidence interval for the left-tail normal probability $\Phi(x; \mu, \sigma)$ based on sample data is given by[1]

$$P\left(\widehat{\Phi}(\widehat{z}) - \delta < \Phi(x; \mu, \sigma) < \widehat{\Phi}(\widehat{z}) + \delta\right) = 1 - \alpha \tag{9.41}$$

---

[1] An alternate confidence interval formulation uses symmetric limits in terms of z which gives asymmetric limits in terms of probability. The two formulations have different but closely related sample size requirements.

9.2. Reliability Demonstration Tests

where the confidence interval half-width is

$$\delta = z_{\alpha/2}\widehat{\sigma}_{\widehat{\Phi}(x)} \qquad (9.42)$$

$$= z_{\alpha/2}\varphi(\widehat{z})\sqrt{\frac{1}{n}\left(1+\frac{1}{2}\widehat{z}^2\right)} \qquad (9.43)$$

and the standard deviation of the $\Phi(x)$ estimate is given by the delta method (see Appendix G.3.10) where $\varphi(\widehat{z})$ is the normal probability density function evaluated at $\widehat{z} = (x - \bar{x})/s$.

If an experiment is to be performed to determine a confidence interval for the normal probability at a specified value of $\widehat{z}$ with specified confidence interval half-width δ, then the required sample size is

$$n = \left(\frac{z_{\alpha/2}\varphi(\widehat{z})}{\delta}\right)^2 \left(1 + \frac{1}{2}\widehat{z}^2\right). \qquad (9.44)$$

Example 9.8 What sample size is required to estimate, with 95% confidence, the 16000 hour failure probability of a product to within 2% if the life distribution is expected to be normal with $\mu \simeq 20000$ and $\sigma \simeq 2000$?

Solution: With $x = 16000$ and $\widehat{z} = (16000 - 20000)/2000 = -2$, the required confidence interval for the 16000 hour failure probability has the form

$$P\left(\widehat{\Phi}(-2) - 0.02 < \Phi(x = 16000; \mu, \sigma) < \widehat{\Phi}(-2) + 0.02\right) = 0.95.$$

From Equation 9.44 the required sample size to obtain this interval is

$$n = \left(\frac{z_{0.025}\varphi(-2)}{0.02}\right)^2 \left(1 + \frac{1}{2}(-2)^2\right)$$

$$= \left(\frac{1.96 \times 0.0540}{0.02}\right)^2 (3)$$

$$= 85.$$

9.2 Reliability Demonstration Tests

Reliability demonstration tests (RDT) are used to confirm that a reliability parameter exceeds a specified minimum or maximum value. The parameters tested may be the location parameter, the reliability or failure rate at a specified time, or a percentile, such as the time associated with a specified reliability. These tests may be interpreted in terms of either confidence intervals or hypothesis tests. They are performed by operating n units for a period t' measured in either hours or cycles of operation. The number of failures x over that period is observed and the demonstration is successful if x does not exceed a critical upper limit r.

In the most common RDT design problems the test goal and other necessary information are provided and either the sample size n or the time-on-test t' must be determined. The algorithms to solve these problems are similar: from the given information including either t' or n, find the failure probability f' at time t' and then find the missing parameter, either n or t', respectively. Such methods and examples are presented here for the exponential, Weibull, and normal reliability models. The exponential is presented even though it is a special case of the Weibull with shape factor $\beta = 1$. The log-normal distribution is omitted, but calculations for it may be performed using the normal distribution methods after applying a log transformation.

In the presentations that follow, the 0 subscript is used to indicate the value of the lower or upper confidence bound on the parameter being studied, or equivalently, the value of the parameter under H_0 in the hypothesis test interpretation. The ' mark indicates the value of a quantity at the end of the demonstration test, that is, at time t'.

9.2.1 Tests for Location Parameters

The goal of an RDT is to show that the unknown location parameter, generically indicated here with the symbol η, exceeds a specified value η_0 with specified confidence $1 - \alpha$; that is, to show that

$$P(\eta_0 < \eta < \infty) = 1 - \alpha. \tag{9.45}$$

Successfully demonstrating this confidence interval is equivalent to rejecting the hypothesis $H_0 : \eta \leq \eta_0$ in favor of $H_A : \eta > \eta_0$. The test is performed by operating n units for specified time t' and observing the number of failures x that occur within that period. The reliability demonstration is successful if the observed number of failures x does not exceed a critical value r.

For specified values of η_0 and α and two of the three test parameters n, t', and r, the unknown parameter of the RDT is determined from the condition

$$\sum_{x=0}^{r} b(x; n, f') \leq \alpha \tag{9.46}$$

where $b(x; n, f)$ is the binomial probability

$$b(x; n, f') = \binom{n}{x} (f')^x (1 - f')^{n-x} \tag{9.47}$$

and $f' = f(t'; \eta_0)$ is the failure probability at time t' when the location parameter is η_0. Table 9.1 gives equations for finding t' and f' in terms of the other parameters.

Equation 9.46 may be solved exactly using a binomial probability calculator or it may be approximated using Larson's nomogram (see Appendix E.1). Another useful approximation for solving Equation 9.46 when n is large and f' is

9.2. Reliability Demonstration Tests

Distribution	Solve for t' given f':	Solve for f' given t':
Exponential	$t' = -\mu_0 ln(1-f')$	$f' = 1 - e^{-t'/\mu_0}$
Weibull	$t' = \eta_0(-ln(1-f'))^{\frac{1}{\beta}}$	$f' = 1 - e^{-(t'/\eta_0)^\beta}$
Normal	$t' = \mu_0 + z_{f'}\sigma$	$f' = \Phi(-\infty < z < z_{f'})$ where $z_{f'} = \frac{t'-\mu_0}{\sigma}$

Table 9.1: Test time (t') and failure probability (f') for location parameter tests.

small is

$$\chi^2_{1-\alpha,2(r+1)} = \frac{1}{2nf'}.$$

Example 9.9 How many units must be tested for 200 hours without any failures to show, with 95% confidence, that the $MTTF$ of a system exceeds 400 hours. The life distribution is exponential and the test is time terminated.

Solution: We must determine the value of n with $r = 0$ failures in $t' = 200$ hours of testing such that

$$P(400 < \mu < \infty) = 0.95.$$

From the f' equation for the exponential distribution from Table 9.1 with $\mu_0 = 400$, the $t' = 200$ hour failure probability is

$$\begin{aligned} f' &= 1 - e^{-t'/\mu_0} \\ &= 1 - e^{-200/400} \\ &= 0.3935. \end{aligned}$$

With $r = 0$ and $\alpha = 0.05$ the smallest value of n that satisfies Equation 9.46 is $n = 6$ because

$$(b(0; 6, 0.3935) = 0.04977) < (\alpha = 0.05).$$

Example 9.10 Determine how long ten units must be life tested with no more than one failure during the test period to demonstrate, with 90% confidence, that the mean life is greater than 1000 hours. Assume that the life distribution is exponential.

Solution: The goal of the experiment is to demonstrate that

$$P(1000 < \mu < \infty) = 0.90.$$

With $n = 10$ and $r = 1$, Equation 9.46 gives

$$\sum_{x=0}^{1} b\left(x; n = 10, f'\right) = 0.10 \qquad (9.48)$$

which is satisfied by $f' = 0.337$. From the t' equation for the exponential distribution from Table 9.1, the required duration of the test in hours is

$$\begin{aligned} t' &= -\mu_0 ln\,(1-f') \\ &= -1000 ln\,(1-0.337) \\ &= 411. \end{aligned}$$

Example 9.11 Determine how long ten units must be life tested with no more than one failure during the test period to demonstrate, with 90% confidence, that the Weibull scale factor is at least 1000 hours. Assume that the Weibull shape factor is known to be $\beta = 2$.

Solution: The design parameters of the RDT are the same as in Example 9.10, so Equation 9.48 still applies and the end-of-test failure probability is $f' = 0.337$. From the t' equation for the Weibull distribution from Table 9.1, the required duration of the test in hours is

$$\begin{aligned} t' &= \eta_0\,(-ln\,(1-f'))^{\frac{1}{\beta}} \\ &= 1000\,(-ln\,(1-0.337))^{\frac{1}{2}} \\ &= 641. \end{aligned}$$

Example 9.12 Determine how long ten units must be life tested with no more than one failure during the test period to demonstrate, with 90% confidence, that the mean life is at least 1000 hours. Assume that the life distribution is normal with $\sigma = 100$.

Solution: The design parameters of the RDT are the same as in Example 9.10, so Equation 9.48 applies and the failure probability is $f' = 0.337$. From the equation for t' from Table 9.1, the required duration of the test in hours is

$$\begin{aligned} t' &= \mu + z_{f'} \sigma \\ &= 10000 + z_{0.337}\,(100) \\ &= 10000 + (-0.42 \times 100) \\ &= 958. \end{aligned}$$

9.2.2 Tests for Specified Reliability

The goal of the reliability demonstration test is to show that the reliability R at a specified time t_0 exceeds a specified minimum value R_0 with specified confidence $1 - \alpha$, that is,

$$P(R_0 < R(t_0) < 1) = 1 - \alpha, \tag{9.49}$$

9.2. Reliability Demonstration Tests

Distribution	t'	f'
Exponential	$t' = t_0 \frac{\ln(1-f')}{\ln(1-f_0)}$	$f' = 1 - (1-f_0)^{t'/t_0}$
Weibull	$t' = t_0 \left(\frac{\ln(1-f')}{\ln(1-f_0)} \right)^{1/\beta}$	$f' = 1 - (1-f_0)^{\left(\frac{t'}{t_0}\right)^\beta}$
Normal	$t' = t_0 + (z_{f_0} + z_{f'})\sigma$	$f' = \Phi(-\infty < z < z_{f'})$ where $z_{f'} = z_{f_0} + \left(\frac{t'-t_0}{\sigma}\right)$

Table 9.2: End of test time (t') and failure probability (f') for reliability tests.

or in terms of the failure probability $f = 1 - R$

$$P(0 < f(t_0) < f_0) = 1 - \alpha. \tag{9.50}$$

Successful demonstration of the confidence interval is equivalent to rejecting the hypothesis $H_0 : R(t_0) = R_0$ or $f(t_0) = f_0$ in favor of $H_A : R(t_0) > R_0$ or $f(t_0) < f_0$. The test is performed by operating n units for time t' and observing the number of failures x that occur within the testing period. The reliability demonstration is successful if the observed number of failures x does not exceed a critical value r.

For specified values of t_0, f_0, and α and two of the three test parameters n, t', and r, the final parameter of the RDT is determined from Equation 9.46. The unknown value of t' or f' is found from the simultaneous solution of the equations for f_0 and f'. The resulting solutions are shown in Table 9.2. When the test is suspended at time t_0, so that $t' = t_0$, it is not necessary to know the Weibull shape factor β.

Example 9.13 How many units must be tested for 400 hours without any failures to demonstrate 90% reliability at 600 hours, with 95% confidence? Assume that the reliability distribution is exponential.

Solution: In terms of the 600 hour failure probability, the goal of the experiment is to demonstrate

$$P(0 < f(600) < 0.10) = 0.95$$

based on a sample of size n tested to $t' = 400$ hours with $r = 0$ failures. From Table 9.2 the equation for f' for the exponential distribution gives

$$\begin{aligned} f' &= 1 - (1-f_0)^{t'/t_0} \\ &= 1 - (1-0.10)^{400/600} \\ &= 0.0678. \end{aligned}$$

With $r = 0$, $f' = 0.0678$, and $\alpha = 0.05$, Equation 9.46 gives

$$b(0; n, 0.0678) \leq 0.05,$$

which is satisfied by $n = 43$.

Example 9.14 How long must ten units be life tested with no more than one failure during the test period to demonstrate, with 80% confidence, that the 3000-hour reliability is at least 90%. Assume that the life distribution is Weibull with $\beta = 1.8$.

Solution: The goal of the experiment is to demonstrate that

$$P(0.90 < R(3000) < 1) = 0.80$$

or in terms of the failure probability

$$P(0 < f(3000) < 0.10) = 0.80.$$

With $n = 10$, $r = 1$, and $\alpha = 0.20$, Equation 9.46 becomes

$$\sum_{x=0}^{1} b(x; 10, f') \leq 0.20,$$

which is satisfied by $f' = 0.271$. From the equation for t' for the Weibull distribution from Table 9.2 with $f_0 = 0.10$ and $t_0 = 3000$, the test time is

$$\begin{aligned} t' &= t_0 \left(\frac{\ln(1-f')}{\ln(1-f_0)} \right)^{1/\beta} \\ &= 3000 \left(\frac{\ln(1-0.271)}{\ln(1-0.10)} \right)^{1/1.8} \\ &= 5523. \end{aligned}$$

Example 9.15 How many units must be tested for 140 hours with no more than one failure to demonstrate that the 100 hour reliability is at least 95% with 90% confidence? Assume that the reliability distribution is normal with $\sigma = 20$.

Solution: The goal of the experiment is to demonstrate

$$P(0 < f(100) < 0.05) = 0.90$$

based on a sample of size n tested to $t' = 140$ hours with no more than $r = 1$ failures. From Table 9.2, the equation for $z_{f'}$ for the normal distribution gives

$$\begin{aligned} z_{f'} &= z_{f_0} + \left(\frac{t' - t_0}{\sigma} \right) \\ &= z_{0.05} + \left(\frac{140 - 100}{20} \right) \\ &= -1.645 + 2.0 \\ &= 0.355, \end{aligned}$$

9.2. Reliability Demonstration Tests

which is satisfied by $f' = \Phi(-\infty < z < 0.355) = 0.639$. With $r = 1$, $f' = 0.639$, and $\alpha = 0.10$, Equation 9.46 becomes

$$\sum_{x=0}^{1} b(x; n, 0.639) \leq 0.10,$$

which is satisfied by $n = 5$.

Example 9.16 How many units must be tested without any failures to t_0 hours to demonstrate 90% reliability at t_0 hours with 95% confidence? Assume that the distribution is Weibull.

Solution: The goal of the experiment is to demonstrate

$$P(0 < f(t_0) < 0.10) = 0.95.$$

With $f_0 = 0.10$ and $t' = t_0$ in the Weibull equation for f' from Table 9.2

$$\begin{aligned} f' &= 1 - (1 - 0.10)^{\left(\frac{t_0}{t_0}\right)^\beta} \\ &= 0.10. \end{aligned}$$

With $f' = 0.10$, $r = 0$, and $\alpha = 0.05$, Equation 9.46 is

$$b(0; n, 0.10) \leq 0.05,$$

which is satisfied by $n = 29$. This is just a case of the rule of three: $n = 3/f_0$.

9.2.3 Tests for Percentiles

The goal of the RDT is to demonstrate that the time associated with a specified failure rate f_0 exceeds a specified value t_0 with confidence $(1 - \alpha)$, or

$$P(t_0 < t_{f_0} < \infty) = 1 - \alpha \qquad (9.51)$$

where time t_{f_0} is the *percentile* associated with f_0. Successful demonstration of Equation 9.51 is equivalent to rejecting the hypothesis $H_0 : t_{f_0} \leq t_0$ in favor of $H_A : t_{f_0} > t_0$. The design of an RDT for a percentile uses the same equations and algorithms as those presented in Section 9.2.2.

Example 9.17 How long must 50 units be tested without any failures to demonstrate that the time at which the first 1% of the population fails exceeds 400 cycles? Assume that the life distribution is exponential and use the 95% confidence level.

Solution: The goal of the experiment is to demonstrate that

$$P(400 < t_{0.01} < \infty) = 0.95.$$

From Equation 9.46 with $r = 0$, $n = 50$, and $\alpha = 0.05$

$$b(0; 50, f') \leq 0.05,$$

which is satisfied by $f' = 0.058$. From the exponential form of t' from Table 9.2 with $t_0 = 400$, the required duration of the test in cycles is

$$\begin{aligned} t' &= t_0 \frac{\ln(1-f')}{\ln(1-f_0)} \\ &= 400 \frac{\ln(1-0.058)}{\ln(1-0.01)} \\ &= 2380. \end{aligned}$$

Example 9.18 How many units must be tested to 30,000 cycles without any failures to demonstrate, with 95% confidence, that the 20,000 cycle reliability is at least 90%? The life distribution is known to be Weibull with $\beta = 3.1$.
Solution: The goal of the experiment is to demonstrate that

$$P(20000 < t_{0.10} < \infty) = 0.95.$$

The equation for f' for the Weibull distribution from Table 9.2 with $f_0 = 0.10$, $t_0 = 20000$, $t' = 30000$, and $\beta = 3.1$ gives

$$\begin{aligned} f' &= 1 - (1-f_0)^{(t'/t_0)^\beta} \\ &= 1 - (1-0.10)^{(30000/20000)^{3.1}} \\ &= 0.3095. \end{aligned}$$

Then, with $r = 0$, $f' = 0.3095$, and $\alpha = 0.05$, Equation 9.46 gives

$$b(0; n, 0.3095) \leq 0.05,$$

which is satisfied by $n = 9$.

9.3 Two-Sample Reliability Tests

Two-sample reliability tests are used to test for differences between two independent reliability distributions. Such tests may be performed for reliability parameters, percentiles, and survival rates at a specified endpoint. This section presents sample size and power calculations for a test for a difference between the means of two exponential distributions and a test for a difference between the survival rates of two populations.

9.3. Two-Sample Reliability Tests

9.3.1 Two-Sample Test for Mean Exponential Life

The hypotheses to be tested are $H_0 : \mu_1 = \mu_2$ versus $H_A : \mu_1 > \mu_2$ where μ_1 and μ_2 are the means of two independent exponential life distributions having the form

$$R(t; \mu_i) = e^{-t/\mu_i} \quad (9.52)$$

where $R(t; \mu_i)$ is the reliability or probability of survival to time t. The test is performed using the F statistic

$$F = \frac{\widehat{\mu}_1}{\widehat{\mu}_2} \quad (9.53)$$

where the estimated mean life $\widehat{\mu}_i$ is determined by the total time on test for n_i units and the observed number of failures x_i:

$$\widehat{\mu}_i = \frac{1}{x_i} \sum_{j=1}^{n_i} t_{ij} \quad (9.54)$$

and the F distribution has $df_1 = 2x_1$ and $df_2 = 2x_2$ degrees of freedom.

For specified values of μ_1 and μ_2 under H_A, the power of the test is given by

$$\pi = P(F_\beta < F < \infty) \quad (9.55)$$

where

$$F_\beta = \frac{\mu_2}{\mu_1} F_{1-\alpha} \quad (9.56)$$

and the F distribution has $df_1 = 2x_1$ and $df_2 = 2x_2$ degrees of freedom where x_1 and x_2 are the number of units that are expected to fail during the test period. If all units are tested to failure, then x_1 and x_2 will be the number of units on test.

Example 9.19 A reliability experiment is to be performed to compare the mean life of two different product designs. Determine the power to reject $H_0 : \mu_1 = \mu_2$ in favor of $H_A : \mu_1 > \mu_2$ when $\mu_1 = 200$ hours and $\mu_2 = 100$ hours using two different strategies: a) $n_1 = n_2 = 30$ units, all tested to failure and b) $n_1 = 40$, $n_2 = 20$, and the test will be suspended when 90% of the units from one of the two designs have failed. Assume that both life distributions are exponential

Solution:

a) With $n_1 = n_2 = 30$ units tested to failure, the F test critical value will be $F_{0.95, 60, 60} = 1.534$ and by Equation 9.55 the power will be

$$\begin{aligned} \pi &= P\left(\left(\frac{100}{200} \times 1.534\right) < F < \infty\right) \\ &= P(0.767 < F < \infty) \\ &= 0.846 \end{aligned}$$

b) Under H_A in the second strategy, the second treatment group has fewer units with lower mean life so they should be exhausted first. The time at which 90% or 18 of these units will have failed is expected to be about $t = -100 \ln(0.1) = 230$ hours. At the same time about $40\left(1 - e^{-230/200}\right) = 27$ of the units from the first treatment group are expected to fail. If the test is suspended then, with $x_1 = 27$ and $x_2 = 18$ the power will be

$$\begin{aligned}\pi &= P\left(\left(\frac{100}{200} \times F_{0.95, 54, 36}\right) < F < \infty\right) \\ &= P(0.842 < F < \infty) \\ &= 0.721.\end{aligned}$$

Under the second strategy, the test will end much earlier (i.e., when the 18th unit with 100 hour mean life fails versus when the 30th unit with 200 hour mean life fails); however, at the penalty of reduced experimental power.

Although the test in this section was presented as a test for a difference between two exponential means ($H_0 : \mu_1 = \mu_2$ versus $H_A : \mu_1 > \mu_2$), the method can also be used to test for other relationships between μ_1 and μ_2 because

$$F = \frac{\mu_2}{\mu_1}\frac{\widehat{\mu_1}}{\widehat{\mu_2}} \tag{9.57}$$

also follows the F distribution with $2x_1$ and $2x_2$ degrees of freedom where the power of the test is still given by Equation 9.55.

9.3.2 Two-Sample Log-Rank Test

The two-sample log-rank test is used to test for a difference in the survival rates between a treatment group and a control group. In most engineering testing situations the samples are put on test at the same time and stay on test until they either fail or the test is suspended; however, the method allows experimental units to enter and/or leave the study at any time. The latter situation, which will not be addressed here, frequently occurs in biomedical studies such as clinical trials.

The hypotheses to be tested are $H_0 : h_1(t) = h_2(t)$ versus $H_A : h_1(t) > h_2(t)$ where $h_1(t)$ and $h_2(t)$ are the time-dependent hazard rates (that is, the instantaneous failure rates given survival to time t) for the control and treatment groups, respectively. The log-rank test does not require knowledge of the specific forms of the hazard rates or the survival distributions except that the hazard ratio $h_2(t)/h_1(t)$ must be constant with respect to time. This condition is called the *proportional hazards assumption*.

The log-rank test hypotheses are usually redefined in terms of the log-hazard ratio, r, which is estimated from survival probabilities $s_1(t)$ and $s_2(t)$ at any common time t under the proportional hazards assumption

$$r = \frac{\ln(s_2(t))}{\ln(s_1(t))}. \tag{9.58}$$

9.3. Two-Sample Reliability Tests

In practice, the log-hazard ratio is usually determined from the end-of-test ($t = t'$) survival probabilities because they are also required to complete power and sample size calculations.

From the definition of the log-hazard ratio under the proportional hazards assumption in Equation 9.58, the log-rank test hypotheses may be written as $H_0 : r = 1$ versus $H_A : r < 1$ where H_A is constructed to reject H_0 when the treatment group's survival rate is significantly greater than the control group's survival rate.

Two popular methods for calculating power and sample size are presented for the log-rank test. The two methods give nearly identical results for the equal-sample-size case but diverge slightly when the sample sizes are not equal. The second method (Lachin's) is preferred in the unequal-sample-size case because it is more conservative.

9.3.2.1 Schoenfeld's Method

The sampling distribution of $\ln(r)$ is asymptotically normal with standard deviation

$$\sigma_{\ln(r)} = \sqrt{\frac{1}{d_1(t')} + \frac{1}{d_2(t')}} \tag{9.59}$$

where the d_i are the number of failures in the two groups. In sample size and power calculations the d_i are usually estimated from the expected end-of-test ($t = t'$) survival probabilities

$$d_i(t') = n_i(1 - s_i(t')) \tag{9.60}$$

where the n_i are the number of units on test and $s_i(t')$ is the ith group's survival probability at the end of the test. When n_1, n_2, $s_1(t')$, and $s_2(t')$ are specified, the power associated with the log-hazard ratio r_A is given by

$$\pi = 1 - \beta = \Phi(-\infty < z < z_\beta) \tag{9.61}$$

where

$$z_\beta = \frac{-\ln(r_A)}{\sigma_{\ln(r)}} - z_\alpha. \tag{9.62}$$

When the sample sizes are equal, the number of units in each group required to obtain power $\pi = 1 - \beta$ when $r = r_A$ is given by

$$n_1 = n_2 = \left(\frac{z_\alpha + z_\beta}{\ln(r_A)}\right)^2 \left(\frac{1}{1 - s_1(t')} + \frac{1}{1 - s_2(t')}\right). \tag{9.63}$$

For unequal sample sizes, it will be necessary to calculate the power for various sample size choices until the desired power is obtained.

Example 9.20 Determine how many units must be included in a study to compare the survival rates of two treatments using the log-rank test if the control treatment is expected to have about 20% survivors at the end of the study and the study should have 90% power to reject H_0 if the experimental treatment has 40% survivors at the end of the study. Assume that the hazard rates are proportional and that the sample sizes will be equal.

Solution: From the expected end-of-study conditions under H_A the log-hazard ratio is estimated to be

$$r_A \simeq \frac{\ln(0.40)}{\ln(0.20)} = 0.5693,$$

so the required sample size is

$$n_1 = n_2 = \left(\frac{z_{0.05} + z_{0.10}}{\ln(0.5693)}\right)^2 \left(\frac{1}{1 - 0.2} + \frac{1}{1 - 0.4}\right) = 79.$$

9.3.2.2 Lachin's Method

Lachin's method is presented in addition to Schoenfeld's method because, although it gives similar results in the equal-sample-size case, it gives more conservative results (i.e., lower power for a given sample size and larger sample size for a specified power) in the unequal-sample-size case.

By Lachin's method, the log-rank test power is given by Equation 9.61, where

$$z_\beta = \frac{1 - r_A}{1 + \left(\frac{n_2}{n_1}\right) r_A} \sqrt{\left(\frac{n_2}{n_1}\right) (d_1(t') + d_2(t'))} - z_\alpha. \tag{9.64}$$

For the unequal-sample-size case, when the sample sizes are unknown but their intended ratio is known, the total number of experimental units required may be determined from

$$n_1 + n_2 = \frac{(z_\alpha + z_\beta)^2 \left(\frac{n_1}{n_2}\right) \left(\frac{1 + \left(\frac{n_2}{n_1}\right) r_A}{1 - r_A}\right)^2}{1 - \frac{s_1(t')}{1 + \frac{n_2}{n_1}} - \frac{s_2(t')}{1 + \frac{n_1}{n_2}}}. \tag{9.65}$$

When the sample sizes will be equal, they are given by

$$n_1 = n_2 = \frac{(z_\alpha + z_\beta)^2}{2 - s_1(t') - s_2(t')} \left(\frac{1 + r_A}{1 - r_A}\right)^2. \tag{9.66}$$

9.4. Interference

Example 9.21 Compare the power of the log-rank test to the power of the two-sample test for exponential mean life for Example 9.19b.

Solution: Because the hazard rate of an exponential distribution is constant, the proportional hazards assumption is satisfied. At 230 hours with $s_1(t') = 2/20 = 0.10$ and $s_2(t') = 13/40 = 0.325$ the hazard ratio under H_A will be

$$r_A = \frac{\ln(0.325)}{\ln(0.10)} = 0.488.$$

With $n_2/n_1 = 2$, $d_1(t') = 18$, and $d_2(t') = 27$, the z_β value from Equation 9.64 is

$$z_\beta = \frac{1 - 0.488}{1 + 2(0.488)}\sqrt{2(18 + 27)} - 1.96 = 0.50.$$

Then the power for the log-rank test is $\pi = \Phi(-\infty < z < 0.50) \simeq 0.69$, which is slightly less than the power for the two-sample exponential test for mean life, which was $\pi = 0.72$. The two-tailed test was used here to match the power obtained in Example 9.19.

Example 9.22 Compare the sample size calculated by Lachin's method to that of Schoenfeld's method in Example 9.20.

Solution: From the information given in the problem statement and Equation 9.66 the sample size by Lachin's method must be

$$n_1 = n_2 = \frac{(1.645 + 1.282)^2}{2 - 0.2 - 0.4}\left(\frac{1 + 0.5693}{1 - 0.5693}\right)^2 = 82,$$

which is in good agreement with Schoenfeld's method, $n = 79$.

9.4 Interference

Interference failures occur when a quality characteristic exceeds a limit where both the quality characteristic and the limit are statistically distributed. Examples of interference failures include a cylindrical shaft that does not fit into its mating hole, a load applied to a component that exceeds its strength, a fuse that does not blow when an electrical device shorts out, and a pressure vessel that ruptures because its pressure relief valve does not open.

Figure 9.1 shows a strength–load interference situation. Even though the nominal strength (S) exceeds the nominal load (L), there is a small probability that the load applied to a component will exceed its strength and the component will fail. The fraction of the components that will fail corresponds to the area in the left tail of the strength minus load $(S - L)$ distribution that is below 0.

In general, if strength and load values are indicated with the symbol x, and if their probability density functions are given by $S(x)$ and $L(x)$, respectively,

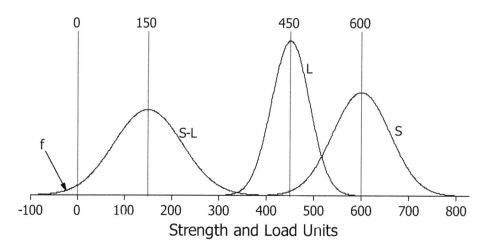

Figure 9.1: Distributions of strength, load, and their difference.

then the probability that the load will exceed a given value of strength x_S is

$$P(x > x_S) = \int_{x_S}^{\infty} L(x)\,dx. \tag{9.67}$$

Then for all possible strengths, the probability of interference failure is given by

$$f = \int_{-\infty}^{\infty} S(x_S) \left(\int_{x_S}^{\infty} L(x_L)\,dx_L \right) dx_S. \tag{9.68}$$

The necessary integrations may be performed to solve for the failure probability when $S(x)$ and $L(x)$ are well defined; however, in many situations the actual interference analysis is performed by resampling from sample strength and load data.

Analyses for normal–normal, exponential–exponential, and Weibull–Weibull interference are presented. To some degree the first two analyses are redundant with the Weibull–Weibull analysis because the exponential distribution is a special case of the Weibull distribution and the Weibull distribution provides a good approximation for the normal distribution. However, the Weibull–Weibull interference analysis requires an approximation that is not necessary in the other two cases, so it should not be used if there is a better alternative.

9.4.1 Normal-Normal Interference

In the special case that the load $L(x)$ and the strength $S(x)$ are both normally distributed, the solution to Equation 9.68 indicates that the distribution of the difference between the strength and load, $x_S - x_L$, is also normal with mean

9.4. Interference

$\mu_{S-L} = \mu_S - \mu_L$ and standard deviation $\sigma_{S-L} = \sqrt{\sigma_S^2 + \sigma_L^2}$. Then the interference failure probability is given by the standard normal probability

$$f = \Phi(-\infty < x < 0; \mu_{S-L}, \sigma_{S-L}) \quad (9.69)$$
$$= \Phi(-\infty < z < z_f) \quad (9.70)$$

where

$$z_f = \frac{\mu_L - \mu_S}{\sqrt{\sigma_S^2 + \sigma_L^2}}. \quad (9.71)$$

When the parameters of either or both S and L are not known and must be estimated from sample data, then Equation 9.69 provides only a point estimate for the failure probability and it becomes necessary to consider confidence intervals for f. In most cases, we require a one-sided upper confidence interval for the failure probability of the form

$$P(0 \le f \le f_U) = 1 - \alpha. \quad (9.72)$$

To determine how f_U depends on the sample data, we must study the sampling distribution of

$$\widehat{z}_f = \frac{\widehat{\mu}_L - \widehat{\mu}_S}{\sqrt{\widehat{\sigma}_S^2 + \widehat{\sigma}_L^2}}. \quad (9.73)$$

The sampling distribution of $\widehat{z}_f$ can be found exactly from the joint distributions of $\widehat{\mu}_L, \widehat{\mu}_S, \widehat{\sigma}_L^2$, and $\widehat{\sigma}_S^2$. However, a large-sample approximation for the distribution of $\widehat{z}_f$ can be made that is sufficient for power and sample size calculations. Equation 9.73 gives an appropriate estimate for the mean of the $\widehat{z}_f$ distribution. The variance of the $\widehat{z}_f$ distribution can be estimated by the delta method:

$$\widehat{\sigma}_{\widehat{z}_f}^2 = \left(\frac{\partial \widehat{z}_f}{\partial \widehat{\mu}_L}\widehat{\sigma}_{\widehat{\mu}_L}\right)^2 + \left(\frac{\partial \widehat{z}_f}{\partial \widehat{\mu}_S}\widehat{\sigma}_{\widehat{\mu}_S}\right)^2 + \left(\frac{\partial \widehat{z}_f}{\partial \widehat{\sigma}_L}\widehat{\sigma}_{\widehat{\sigma}_L}\right)^2 + \left(\frac{\partial \widehat{z}_f}{\partial \widehat{\sigma}_S}\widehat{\sigma}_{\widehat{\sigma}_S}\right)^2$$
$$= \frac{1}{\widehat{\sigma}_L^2 + \widehat{\sigma}_S^2}\left(\frac{\widehat{\sigma}_L^2}{n_L} + \frac{\widehat{\sigma}_S^2}{n_S} + \frac{1}{2}\left(\frac{\widehat{\mu}_L - \widehat{\mu}_S}{\widehat{\sigma}_L^2 + \widehat{\sigma}_S^2}\right)^2 \left(\frac{\widehat{\sigma}_L^4}{n_L} + \frac{\widehat{\sigma}_S^4}{n_S}\right)\right) \quad (9.74)$$

where n_L and n_S are the L and S sample sizes, respectively. When $n_L > 30$ and $n_S > 30$ the sampling distribution of $\widehat{z}_f$ is approximately normal so the approximate one-sided upper $(1-\alpha)100\%$ confidence bound for f in Equation 9.72 is given by

$$\widehat{f}_U = \Phi(-\infty < z < \widehat{z}_{f_U}) \quad (9.75)$$

where

$$\widehat{z}_{f_U} = \widehat{z}_f + z_\alpha \widehat{\sigma}_{\widehat{z}_f}. \quad (9.76)$$

This is all complicated enough to justify an example before continuing on to sample size considerations.[2]

---

[2]Lloyd and Lipow [41] present a similar confidence interval but they assume that the sampling distribution of $\widehat{f}$ is normal instead of $\widehat{z}_f$ as I've done here. The two methods are in good agreement, but the normality assumption seems to be better satisfied for $\widehat{z}_f$ than for $\widehat{f}$ when the sample size is small.

Example 9.23 A random sample of component strengths gave $n_S = 100$, $\hat{\mu}_S = 600$, and $\hat{\sigma}_S = 60$ and a random sample of loads gave $n_L = 36$, $\hat{\mu}_L = 450$, and $\hat{\sigma}_L = 40$. Both distributions are known to be normal.[3] Determine the 90% upper confidence limit for the interference failure rate.

Solution: The point estimate for $\hat{z}_f$ is given by Equation 9.73:

$$\hat{z}_f = \frac{450 - 600}{\sqrt{40^2 + 60^2}} = -2.08$$

and the corresponding point estimate for the interference failure rate is

$$\hat{f} = \Phi(-\infty < z < -2.08) = 0.0188.$$

The approximate standard deviation of the $\hat{z}_f$ distribution is given by Equation 9.74:

$$\hat{\sigma}_{\hat{z}_f} = \sqrt{\frac{1}{40^2 + 60^2}\left(\frac{40^2}{36} + \frac{60^2}{100} + \frac{1}{2}\left(\frac{450-600}{40^2+60^2}\right)^2\left(\frac{40^4}{36} + \frac{60^4}{100}\right)\right)} = 0.178.$$

Then, from Equation 9.76 with $z_{0.10} = 1.282$

$$\hat{z}_{f_U} = -2.08 + 1.282 \times 0.178$$
$$= -1.85,$$

so from Equation 9.75 the 90% upper confidence limit for the interference failure probability is

$$\hat{f}_U = \Phi(-\infty < z < -1.85)$$
$$= 0.032.$$

That is, on the basis of the sample data, we can claim that the one-sided upper 90% confidence interval for the interference failure rate is

$$P(0 < f < 0.032) = 0.90.$$

The calculation of sample size for an interference study to demonstrate a specified upper confidence limit for the interference failure rate requires initial estimates of $\mu_L, \mu_S, \sigma_L,$ and σ_S. Then the values of n_L and n_S that meet the confidence limit goal can be determined by iteration. This procedure is tedious, but it works. To consider a simpler but common situation, suppose that one of the two distributions is already well known and it is necessary to calculate only the sample size for an experiment to study the unknown distribution. In this case,

---
[3] These are the conditions shown in Figure 9.1.

9.4. Interference

and if we arbitrarily assume that the load distribution L is known (that is, n_L is effectively infinite) and the strength distribution S is unknown, then Equation 9.74 for the $\hat{z}_f$ sampling variance reduces to

$$\hat{\sigma}_{\hat{z}_f}^2 = \frac{1}{\sigma_L^2 + \hat{\sigma}_S^2}\left(\frac{\hat{\sigma}_S^2}{n_S} + \frac{1}{2}\left(\frac{\mu_L - \hat{\mu}_S}{\sigma_L^2 + \hat{\sigma}_S^2}\right)^2\left(\frac{\hat{\sigma}_S^4}{n_S}\right)\right)$$

$$= \frac{1}{n_S}\left(\frac{\hat{\sigma}_S^2}{\sigma_L^2 + \hat{\sigma}_S^2}\right)\left(1 + \frac{\hat{\sigma}_S^2}{2}\left(\frac{\mu_L - \hat{\mu}_S}{\sigma_L^2 + \hat{\sigma}_S^2}\right)^2\right). \quad (9.77)$$

Then, to demonstrate a one-sided upper $(1-\alpha)\,100\%$ confidence limit for f, $\hat{f}_U$, which has a corresponding $\hat{z}_{f_U}$ value given by Equation 9.76, the sample size n_S must be

$$n_S = \left(\frac{z_\alpha}{\hat{z}_{f_U} - \hat{z}_f}\right)^2 \left(\frac{\hat{\sigma}_S^2}{\sigma_L^2 + \hat{\sigma}_S^2}\right)\left(1 + \frac{\hat{\sigma}_S^2}{2}\left(\frac{\mu_L - \hat{\mu}_S}{\sigma_L^2 + \hat{\sigma}_S^2}\right)^2\right). \quad (9.78)$$

Remember that this sample size is valid only in the large-sample limit.

Example 9.24 What sample size is required to demonstrate that the interference failure probability is less than 0.1% with 90% confidence if the strength distribution is known to be normal with $\mu_S = 20$ and $\sigma_S = 2$ and the load distribution is expected to be normal with $\hat{\mu}_L = 13$ and $\hat{\sigma}_L = 1$?

Solution: The point estimate for the interference failure probability determined from the S parameters and the L parameter estimates is

$$\hat{f} = \Phi(-\infty < z < \hat{z}_f)$$

$$= \Phi\left(-\infty < z < \frac{13-20}{\sqrt{1^2+2^2}}\right)$$

$$= \Phi(-\infty < z < -3.13)$$

$$= 0.000874.$$

The sample size required to study the load distribution is given by swapping the relevant S and L subscripts in Equation 9.78:

$$n_L = \left(\frac{z_\alpha}{\hat{z}_{f_U} - \hat{z}_f}\right)^2 \left(\frac{\hat{\sigma}_L^2}{\hat{\sigma}_L^2 + \sigma_S^2}\right)\left(1 + \frac{\hat{\sigma}_L^2}{2}\left(\frac{\hat{\mu}_L - \mu_S}{\hat{\sigma}_L^2 + \sigma_S^2}\right)^2\right) \quad (9.79)$$

$$= \left(\frac{z_{0.10}}{z_{0.001} - z_{0.000874}}\right)^2 \left(\frac{1^2}{1^2 + 2^2}\right)\left(1 + \frac{1^2}{2}\left(\frac{13-20}{1^2+2^2}\right)^2\right)$$

$$= \left(\frac{1.282}{-3.09 - (-3.13)}\right)^2 \left(\frac{1^2}{1^2+2^2}\right)\left(1 + \frac{1^2}{2}\left(\frac{13-20}{1^2+2^2}\right)^2\right)$$

$$= 407.$$

9.4.2 Exponential–Exponential Interference

When the strength and load are both exponentially distributed with probability density functions of the form

$$\xi(x;\theta) = \frac{1}{\theta}e^{-x/\theta} \qquad (9.80)$$

where θ is the mean and $\theta_S > \theta_L$, then the interference failure rate (f) is

$$\begin{aligned} f &= \int_0^\infty \xi(x_L;\theta_L) \left(\int_0^{x_L} \xi(x_S;\theta_S)\,dx_S \right) dx_L \\ &= \int_0^\infty \frac{1}{\theta_L} e^{-x_L/\theta_L} \left(\int_0^{x_L} \frac{1}{\theta_S} e^{-x_S/\theta_S}\,dx_S \right) dx_L \\ &= \frac{\theta_L}{\theta_S + \theta_L}. \end{aligned} \qquad (9.81)$$

When both θ_L and θ_S are unknown, the sampling distribution of f may be estimated by

$$\widehat{f} = \frac{\widehat{\theta}_L}{\widehat{\theta}_S + \widehat{\theta}_L} \qquad (9.82)$$

where the θ estimates are obtained from experimental results with

$$\widehat{\theta} = \frac{1}{n}\sum_{i=1}^n x_i. \qquad (9.83)$$

For large n_L and n_S, the sampling distribution of f is approximately normal with mean given by Equation 9.82 and standard deviation

$$\begin{aligned} \widehat{\sigma}_{\widehat{f}} &= \sqrt{\left(\frac{\partial \widehat{f}}{\partial \widehat{\theta}_L}\widehat{\sigma}_{\widehat{\theta}_L}\right)^2 + \left(\frac{\partial \widehat{f}}{\partial \widehat{\theta}_S}\widehat{\sigma}_{\widehat{\theta}_S}\right)^2} \\ &= \frac{\widehat{\theta}_L \widehat{\theta}_S}{\left(\widehat{\theta}_L + \widehat{\theta}_S\right)^2}\sqrt{\frac{1}{n_L}+\frac{1}{n_S}} \\ &= \widehat{f}(1-\widehat{f})\sqrt{\frac{1}{n_L}+\frac{1}{n_S}}. \end{aligned} \qquad (9.84)$$

The approximate one-sided upper confidence limit for f is given by

$$\Phi\left(0 < f < \widehat{f}_U\right) = 1 - \alpha \qquad (9.85)$$

where

$$\begin{aligned} \widehat{f}_U &= \widehat{f} + z_\alpha \widehat{\sigma}_{\widehat{f}} \\ &= \widehat{f} + z_\alpha \widehat{f}(1-\widehat{f})\sqrt{\frac{1}{n_L}+\frac{1}{n_S}}. \end{aligned} \qquad (9.86)$$

9.4. Interference

If the upper confidence limit is expressed relative to $\widehat{f}$, that is, in the form

$$\Phi\left(0 < f < \widehat{f}(1+\delta)\right) = 1 - \alpha, \tag{9.87}$$

then

$$\delta = z_\alpha \left(1 - \widehat{f}\right)\sqrt{\frac{1}{n_L} + \frac{1}{n_S}}. \tag{9.88}$$

When the sample sizes are equal

$$n_L = n_S = 2\left(\frac{z_\alpha\left(1 - \widehat{f}\right)}{\delta}\right)^2. \tag{9.89}$$

Because the design goal is usually $f = 0$, the sample size is approximately

$$n \simeq 2\left(\frac{z_\alpha}{\delta}\right)^2. \tag{9.90}$$

Remember that this sample size is valid only in the large-sample limit.

Example 9.25 What sample size is required to determine the 95% two-sided confidence interval for the exponential-exponential interference failure rate if the confidence limits must be within 50% of the predicted mean failure rate?
Solution: The goal of the experiment is to obtain a confidence interval for the exponential-exponential interference failure rate f of the form

$$\Phi\left(0.50\widehat{f} < f < 1.50\widehat{f}\right) = 0.95.$$

With $z_{0.025} = 1.96$ and $\delta = 0.50$ in Equation 9.90, the required equal sample sizes are

$$n_L = n_S = 2\left(\frac{1.96}{0.50}\right)^2 = 31.$$

9.4.3 Weibull-Weibull Interference

When the strength and load are both Weibull-distributed with probability density functions of the form

$$\xi(x; \eta, \beta) = \beta \frac{x^{\beta-1}}{\eta^\beta} e^{-\left(\frac{x}{\eta}\right)^\beta} \tag{9.91}$$

where η is the scale factor and β is the shape factor, then the strength/load interference failure rate (f) is given by

$$f = \int_0^\infty \xi(x_L; \eta_L, \beta_L) \left(\int_0^{x_L} \xi(x_S; \eta_S, \beta_S) \, dx_S \right) dx_L$$

$$\simeq \left(\frac{\eta_L}{\eta_S} \right)^{\beta_S} \Gamma \left(1 + \frac{\beta_S}{\beta_L} \right) \tag{9.92}$$

where $\Gamma()$ is the gamma function.[4] The approximation is satisfied when

$$\left(\frac{\eta_L}{\eta_S} \right)^{\beta_S} \ll 1, \tag{9.93}$$

which corresponds to small f - the usual condition of interest.

Under the assumption that β_L and β_S are known, for large samples, the sampling distribution of f is approximately normal with mean $\widehat{f}$ given by substituting the scale parameter estimates into Equation 9.92 and approximate standard deviation given by the delta method

$$\widehat{\sigma}_{\widehat{f}} = \sqrt{\left(\frac{\partial \widehat{f}}{\partial \widehat{\eta}_L} \widehat{\sigma}_{\widehat{\eta}_L} \right)^2 + \left(\frac{\partial \widehat{f}}{\partial \widehat{\eta}_S} \widehat{\sigma}_{\widehat{\eta}_S} \right)^2}$$

$$= \sqrt{\left(\frac{\beta_S}{\widehat{\eta}_L} \left(\frac{\widehat{\eta}_L}{\widehat{\eta}_S} \right)^{\beta_S} \widehat{\sigma}_{\widehat{\eta}_L} \right)^2 + \left(-\frac{\beta_S}{\widehat{\eta}_S} \left(\frac{\widehat{\eta}_L}{\widehat{\eta}_S} \right)^{\beta_S} \widehat{\sigma}_{\widehat{\eta}_S} \right)^2}$$

$$= \beta_S \left(\frac{\widehat{\eta}_L}{\widehat{\eta}_S} \right)^{\beta_S} \sqrt{\left(\frac{\widehat{\sigma}_{\widehat{\eta}_L}}{\widehat{\eta}_L} \right)^2 + \left(\frac{\widehat{\sigma}_{\widehat{\eta}_S}}{\widehat{\eta}_S} \right)^2}. \tag{9.94}$$

From Section 9.1.2.1 the scale parameter estimate's standard error is

$$\widehat{\sigma}_{\widehat{\eta}} \simeq \frac{\widehat{\eta}}{\beta \sqrt{n}}, \tag{9.95}$$

so $\widehat{\sigma}_{\widehat{f}}$ is approximately

$$\widehat{\sigma}_{\widehat{f}} \simeq \beta_S \left(\frac{\widehat{\eta}_L}{\widehat{\eta}_S} \right)^{\beta_S} \sqrt{\frac{1}{n_L \beta_L^2} + \frac{1}{n_S \beta_S^2}}. \tag{9.96}$$

The approximate one-sided upper confidence limit for f is given by

$$\Phi \left(0 < f < \widehat{f}_U \right) = 1 - \alpha \tag{9.97}$$

---

[4]The factorial function is related to the gamma function by $n! = \Gamma(n-1)$, but the factorial function operates on integers where the gamma function is continuous.

9.4. Interference

where

$$\hat{f}_U = \hat{f} + z_\alpha \hat{\sigma}_{\hat{f}}$$

$$= \left(\frac{\hat{\eta}_L}{\hat{\eta}_S}\right)^{\beta_S} \Gamma\left(1 + \frac{\beta_S}{\beta_L}\right) + z_\alpha \beta_S \left(\frac{\hat{\eta}_L}{\hat{\eta}_S}\right)^{\beta_S} \sqrt{\frac{1}{n_L \beta_L^2} + \frac{1}{n_S \beta_S^2}}$$

$$= \hat{f}\left(1 + \frac{z_\alpha \beta_S}{\Gamma\left(1 + \frac{\beta_S}{\beta_L}\right)} \sqrt{\frac{1}{n_L \beta_L^2} + \frac{1}{n_S \beta_S^2}}\right). \quad (9.98)$$

This upper confidence limit is expressed relative to $\hat{f}$ in the form $\hat{f}_U = \hat{f}(1 + \delta)$ where

$$\delta = \frac{z_\alpha \beta_S}{\Gamma\left(1 + \frac{\beta_S}{\beta_L}\right)} \sqrt{\frac{1}{n_L \beta_L^2} + \frac{1}{n_S \beta_S^2}}. \quad (9.99)$$

For a specified value of the relative precision of the estimate δ in the equal-sample-size case ($n_L = n_S = n$), this equation can be solved for the sample size to obtain

$$n = \left(\frac{z_\alpha}{\delta \Gamma\left(1 + \frac{\beta_S}{\beta_L}\right)}\right)^2 \left(1 + \frac{\beta_S^2}{\beta_L^2}\right). \quad (9.100)$$

Example 9.26 How many measurements of mating components in a device must be taken to demonstrate, with 95% confidence, that their true interference failure rate does not exceed the observed failure rate by 20% if the two distributions are known to be Weibull with $\beta_S = 2.5$ and $\beta_L = 1.5$?

Solution: The goal of the experiment is to acquire sufficient information to demonstrate the following one-sided upper confidence interval for the interference failure rate f:

$$P\left(0 < f < \hat{f}(1 + 0.2)\right) = 0.95.$$

With $\delta = 0.2$ and $\alpha = 0.05$ in Equation 9.100, we obtain the sample size

$$n = \left(\frac{1.645}{0.2 \times \Gamma\left(1 + \frac{2.5}{1.5}\right)}\right)^2 \left(1 + \frac{2.5^2}{1.5^2}\right)$$
$$= 113.$$

Chapter 10

Statistical Quality Control

The purpose of this chapter is to present sample size and power calculations for the fundamental methods of statistical quality control including

- statistical process control (SPC).
- process capability.
- tolerance intervals.
- acceptance sampling.
- gage repeatability and reproducibility (gage R&R) studies.

In the case of acceptance sampling, where the technology is well documented in textbooks and published sampling standards, the analyses presented here are superficial — limited to single sampling plans for attributes and variables.

10.1 Statistical Process Control

Two issues in the design and operation of control charts are presented in this section:

- The design of rules used to interpret patterns of points on control charts.
- The selection of sample size for attribute and variable control charts.

10.1.1 Control Chart Run Rules

A statistically valid control chart run rule must satisfy three conditions:

1. The rule must be easy to recognize.

2. The rule must have a low probability of occurring when the process is in control, that is, it must have a low type I or false alarm error rate.

3. The rule must have a high probability (or power) of occurring when the process is out of control, that is, it must have a low type II or missed alarm error rate.

When the first condition is satisfied, usually the third condition is also satisfied. This leaves the second condition, low type I error rate, as the most discriminating condition of a good control chart rule.

When more than one run rule is used to interpret a control chart, the combined type I error rate of the family of rules must be acceptably low. If α_i is the type I error rate for the ith of k run rules, then the overall type I error rate for all of the rules is

$$\begin{aligned}\alpha_{family} &= 1 - \prod_{i=1}^{k}(1-\alpha_i) \\ &\simeq \sum_{i=1}^{k}\alpha_i.\end{aligned} \quad (10.1)$$

If the family of run rules is determined such that the α_i are all approximately equal, then

$$\alpha_{family} \simeq k\alpha_i. \quad (10.2)$$

These approximations for α_{family} assume that the rules are independent of each other, which is not strictly the case; however, the approximations are still useful as guidelines for good practice in run rule design.

Just as the type I family error rate increases as more and more run rules are used to interpret a control chart, the overall type I error rate increases further when several simultaneous charts are kept. If the same family of run rules is used to interpret m simultaneous charts, then the overall type I error rate α_{FAMILY} for all m charts, each using the same k run rules, is

$$\begin{aligned}\alpha_{FAMILY} &= 1 - (1 - \alpha_{family})^m \\ &\simeq m\alpha_{family}\end{aligned} \quad (10.3)$$

and if the k rules all have approximately the same α_i then

$$\alpha_{FAMILY} \simeq m\alpha_{family} \simeq mk\alpha_i. \quad (10.4)$$

This result suggests several important control chart management guidelines:

- Keep only the charts that are necessary.
- Use only run rules that are effective.
- Design run rules with sufficiently low type I error rates to keep α_{FAMILY} acceptably low.

10.1. Statistical Process Control

Example 10.1 Evaluate the following control chart run rule: A process is judged to be out of control if at least two of three consecutive observations falls beyond the same 2σ limit on the chart.

Solution: The rule is easy to identify on the chart, so it satisfies the first condition for a valid run rule. If the process is in control and the distribution of the statistic (call it w) is approximately normal, then the probability that any point on the chart falls above $\mu_w + 2\sigma_w$ is $p = \Phi(2 < z < \infty) = 0.023$. The probability that at least $x = 2$ of $n = 3$ consecutive points fall above that limit is given by the binomial probability

$$\sum_{x=2}^{3} b(x; n=3, p=0.023) = 0.0016.$$

Because this pattern could also appear on the bottom half of the chart, the type I error rate for this rule is $\alpha = 2(0.0016) = 0.0032$, which is acceptably low, so the rule meets the second requirement for a valid control chart rule. If the process mean shifted to $\mu_w + 2\sigma_w$, then the probability that an observation would fall beyond $\mu_w + 2\sigma_w$ is $p = 0.5$ and the corresponding power of the rule is

$$\pi = \sum_{x=2}^{3} b(x; n=3, p=0.5) = 0.5.$$

This meets the third requirement of a valid control chart rule. Because all three conditions are satisfied, that is: 1) the rule is easy to recognize, 2) it has a low type I error rate, and 3) it has good power to detect shifts in the process, then it is a valid control chart run rule.

Example 10.2 One of the weaknesses of defects charts when the sampling unit is small is that it is not possible to declare a process to be out of control on the lower side of the chart with a single observation. Evaluate the following special run rule for defects charts: If a defects chart's sampling unit size is sufficient to deliver $\lambda \geq 3$, then the process is out of control if two consecutive sampling units have 0 defects.

Solution: The chart obviously meets the first and third conditions for valid control chart run rules, but it is not clear if the second condition (low type I error rate) is satisfied. If the mean defect rate is $\lambda = 3$, then the probability of a sampling unit having 0 defects when the process is in control ($H_0 : \lambda = 3$) is $Poisson(x = 0; \lambda = 3) = 0.05$, a rather common occurrence. Under the same conditions, the probability of observing two consecutive zeros is $b(x = 2; n = 2, p = 0.05) = 0.0025$, but this is just the type I error rate for the rule. Because $\alpha = 0.0025$ is acceptably low, the rule meets all three conditions for a valid control chart run rule.

Example 10.3 When Walter Shewhart invented control charts, he expected that an operator would be using about four run rules to interpret at most three control charts. If each of Shewhart's run rules had $\alpha_i \simeq 0.004$, what is the expected overall type I error rate?
Solution: From Equation 10.4

$$\alpha_{FAMILY} \simeq 3 \times 4 \times 0.004 = 0.048.$$

That is, with three simultaneous charts and four run rules, Shewhart expected about 5% of the sampling intervals to result in type I errors.

10.1.2 Power and Sample Size for Control Charts

Power and/or sample size calculations are presented for control charts for defectives, defects, and $\bar{x}$ charts.

10.1.2.1 Defectives Charts

The following sample size calculations for defectives charts assume that the only out-of-control rule used is one point beyond the usual $\pm 3\sigma$ defective chart control limits given by

$$UCL/LCL = np \pm 3\sqrt{np(1-p)}. \tag{10.5}$$

The use of additional rules will increase the sensitivity of the family of rules to small changes in the process faction defective and increase the type I error rate. In all sample size calculations, the value of the process fraction defective must be known or estimated.

- The sample size required to obtain a positive lower control limit on a defectives chart is determined by setting $LCL = 0$ and solving for n. Then any n greater than this critical value will have $LCL > 0$. The benefit of a positive LCL it makes it possible to identify an out of control state on a single sample. The necessary sample size is

$$n > \frac{9(1-p)}{p}. \tag{10.6}$$

Because np is also the defectives chart center line, this sample size condition is satisfied when $(CL = np) > 9$.

- The sample size required to deliver an average run length of $ARL = 2$ (or $\beta = 0.5$) for a specified shift in the defective rate from p to $p+\delta$ requires $\delta = 3\sqrt{np(1-p)}$ from which the following sample size condition is obtained:

$$n = \frac{9p(1-p)}{\delta^2}. \tag{10.7}$$

10.1. Statistical Process Control

- Defective charts with very small np will have many zero or near-zero observations which makes the lower half of the chart difficult to interpret. The sample size required to limit the fraction of the observations on a defectives chart that fall at $x = 0$ to some acceptably low value α can be determined from the binomial probability

$$b(0; n, p) = \alpha. \tag{10.8}$$

Unless appropriate software or Larson's nomogram are available, this expression is inconvenient for determining the sample size, but from the χ^2 distribution form of the Poisson approximation to the binomial distribution the approximate n value is

$$n = \frac{\chi^2_{1-\alpha, 2}}{2p}. \tag{10.9}$$

For the special case of $\alpha = 0.05$, which limits the number of zero-defective samples to about 5%, Equation 10.9 simplifies to the rule of three:

$$n = \frac{3}{p}. \tag{10.10}$$

Because np is also the defectives chart center line, this sample size condition is satisfied when $(CL = np) > 3$.

Example 10.4 What is the minimum sample size required to have a positive lower control limit on a defectives chart if the process fraction defective is expected to be $p = 0.01$?
Solution: From Equation 10.6 the required sample size is

$$n > \left(\frac{9(1 - 0.01)}{0.01}\right)$$
$$> 891.$$

10.1.2.2 Defects Charts

The sample size calculations for defects charts are analogous to those for defectives charts except that the defects charts are based on the Poisson distribution instead of the binomial distribution. The following calculations assume that the only out-of-control rule used is one point beyond the usual $\pm 3\sigma$ defect chart control limits given by

$$UCL/LCL = \lambda \pm 3\sqrt{\lambda} \tag{10.11}$$

where λ is the Poisson mean defect rate per sampling unit. If additional rules are used the defects chart will be more sensitive to shifts in the process mean and the family type I error rate will increase.

- The sampling unit size required to obtain a positive lower control limit on a defects chart is determined by setting $LCL = 0$ and solving for the size of the sampling unit. Then any sampling unit size greater than this critical value will have $LCL > 0$. The benefit of a positive LCL is that it makes it possible to identify an out of control state on a single sample. The $LCL > 0$ condition leads to the following condition, which constrains the sampling unit size:

$$\lambda > 9. \tag{10.12}$$

That is, the size of the sampling unit must be chosen such that the mean number of defects observed per sampling unit is greater than nine. This condition is often considered to be too conservative, that is, it gives an excessively large sample size.

- The sample size required to limit the fraction of the observations on a defects chart that fall at $x = 0$ to α is given by the condition

$$Poisson\,(x = 0; \lambda) = \alpha. \tag{10.13}$$

From the χ^2 distribution form of the Poisson distribution, the required sampling unit size must meet the condition

$$\lambda = \frac{1}{2}\chi^2_{1-\alpha,2}. \tag{10.14}$$

Example 10.5 What is the smallest sampling unit size for a defects chart that will deliver no more than about 5% zero-defect observations when the process delivers 0.6 defects per unit?

Solution: From Equation 10.13 the mean number of defects per sampling unit is the value of λ that satisfies the condition

$$Poisson\,(x = 0; \lambda) = 0.05, \tag{10.15}$$

which is $\lambda = 3$. Consequently, the sampling unit size must be $3/0.6 = 5$ units.

10.1.2.3 $\bar{x}$ Charts

The upper and lower control limits for $\bar{x}$ charts have the form

$$UCL/LCL = \hat{\mu}_x \pm 3\hat{\sigma}_{\bar{x}} \tag{10.16}$$

$$= \hat{\mu}_x \pm \frac{3\hat{\sigma}_x}{\sqrt{n}} \tag{10.17}$$

where $\hat{\mu}_x$ and $\hat{\sigma}_x$ are usually determined from historical or preliminary data.[1] Under the assumption that the distribution of $\bar{x}$ is normal, the power to reject

---

[1] $\hat{\mu}_{\bar{x}}$ is usually determined from $\bar{\bar{x}}$ and $3\hat{\sigma}_{\bar{x}}$ is usually determined from one of $A\sigma$, $A_2\bar{R}$, or $A_3\bar{s}$.

10.1. Statistical Process Control

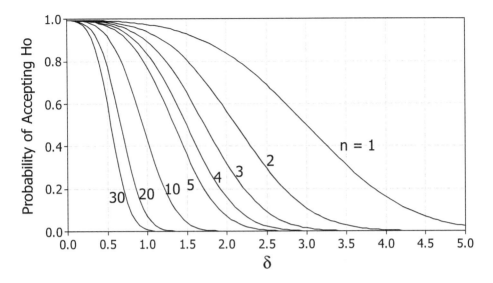

Figure 10.1: Operating characteristic curves for $\bar{x}$ charts.

$H_0: \mu_x = \mu_0$ (where μ_0 is the center line of the chart) when H_0 is false is

$$\pi = 1 - \beta = 1 - \Phi\left(LCL < \bar{x} < UCL; \mu_x, \sigma_{\bar{x}}\right). \tag{10.18}$$

It may be easier to communicate the concept of power to a statistically naive audience in terms of the average run length (ARL) – the expected number of samples that must be drawn after a process shift occurs to detect the shift. ARL is related to π and β by

$$\begin{aligned} ARL &= \sum_{L=1}^{\infty} L \times E(L) \\ &= \sum_{L=0}^{\infty} L\beta^{L-1}(1-\beta) \\ &= \frac{1}{1-\beta} \tag{10.19} \\ &= \frac{1}{\pi} \tag{10.20} \end{aligned}$$

where L is the length of the run and $E(L)$ is its expected value.

Figures 10.1 and 10.2 show the operating characteristic and average run length curves for $\bar{x}$ charts with various sample sizes for shifts from μ_0 to $\mu_0 \pm \delta\sigma_x$ using the single rule that the process is out of control if a point falls outside of the control limits at $UCL/LCL = \mu \pm 3\sigma_{\bar{x}}$.

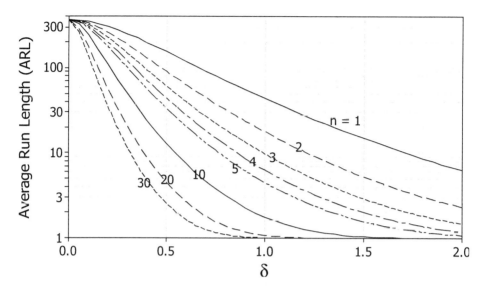

Figure 10.2: Average run length curves for $\bar{x}$ charts.

Example 10.6 Calculate the power to reject $H_0 : \mu = 30$ when $\mu = 32$ if an $\bar{x}$ chart is kept using $n = 4$ and $\sigma_x = 2$. Also determine the corresponding ARL.
Solution: The shift in the mean relative to $\sigma_{\bar{x}}$ is $\delta = 1$, so from Figure 10.1 the power is approximately $\pi = 1 - \beta \simeq 0.15$ and from Figure 10.2 the average run length is approximately $ARL \simeq 6$.

The $\bar{x}$ chart control limits will fall at

$$UCL/LCL = 30 \pm \frac{3 \times 2}{\sqrt{4}} = 33/27.$$

Assuming that the only out-of-control run rule used is one point beyond three sigma limits, the power is given by Equation 10.18:

$$\begin{aligned}\pi &= 1 - \Phi\left(27 < \bar{x} < 33; \mu_x = 32, \sigma_{\bar{x}} = \frac{2}{\sqrt{4}} = 1\right) \\ &= 1 - \Phi(-5 < z < 1) \\ &= 0.16.\end{aligned}$$

Under the same conditions, the average number of subgroups that will have to be drawn to detect a shift from $\mu = 30$ to $\mu = 32$ is

$$ARL = \frac{1}{0.16} = 6.3.$$

10.2 Process Capability

Process capability analysis is used to compare the mean and standard deviation of a process to its specification limits. The most commonly used process capability parameters are: $c_p = (USL - LSL)/6\sigma$ and $c_{pk} = |NSL - \mu|/3\sigma$ where NSL is the nearest specification limit to the process mean μ. The values of μ and σ are usually estimated from control charts.

A pair of c_p and c_{pk} values is only a substitute for the true defective rate of a process p because

$$\begin{aligned} 1 - p &= \Phi(LSL < x < USL; \mu, \sigma) \\ &= \Phi(6c_p - 3c_{pk} < z < 3c_{pk}) \end{aligned} \quad (10.21)$$

where $z = (x - \mu)/\sigma$. Obviously, unless the distribution of the characteristic being measured is in statistical control and normally distributed, estimates for c_p and c_{pk} from sample data are nearly meaningless and nonparametric methods of estimating the defective rate directly might be more appropriate.

10.2.1 Confidence Intervals for c_p and c_{pk}

10.2.1.1 Confidence Interval for c_p

Because c_p is related to σ by a simple mathematical transform, a confidence interval for c_p can be derived from the confidence interval for σ (Equation 3.4) which gives

$$P\left(\widehat{c}_p \sqrt{\frac{\chi^2_{\alpha/2}}{n-1}} < c_p < \widehat{c}_p \sqrt{\frac{\chi^2_{1-\alpha/2}}{n-1}}\right) = 1 - \alpha \quad (10.22)$$

where the χ^2 distribution has $df = n - 1$ degrees of freedom. When $n > 100$, which is usually the situation in process capability studies, an approximate confidence interval for c_p is

$$P\left(\widehat{c}_p \left(1 - \frac{z_{\alpha/2}}{\sqrt{2n}}\right) < c_p < \widehat{c}_p \left(1 + \frac{z_{\alpha/2}}{\sqrt{2n}}\right)\right) = 1 - \alpha. \quad (10.23)$$

If the goal of an experiment is to provide a confidence interval for c_p with specified relative uncertainty δ, that is, a confidence interval of the form

$$P(\widehat{c}_p(1-\delta) < c_p < \widehat{c}_p(1+\delta)) = 1 - \alpha, \quad (10.24)$$

then by comparison to Equation 10.23 we have

$$\delta = \frac{z_{\alpha/2}}{\sqrt{2n}} \quad (10.25)$$

and the required sample size is

$$n = \frac{1}{2}\left(\frac{z_{\alpha/2}}{\delta}\right)^2. \quad (10.26)$$

Example 10.7 What sample size is required to determine c_p to within 10% of its true value with 90% confidence?
Solution: With $\delta = 0.10$ and $\alpha = 0.10$ in Equation 10.26 the required sample size is
$$n = \frac{1}{2}\left(\frac{1.645}{0.10}\right)^2 = 136.$$

10.2.1.2 Confidence Interval for c_{pk}

When n is large, the sampling distribution of c_{pk} is approximately normal with approximate standard deviation given by the delta method (see Appendix G.3.15):

$$\hat{\sigma}_{c_{pk}} = \hat{c}_{pk}\sqrt{\frac{1}{n}\left(\frac{1}{9\hat{c}_{pk}^2} + \frac{1}{2}\right)}. \qquad (10.27)$$

Then an approximate $(1-\alpha)\,100\%$ confidence interval for c_{pk} is

$$P\left(\hat{c}_{pk}\left(1-\delta\right) < c_{pk} < \hat{c}_{pk}\left(1+\delta\right)\right) = 1-\alpha, \qquad (10.28)$$

where the confidence interval's relative half-width is

$$\delta = z_{\alpha/2}\sqrt{\frac{1}{n}\left(\frac{1}{9\hat{c}_{pk}^2} + \frac{1}{2}\right)}. \qquad (10.29)$$

The approximate sample size required to estimate c_{pk} with specified relative precision is

$$n \simeq \left(\frac{z_{\alpha/2}}{\delta}\right)^2\left(\frac{1}{9\hat{c}_{pk}^2} + \frac{1}{2}\right) \qquad (10.30)$$

where some expected value of $\hat{c}_{pk}$ will have to be chosen to complete the calculation. As c_{pk} gets very large, Equation 10.30 may be approximated by

$$n \simeq \frac{1}{2}\left(\frac{z_{\alpha/2}}{\delta}\right)^2. \qquad (10.31)$$

This approximation serves as a lower limit for the sample size required to estimate c_{pk} and is the same as the sample size required for estimating c_p.

Example 10.8 What sample size is required to estimate c_{pk} to within 5% of its true value with 90% confidence if $c_{pk} = 1.0$ is expected?
Solution: From Equation 10.30 with $\delta = 0.05$ and $\alpha = 0.05$ the required sample size is
$$n \simeq \left(\frac{1.645}{0.05}\right)^2\left(\frac{1}{9\,(1.0)^2} + \frac{1}{2}\right) = 662.$$

10.2. Process Capability

Example 10.9 What sample size is required to estimate c_{pk} to within 5% of its true value with 90% confidence if c_{pk} is expected to be very large?
Solution: From Equation 10.31 with $\delta = 0.05$ and $\alpha = 0.05$ the required sample size is
$$n \simeq \frac{1}{2}\left(\frac{1.645}{0.05}\right)^2 = 541.$$

10.2.2 Tests for c_p and c_{pk}

10.2.2.1 Test for c_p

The hypotheses to be tested are $H_0 : c_p = (c_p)_0$ versus $H_A : c_p > (c_p)_0$ where $(c_p)_0$ is a specified value, often 1.33 or 2.0. By a variable transformation of Equation 3.17, the sample size n required to provide power $\pi = 1 - \beta$ to reject H_0 when in fact $c_p = (c_p)_1$ is given by the smallest value of n which satisfies the condition

$$\frac{(c_p)_1}{(c_p)_0} \leq \sqrt{\frac{\chi^2_{1-\beta}}{\chi^2_\alpha}}. \tag{10.32}$$

When the sample size is expected to be very large, which is usually the case for process capability studies, the normal approximation for Equation 10.32 (see Section 3.1.2.2) gives the much more convenient approximate sample size

$$n = \frac{1}{2}\left(\frac{z_\alpha + z_\beta}{\ln\left(\frac{(c_p)_1}{(c_p)_0}\right)}\right)^2. \tag{10.33}$$

Example 10.10 Determine the sample size required to reject $H_0 : c_p = 1.33$ in favor of $H_A : c_p > 1.33$ with 90% power when $c_p = 1.5$.
Solution: With $(c_p)_0 = 1.33$, $(c_p)_1 = 1.5$, $\alpha = 0.05$, and $\beta = 0.10$ in Equation 10.33, the required sample size is
$$n \simeq \frac{1}{2}\left(\frac{1.645 + 1.282}{\ln\left(\frac{1.5}{1.33}\right)}\right)^2 = 297.$$

10.2.2.2 Test for c_{pk}

The hypotheses to be tested are $H_0 : c_{pk} = (c_{pk})_0$ versus $H_A : c_{pk} > (c_{pk})_0$ where $(c_{pk})_0$ is a specified value. When the sample size is very large, the distribution of $\hat{c}_{pk}$ is approximately normal with standard error given by Equation 10.27. At the critical value $(c_{pk})_{A/R}$ that defines the accept/reject boundary, the type I and type II error conditions are related by

$$(c_{pk})_{A/R} = (c_{pk})_0 + z_\alpha \widehat{\sigma}_{(c_{pk})_0} = (c_{pk})_1 - z_\beta \widehat{\sigma}_{(c_{pk})_1} \qquad (10.34)$$

where the power is $\pi = 1 - \beta$ to reject H_0 when in fact $c_{pk} = (c_{pk})_1$. This equation is somewhat complicated by the fact that $\sigma_{c_{pk}}$ depends on the value of c_{pk}; however, from the form of $\sigma_{c_{pk}}$ in Equation 10.27 it leads to the following sample size requirement:

$$n = \left(\frac{z_\alpha (c_{pk})_0 \sqrt{\frac{1}{9(c_{pk})_0^2} + \frac{1}{2}} + z_\beta (c_{pk})_1 \sqrt{\frac{1}{9(c_{pk})_1^2} + \frac{1}{2}}}{(c_{pk})_1 - (c_{pk})_0} \right)^2. \qquad (10.35)$$

When $(c_{pk})_0$ and $(c_{pk})_1$ are both large, say, greater than or equal to 1.33, this sample size condition may be approximated by

$$n \simeq \frac{1}{2} \left(\frac{z_\alpha (c_{pk})_0 + z_\beta (c_{pk})_1}{(c_{pk})_1 - (c_{pk})_0} \right)^2. \qquad (10.36)$$

This approximation underestimates the sample size given by Equation 10.35, but it is usually sufficiently accurate considering the many sources of uncertainty present when planning a process capability study.

Example 10.11 Determine the sample size required to reject $H_0 : c_{pk} = 1.33$ in favor of $H_A : c_{pk} > 1.33$ with 90% power when $c_{pk} = 1.5$.
Solution: With $(c_{pk})_0 = 1.33$, $(c_{pk})_1 = 1.5$, $\alpha = 0.05$, and $\beta = 0.10$ in Equation 10.35, the sample size required to reject H_0 is

$$n = \left(\frac{1.645 (1.33) \sqrt{\frac{1}{9 \times 1.33^2} + \frac{1}{2}} + 1.282 (1.5) \sqrt{\frac{1}{9 \times 1.5^2} + \frac{1}{2}}}{1.5 - 1.33} \right)^2$$

$$= 326.$$

As expected, this value is comparable to the $n = 297$ sample size required for the test of c_p determined in Example 10.10 for similar conditions.

Example 10.12 Determine the sample size for Example 10.11 using the large sample approximation and compare the result to the original sample size.
Solution: From the information given in the original problem statement and Equation 10.36 the approximate sample size is

$$n \simeq \frac{1}{2} \left(\frac{1.645 (1.33) + 1.282 (1.5)}{1.5 - 1.33} \right)^2$$

$$\simeq 292.$$

This value is about 10% lower than the more accurate value calculated in Example 10.11.

10.3 Tolerance Intervals

Tolerance intervals are constructed from sample data to set an upper limit for the fraction defective of a quality characteristic with some specified degree of confidence. The complement of the fraction defective is often referred to as the yield or the coverage factor. The yield (Y), defective rate (p), and the tolerance limits (UTL/LTL) are related by

$$Y = 1-p$$
$$= P(LTL \leq x \leq UTL). \qquad (10.37)$$

The symbols Y and p are used here as parameters. Subscripts U and L will be used to indicate the upper and lower confidence limits on Y and p. For example, p_U will be used to indicate the upper confidence limit for the fraction defective, as in $P(0 < p < p_U) = 1 - \alpha$, and $Y_L = 1 - p_U$ will be used to indicate the corresponding lower confidence limit for the yield, as in $P(Y_L < Y < 1) = 1 - \alpha$.

10.3.1 Nonparametric Tolerance Intervals

Sample size calculations are presented for studies to determine one-sided and two-sided nonparametric tolerance limits.

10.3.1.1 Two-Sided Nonparametric Tolerance Intervals

When a random sample of size n is drawn from a single stable population of continuous measurement values (x), the interval given by the minimum and maximum values in the sample, x_{min} and x_{max}, respectively, has probability $1 - \alpha$ of containing at least $Y_L = 1 - p_U$ of the population, that is

$$P(Y_L < P(x_{min} \leq x \leq x_{max}) < 1) = 1 - \alpha, \qquad (10.38)$$

where α, p_U, and n are related by the cumulative binomial probability

$$\alpha = \sum_{i=0}^{1} b(i; n, p_U). \qquad (10.39)$$

Because no assumptions about the shape of the x distribution have been made, the interval $x_{min} \leq x \leq x_{max}$ is called a *nonparametric tolerance interval* for x.

Given two of the three values of α, p_U, and n, the third value can be found from Equation 10.39. For specified values of p_U and α, Equation 10.39 is transcendental; however, n may be approximated using Larson's nomogram (see Appendix E.1) or with

$$n \simeq \frac{\chi^2_{1-\alpha,4}}{2p_U}. \qquad (10.40)$$

Example 10.13 What sample size is required to be 95% confident that at least 99% of a population of continuous measurement values falls within the extreme values of the sample?

Solution: With $\alpha = 0.05$ and $p_U = 0.01$ the required sample size is approximately

$$n \simeq \frac{\chi^2_{0.95,4}}{2 \times 0.01}$$
$$\simeq 475.$$

Further iterations indicate that the smallest value of n for which $\alpha \leq 0.05$ is $n = 473$, which leads to the following nonparametric tolerance interval for x:

$$P(0.99 < P(x_{min} \leq x \leq x_{max}) < 1) = 0.9502.$$

10.3.1.2 One-Sided Nonparametric Tolerance Intervals

When a random sample of size n is drawn from a single stable population of continuous measurement values, the one-sided interval given by the minimum or maximum value in the sample, x_{min} and x_{max}, respectively, has probability $1 - \alpha$ of containing at least $Y_L = 1 - p_U$ of the population of values. That is, for a one-sided lower specification limit on x

$$P(Y_L < P(x_{min} \leq x \leq \infty) < 1) = 1 - \alpha \qquad (10.41)$$

and for the one-sided upper limit case

$$P(Y_L < P(-\infty \leq x \leq x_{max}) < 1) = 1 - \alpha \qquad (10.42)$$

where in both cases α, p_U, and n are related by the binomial probability

$$\alpha = b(0; n, p_U). \qquad (10.43)$$

Then for specified values of α and p_U, the one-sided tolerance interval can be determined from Equation 10.43. When n is expected to be large, it may be approximated from

$$n = \frac{\chi^2_{1-\alpha,2}}{2p_U}. \qquad (10.44)$$

Example 10.14 What sample size n is required to be 95% confident that at least 99% of a population of continuous measurement values falls below the maximum value of the sample?

Solution: With $\alpha = 0.05$ and $p_U = 0.01$ the required sample size is

$$n \simeq \frac{\chi^2_{0.95,2}}{2 \times 0.01}$$
$$\simeq 300.$$

10.3.2 Normal Tolerance Intervals

Sample size methods are presented for studies to determine one-sided and two-sided normal tolerance intervals.

10.3.2.1 Two-Sided Normal Tolerance Intervals

When a random sample of size n is drawn from a normal population, the two-sided tolerance interval $LTL \leq x \leq UTL$ has probability $1 - \alpha$ of containing at least $Y_L = 1 - p_U$ of the population of values, that is

$$P(Y_L < \Phi(LTL \leq x \leq UTL) < 1) = 1 - \alpha \qquad (10.45)$$

where the upper and lower tolerance limits are given by

$$UTL/LTL = \bar{x} \pm k_2 s \qquad (10.46)$$

and k_2, the tolerance interval half-width, is given in Appendix E.7 for $\alpha = 0.05$ and $Y_L = 0.99$ and 0.999. The same table may be used to determine the sample size required to obtain the desired tolerance interval half-width for specified values of α and Y_L.

If a table of k_2 values is not available, k_2 may be approximated with

$$k_2 \simeq z_{p_U/2} \sqrt{1 + \frac{1}{n}} \sqrt{\frac{n-1}{\chi^2_{\alpha,n-1}}}. \qquad (10.47)$$

This equation cannot be solved for n and approximations to it have too much error, so the sample size required to obtain a desired k_2 value must be obtained by iteration.

Example 10.15 Determine the sample size required to obtain a 95% confidence two-sided 99% coverage normal distribution tolerance interval with tolerance limits $UTL/LTL = \bar{x} \pm 3.5s$.

Solution: The desired tolerance interval has the form

$$P(0.99 \leq \Phi(\bar{x} - 3.5s \leq x \leq \bar{x} + 3.5s) \leq 1) = 0.95.$$

From Appendix E.7, a sample of size $n = 25$ gives $k_2 = 3.46$. A spreadsheet (not shown) was set up to calculate k_2 as a function of n using Equation 10.47 with $p = 0.01$ and $\alpha = 0.05$. The spreadsheet indicated that the sample size $n = 24$ delivers $k_2 = 3.485$ and that $n = 23$ delivers $k_2 = 3.514$, so $n = 24$ should be used to be conservative. These approximate k_2 values differ from the exact values given in Appendix E.7 in the thousandths place.

10.3.2.2 One-Sided Normal Tolerance Intervals

The one-sided lower tolerance interval for a normal population with coverage $Y_L = 1 - p_U$ and one-sided confidence level $1 - \alpha$ is given by

$$P(Y_L < \Phi(LTL \leq x \leq \infty) < 1) = 1 - \alpha \tag{10.48}$$

where the lower tolerance limit is calculated from the sample mean $\bar{x}$ and sample standard deviation s by

$$LTL = \bar{x} - k_1 s \tag{10.49}$$

and k_1 is given by

$$k_1 = \frac{t_{1-\alpha, n-1, \phi}}{\sqrt{n}} \tag{10.50}$$

where n is the sample size and the noncentral Student's t distribution noncentrality parameter is

$$\phi = z_p \sqrt{n}. \tag{10.51}$$

The analysis for a one-sided upper tolerance limit is analogous to that for the one-sided lower limit.

Equation 10.50 is transcendental in n, so the sample size required to obtain the desired k_1 value for specified coverage and confidence level is most easily determined using a table of published k_1 values. Values of k_1 as a function of n are given in Appendix E.7 for $\alpha = 0.05$ and $Y_L = 0.99$ and 0.999.

Example 10.16 Determine the sample size required to obtain a 95% confidence 99% coverage normal distribution tolerance interval with one-sided upper tolerance limit $UTL = \bar{x} + 3s$.

Solution: The required interval has the form

$$P(0.99 < \Phi(-\infty < x \leq UTL) < 1) = 0.95 \tag{10.52}$$

where $UTL = \bar{x} + k_1 s$ with $k_1 = 3$. From Table E.7 of Appendix E with $\alpha = 0.05$ and $Y = 0.99$, the required sample size is $n = 35$.

10.4 Acceptance Sampling

Acceptance sampling is too broad a topic to be covered here in any detail, so only the basic single sampling methods for attributes and variables are presented. The following topics are covered for attribute (pass/fail) sampling: type A (finite lot) sampling plans, type B (infinite lot) sampling plans, LTPD rectifying inspection plans, and AOQL rectifying inspection plans. Single sampling for variables data is covered for the known and unknown standard deviation cases. The basic structure of and the sample size to lot size relationships of the ANSI/ASQ Z1.4 and Z1.9 standards are described, but see the references for more details on the design and use of those standards. Schilling [59] is an excellent reference for all of the topics in this section.

10.4.1 Single Sampling Plans for Attributes

Acceptance sampling for attributes is a straightforward application of the one-sample hypothesis test of a proportion: $H_0 : p = p_0$ versus $H_A : p > p_0$ where p is the fraction defective.[2] The test is performed by counting the number of defectives d observed in a random sample of size n. The decision to accept or reject H_0 is made by comparing d to a critical value called the acceptance number c. If $0 \leq d \leq c$ then H_0 and the lot are accepted; otherwise $d > c$ and H_0 and the lot are rejected. A rejected lot might be returned to its supplier or the lot might be 100% inspected.

The performance of a sampling plan as a function of product quality is presented in the form of an operating characteristic or OC curve - a plot of the probability of accepting H_0 versus the process fraction defective, that is, of P_A versus p.[3] There are two types of OC curves for single sampling attributes plans that are distinguished by the nature of the lot:

- Type A: When the lot is unique, the OC curve is given by the hypergeometric distribution:

$$P_A = \sum_{d=0}^{c} h(d; D, N, n) \qquad (10.53)$$

where N is the lot size, D is the number of defectives in the lot, and $p = D/N$ is the lot fraction defective.

- Type B: When the lot comes from a stream of similar lots, then the OC curve for the sampling plan is given by the binomial distribution:

$$P_A = \sum_{d=0}^{c} b(d; n, p) \qquad (10.54)$$

where p is called the *process fraction defective*.

If the sample size n is relatively small compared to the lot size N in a Type A situation $\left(n < \frac{N}{10}\right)$, then the hypergeometric distribution can be safely approximated with the binomial distribution and the Type A OC curve can be approximated safely with the corresponding Type B OC curve.

Two special points on an OC curve are often used to determine a sampling plan. By definition, the probability of a type I error (rejecting a good lot) is α when the fraction defective is equal to an acceptably low value called the *acceptable quality level* (AQL) and the probability of a type II error (accepting a bad lot) is β when the fraction defective is equal to a rejectably high level called the *rejectable quality level* (RQL). The corresponding points on the OC curve are shown in Figure 10.3.

---

[2] Some acceptance sampling systems like ANSI/ASQ Z1.4 formulate the hypotheses as shown here whereas other systems reverse H_0 and H_A.

[3] Under H_A, P_A is the type II error rate and the complement of P_A is the power.

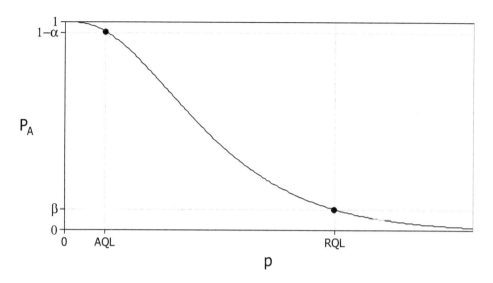

Figure 10.3: Operating characteristic curve with $(p, P_A) = (AQL, 1 - \alpha)$ and $(p, P_A) = (RQL, \beta)$.

There are three common methods used for designing sampling plans:

1. Calculate n and c for specified values of $(p = AQL, P_A = 1 - \alpha)$ and $(p = RQL, P_A = \beta)$.

2. Calculate n for $c = 0$ with specified $(p = AQL, P_A = 1 - \alpha)$ values.

3. Calculate n for $c = 0$ with specified $(p = RQL, P_A = \beta)$ values.

Applications of these methods are presented in the following sections.

10.4.1.1 Design of Type B Plans

When the first sampling plan design method is used in a Type B situation, the two (p, P_A) conditions give two equations corresponding to two points on the OC curve. Simultaneous solution of these two equations, which may be performed exactly using appropriate software or approximately with Larson's nomogram for the cumulative binomial distribution, leads to unique values of n and c. These methods are described in Section 4.1.2.

When software or Larson's nomogram are not available, then approximate values for n and c may be found if the Poisson approximations to the two binomial distributions are justified, where the Poisson distributions are expressed in their χ^2 forms. Then the approximate value of c may be determined from the

10.4. Acceptance Sampling

	α/β			
c	0.05/0.10	0.05/0.05	0.01/0.10	0.01/0.05
0	44.89	58.4	229.1	298.1
1	10.95	13.35	26.18	31.93
2	6.51	7.70	12.21	14.44
3	4.89	5.67	8.12	9.42
4	4.06	4.65	6.25	7.16
5	3.55	4.02	5.20	5.89
6	3.21	3.60	4.52	5.08
7	2.96	3.30	4.05	4.52
8	2.77	3.07	3.70	4.12
9	2.62	2.89	3.44	3.80
10	2.50	2.75	3.23	3.56

Table 10.1: RQL/AQL values for finding acceptance numbers.

smallest value of c that meets the condition

$$\frac{RQL}{AQL} \geq \frac{\chi^2_{1-\beta,2(c+1)}}{\chi^2_{\alpha,2(c+1)}}. \tag{10.55}$$

While this condition appears to be tedious to solve, it is actually quite easy using an appropriate χ^2 table or spreadsheet by scanning down the $\chi^2_{1-\beta}$ and χ^2_α columns as functions of degrees of freedom. Table 10.1 provides a convenient table of RQL/AQL values for this method.

After c has been determined from Table 10.1, the value of n can be approximated from either of the (p, P_A) conditions using

$$n \simeq \frac{\chi^2_{1-P_A,2(c+1)}}{2p}. \tag{10.56}$$

This sample size approximation is more accurate for large n than small n and should be confirmed using the exact binomial calculations.

Whether exact or approximate methods are used, it is usually necessary to select a sampling plan that deviates slightly from the original specifications because n and c are integers. Such deviations should be adopted to make the sampling plan more conservative, that is, to decrease either α, or β, or both of them.

The second and third methods for designing sampling plans are the popular zero acceptance number ($c = 0$) plans. Because they involve only one equation and one unknown, they are easier cases to solve for the sample size. A good first approximation for the sample size for the specified (p, P_A) condition is given by Equation 10.56 with $c = 0$.

Example 10.17 Design the single sampling plan for attributes that will accept 95% of lots when the process fraction defective is 1% and accept only 10% of lots when the process fraction defective is 5%.

Solution: From the problem statement, the lots are coming from a continuous process, so the sampling plan will be Type B with points on the OC curve at $(AQL, 1 - \alpha) = (0.01, 0.95)$ and $(RQL, \beta) = (0.05, 0.10)$. From Table 10.1 with $RQL/AQL = 0.05/0.01 = 5.0$, the acceptance number must be $c = 3$. Then, from the RQL condition, the required sample size is approximately

$$n \simeq \frac{\chi^2_{0.90,8}}{2(0.05)}$$
$$\simeq \frac{13.36}{2 \times 0.05}$$
$$\simeq 134.$$

The exact sampling plan that meets the specifications in the problem statement is $n = 132$ and $c = 3$.

Example 10.18 Find the $c = 0$ plans that meet a) the AQL requirement and b) the RQL requirement from Example 10.17. Plot the three OC curves on the same graph.

Solution:

a) The sample size for the $c = 0$ plan that meets the AQL requirement $(p, P_A) = (0.01, 0.95)$ is approximately

$$n \simeq \frac{\chi^2_{\alpha,2}}{2 \times AQL}$$
$$\simeq \frac{0.1026}{2 \times 0.01}$$
$$\simeq 6.$$

A few binomial calculations indicate that the exact sample size is $n = 5$ because $(b(0; 5, 0.01) = 0.951) > (1 - \alpha = 0.95)$.

b) The sample size for the $c = 0$ plan that meets the RQL requirement $(p, P_A) = (0.05, 0.10)$ is approximately

$$n \simeq \frac{\chi^2_{1-\beta,2}}{2 \times RQL}$$
$$\simeq \frac{4.61}{2 \times 0.05}$$
$$\simeq 46.$$

The exact sample size is $n = 45$ because $(b(0; 45, 0.05) = 0.099) < (\beta = 0.10)$. Figure 10.4 shows the three OC curves.

10.4. Acceptance Sampling

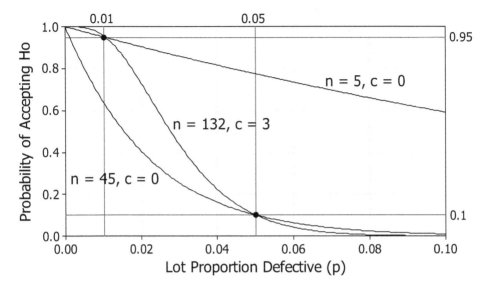

Figure 10.4: Operating characteristic curves for three sampling plans.

10.4.1.2 Design of Type A Plans

Type A (hypergeometric) single sampling plans are calculated in a manner analogous to that used for Type B (binomial) plans using the following equations based on the hypergeometric distribution:

$$\sum_{x=0}^{c} h(x; D_1, N, n) \geq 1 - \alpha \tag{10.57}$$

where lots with D_1 defectives are considered to be acceptable and

$$\sum_{x=0}^{c} h(x; D_2, N, n) < \beta \tag{10.58}$$

where lots with D_2 defectives are considered to be rejectable. The exact solution of these equations is tedious to obtain, so it is best done using appropriate software; however, most practical situations may be solved approximately using the Type B (binomial) methods. While the approximate methods are not accurate enough, they still provide a convenient starting point for iterations toward an exact solution.

There is an important special application of the Type A (hypergeometric) method that deserves to be discussed. It arises when a 100% inspection operation is replaced with a $c = 0$ plan that either 1) accepts the lot when a critical sample size is reached, or 2) rejects the lot when the first defective unit is found. When designed correctly, these zero defects plans are a safe and cost-effective alternative to 100% inspection.

As in the last section, the Type A OC curve given by Equation 10.53 can be approximated with a binomial Type B OC curve when $n/N < 0.10$; however, when n is not necessarily small compared to N but $D \ll N$, then the hypergeometric distribution can be approximated with the rare-event binomial approximation given by

$$h(x; D, N, n) \simeq b\left(x; D, \frac{n}{N}\right)$$
$$\simeq \binom{D}{x}\left(\frac{n}{N}\right)^x \left(1 - \frac{n}{N}\right)^{D-x}. \tag{10.59}$$

With this approximation applied to the special case of a $c = 0$ plan, the Type A OC curve is given by

$$P_A = h(0; D, N, n)$$
$$\simeq \left(1 - \frac{n}{N}\right)^D. \tag{10.60}$$

Then for specified values of (D, P_A), the fraction of the lot that must be inspected is

$$\frac{n}{N} \simeq 1 - P_A^{1/D}. \tag{10.61}$$

Example 10.19 Determine the sampling plan for lots of size $N = 50$ that will accept 95% of the lots with $D \leq 1$ defectives and reject 90% of the lots with $D \geq 5$ defectives.

Solution: The sampling plan must meet the simultaneous conditions given by Equation 10.57 with $D_1 = 1$ and $\alpha \leq 0.05$:

$$\sum_{x=0}^{c} h(x; D_1 = 1, N = 50, n) \geq 0.95 \tag{10.62}$$

and Equation 10.58 with $D_2 = 5$ and $\beta \leq 0.10$:

$$\sum_{x=0}^{c} h(x; D_2 = 5, N = 50, n) < 0.10. \tag{10.63}$$

The acceptance number c is not specified, so different values of c must be considered. The approximate sample size for the $c = 0$ sampling plan to meet the condition in Equation 10.63 is given by Equation 10.61:

$$n \simeq 50\left(1 - 0.10^{1/5}\right) = 19;$$

however, the condition in Equation 10.62 is not satisfied because

$$(h(x = 0; D_1 = 1, N = 50, n = 19) = 0.525) \not\geq 0.95.$$

10.4. Acceptance Sampling

Iterations with a hypergeometric calculator show that with $c = 1$ Equation 10.63 is satisfied when $n = 29$ because

$$\left(\sum_{x=0}^{1} h(x; D_1 = 5, N = 50, n = 28) = 0.109 \right) \not\leq 0.10$$

$$\left(\sum_{x=0}^{1} h(x; D_1 = 5, N = 50, n = 29) = 0.092 \right) \leq 0.10$$

and Equation 10.62 is satisfied because

$$\left(\sum_{x=0}^{1} h(x; D_1 = 1, N = 50, n = 29) = 1 \right) \geq 0.95.$$

The sampling plan that meets the requirements is $n = 29$ with $c = 1$.

Example 10.20 A 100% inspection process for large lots is to be replaced with a $c = 0$ sampling plan. What fraction of each lot must be inspected if lots that contain five or more defectives must be rejected 90% of the time?

Solution: From Equation 10.61 with $D = 5$ and $P_A = 0.10$, the fraction of each lot that must be inspected is

$$\frac{n}{N} \simeq 1 - 0.10^{1/5}$$
$$\simeq 0.37.$$

Example 10.21 The calculated sample size in Example 4.2 was quite large compared to the lot size, which violates the small-sample approximation assumption. Repeat that example using the rare-event approximation method.

Solution: The solution in the example indicated that 30% of the lot needed to be inspected. From Equation 10.61, which takes the relatively large sample size into account, the fraction of the lot that has to be inspected is more accurately

$$\frac{n}{N} \simeq 1 - 0.05^{1/10}$$
$$\simeq 0.259.$$

10.4.1.3 ANSI/ASQ Z1.4

The purpose of this section is to review the ANSI/ASQ Z1.4 acceptance sampling standard for attribute inspection. See the standard itself or one of the many excellent references if you require more information about its use.

Section 10.4.1.1 described how to find the unique sample size n and acceptance number c that forces a sampling plan's OC curve through two preselected points, one corresponding to $(p_0 = AQL, 1 - \alpha)$ and the other corresponding to $(p_1 = RQL, \beta)$. It will be shown in this section that ANSI/ASQ Z1.4 (and the related standard ANSI/ASQ Z1.9, in Section 10.4.3.3) uses a different strategy for selecting sampling plans. Sampling plans in ANSI/ASQ Z1.4 are designed to provide a high probability of accepting lots of AQL quality using a sample size that is determined from the lot size. This approach gives the individual sampling plans in ANSI/ASQ Z1.4 inconsistent consumer protection. Instead of relying on a single sampling plan, however, ANSI/ASQ Z1.4 uses a combination of three sampling plans called *normal, tightened,* and *reduced* plans that are linked together by switching rules.

Sampling always begins with normal inspection, shifts to tightened inspection when too many lots have been rejected, and shifts to reduced inspection when many consecutive lots have been accepted. The OC curve for this system of plans is a combination of the OC curves from the individual plans, which provides economical sampling, a high probability of accepting lots of AQL quality, and good protection for the consumer against lots of poor quality.

By definition, a *sampling plan* consists of a sample size and an acceptance condition that together determine the sampling plan's operating characteristic (OC) curve. A *sampling scheme* consists of two or more sampling plans that are used together under different quality conditions. The sampling plans in a sampling scheme are linked together by *switching rules* that determine which sampling plan within a sampling scheme to use based on recent quality conditions. A *sampling system* is a collection of sampling schemes.

The ANSI/ASQ Z1.4 standard is a sampling system for attribute inspection for controlling the defective rate in lots. ANSI/ASQ Z1.4 has

- seven inspection levels: four special levels (S1, S2, S3, S4) and three general levels (I, II, and III).

- single, double, and multiple sampling plans.

- sampling schemes with normal, tightened, and reduced sampling plans linked by switching rules.[4]

ANSI/ASQ Z1.4 sampling schemes are indexed by lot size (N), inspection level (which together determine a sampling plan code letter), and acceptable quality level (AQL). The standard includes discrete AQLs ranging from 0.01% to 10%, with a multiplicative factor of about $10^{1/5} = 1.585$ between adjacent AQLs. The probability of accepting lots of AQL quality increases monotonically from 89% for code letter B to 99.5% for code letter P. The standard includes

- tables for sample size and acceptance number for the normal, tightened, and reduced sampling plans within each sampling scheme.

---

[4]The individual sampling plans within a sampling scheme are not intended to be used without the switching rules.

10.4. Acceptance Sampling

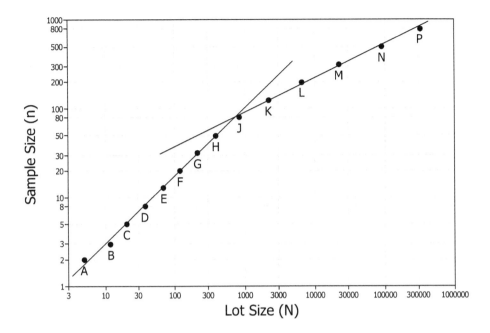

Figure 10.5: Sample size to lot size relationship for ANSI/ASQ Z1.4 single sampling plans, level II, normal inspection.

- OC curves for the normal, tightened, and reduced sampling plans within each sampling scheme. The OC curves are calculated using Equation 10.54.
- OC curves for sampling schemes using the switching rules.

Figure 10.5 shows a plot of $\log(n)$ versus $\log(\overline{N})$ for single sampling, normal inspection, where $\overline{N}$ is the mean lot size for the indicated lot size code letter. There is a similar relationship between sample size and lot size for the other plans in the standard.

Example 10.22 Create the OC curves for normal, tightened, and reduced inspection under ANSI/ASQ Z1.4 using general inspection level II, single sampling, $N = 1000$, and $AQL = 1\%$.

Solution: The sampling plans determined for code letter J from the standard were normal $(n = 200, c = 5)$, tightened $(n = 200, c = 3)$, and reduced $(n = 80, c = 2)$. The operating characteristic curves were calculated using Equation 10.54 and are shown in Figure 10.6. For reference, the figure also shows the OC curve for the sampling plan determined for the same conditions using the Squeglia [64] zero acceptance number sampling standard, which is often used instead of ANSI/ASQ Z1.4.

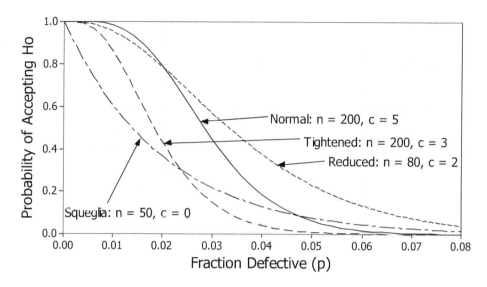

Figure 10.6: OC curves for normal, tightened, and reduced inspection under ANSI/ASQ Z1.4 Level II, single sampling, $N = 4000$, $AQL = 1\%$.

10.4.2 Rectifying Inspection for Attributes

In rectifying inspection, rejected lots are 100% inspected and defective parts are either removed from the lot or are replaced with good parts. Because the number of parts inspected for a lot is either the sample size n or the lot size N, a common goal of all rectifying inspection plans is to minimize the average total inspection (ATI) that must fall between n and N. The rectifying inspection sampling plan that minimizes the ATI is also a function of the process fraction defective.

There are two types of rectifying inspection plans: lot tolerance percent defective (LTPD) plans and average outgoing quality limit (AOQL) plans. LTPD plans are designed to provide specified protection against defective lots for each lot inspected. AOQL plans are designed to control the average postinspection defective rate in a series of similar lots. LTPD plans require larger sample sizes than AOQL plans.

The best known source of tabulated rectifying inspection plans is Dodge and Romig [17]. The following algorithms approximate the published Dodge-Romig rectifying inspection single sampling plans for LTPD and AOQL. The Dodge-Romig plans provide minimum sample size sampling plans for six contiguous intervals of process fraction defective, but the algorithms presented here treat the process fraction defective as a continuous variable, so sampling plans determined by the two methods might differ slightly.

10.4.2.1 LTPD Plans

Lot tolerance percent defective (LTPD) plans are designed to control the quality of individual lots, so the probability of accepting a lot P_A is governed by the cumulative hypergeometric probability distribution

$$P_A = \sum_{x=0}^{c} h(x; D, N, n) \qquad (10.64)$$

where x is the number of defectives in the sample of size n, D is the number of defectives in the lot of size N, and the acceptance number c is the maximum number of defectives allowed in the sample to accept the lot. By definition $LTPD$ is the lot fraction defective that should be accepted with low probability β, that is, the point $(p = LTPD, P_A = \beta)$ is a point on the sampling plan's OC curve. β is typically taken to be a small value, like 10%, so a corresponding lot of $p = LTPD$ quality is undesirable. $LTPD$ and β are related by a special case of Equation 10.64:

$$\beta = \sum_{x=0}^{c} h(x; D = N \times LTPD, N, n). \qquad (10.65)$$

When n is small compared to N and the fraction defective is small, the sample size can be approximated using Equation 10.56 with $p = LTPD$ and $P_A = 1 - \beta$, giving

$$n \simeq \frac{\chi^2_{1-\beta, 2(c+1)}}{2 \times LTPD}. \qquad (10.66)$$

There are many combinations of n and c that satisfy Equation 10.65, but the preferred sampling plan is the one that minimizes the average total inspection (ATI) for the historical lot fraction defective p_0, where ATI is given by the expectation value of the sample size:

$$ATI = nP_A + N(1 - P_A). \qquad (10.67)$$

ATI is transcendental in c, so the easiest way to find the optimal sampling plan is to calculate ATI for each c using a spreadsheet and then to identify the sampling plan with the smallest ATI.

Example 10.23 Determine the optimal rectifying inspection sampling plan for $LTPD = 0.04$ with $\beta = 0.10$ when the lot size is $N = 2500$ and the historical process fraction defective is $p = 0.01$.
Solution: A spreadsheet was used to solve Equations 10.66 and 10.67 as a function of acceptance number c as shown in Table 10.2. For the specified conditions, the sampling plan that minimizes ATI when $p = 0.01$ is $n = 232$ with $c = 5$. By comparison, the Dodge-Romig LTPD tables indicate a sampling plan with $n = 230$ and $c = 5$, which is in excellent agreement with the calculated plan.

c	$2(c+1)$	$\chi^2_{0.90, 2(c+1)}$	n	$P_A\|_{p=0.01}$	ATI
2	6	10.64	134	0.8486	492.2
3	8	13.36	168	0.9108	376.1
4	10	15.99	200	0.9483	319.0
5	**12**	**18.55**	**232**	**0.9697**	**300.8**
6	14	21.06	264	0.9821	304.1
7	16	23.54	295	0.9895	318.0
8	18	25.99	325	0.9940	338.1

Table 10.2: Solution for the $LTPD$ sampling plan that minimizes ATI.

As the LTPD plan sample size becomes larger than about 20% of the lot size, the plan obtained by the approximate method diverges from the exact optimal plan. Most of these situations are satisfied with the $c = 0$ plans for which the exact hypergeometric distribution can be approximated by the large-sample binomial approximation

$$\begin{aligned} P_A &= h(x=0; D, N, n) \\ &\simeq b\left(x=0; D, \frac{n}{N}\right) \\ &\simeq \left(1 - \frac{n}{N}\right)^D. \end{aligned} \tag{10.68}$$

Then, by the definition of $LTPD$

$$\beta \simeq \left(1 - \frac{n}{N}\right)^{N \times LTPD}, \tag{10.69}$$

which can be solved to obtain the following sample size condition for the large-sample $c = 0$ plans:

$$n \simeq N\left(1 - \beta^{\frac{1}{N \times LTPD}}\right). \tag{10.70}$$

For these special plans the sample size does not depend on the process fraction defective as it did for the small-sample plans.

Example 10.24 Find the rectifying inspection plan with $LTPD = 0.04$ and $\beta = 0.1$ for a lot size of $N = 50$.
Solution: For the given conditions, the spreadsheet method gives $n = 58$, which exceeds the lot size. From Equation 10.70 the sample size required for the $c = 0$ plan is

$$\begin{aligned} n &\simeq 50\left(1 - 0.1^{\frac{1}{50 \times 0.04}}\right) \\ &\simeq 35. \end{aligned}$$

By comparison, the corresponding Dodge-Romig plan calls for $n = 34$ and is independent of the historical fraction defective.

10.4.2.2 AOQL Plans

If the incoming lot fraction defective in a rectifying inspection sampling plan is p, then the post-inspection fraction defective or *average outgoing quality (AOQ)* is given by

$$AOQ = \left(\frac{N-n}{N}\right) p P_A + 0 \left(1 - P_A\right)$$
$$= \left(\frac{N-n}{N}\right) p P_A \qquad (10.71)$$

where N is the lot size, n is the sample size, and P_A is the probability of accepting the lot. The distribution of P_A could be hypergeometric or binomial, depending on the circumstances, but in either case, if p is small and $n \ll N$, then P_A can be approximated by the Poisson distribution.

AOQ has a maximum value, called the *average outgoing quality limit (AOQL)*, when p has the critical value

$$p_c = \frac{\chi_c^2}{2n} \qquad (10.72)$$

where χ_c^2 satisfies

$$\chi_c^2 pdf\left(\chi_c^2; \nu\right) = 1 - cdf\left(\chi_c^2; \nu\right) \qquad (10.73)$$

where $pdf\,()$ and $cdf\,()$ are the χ^2 probability density and cumulative distribution functions, respectively, with $\nu = 2\left(c+1\right)$ degrees of freedom where c is the usual acceptance number.[5] The χ_c^2 values, which do not depend on n, are given for $c = 0$ to 10 in Table 10.3.[6]

The value of $AOQL$ is given by the condition

$$AOQL = AOQ|_{p=p_c}$$
$$= A_c \left(\frac{N-n}{nN}\right) \qquad (10.74)$$

where

$$A_c = \frac{1}{2}\chi_c^2 \left(1 - cdf\left(\chi_c^2; \nu\right)\right). \qquad (10.75)$$

The A_c values depend on c and are also shown in Table 10.3. From Equation 10.74 with specified values of c and $AOQL$ the required sample size is

$$n = \frac{1}{\frac{AOQL}{A_c} + \frac{1}{N}}. \qquad (10.76)$$

---

[5]The c subscript on χ_c^2 indicates the relationship with p_c and does not indicate an area under the χ^2 distribution or the degrees of freedom.

[6]The approximate A_c values in Table 10.3 differ slightly from the values given by Duncan [19], but the n and c solutions obtained by the two methods are nearly identical.

c	χ_c^2	A_c
0	2.00	0.368
1	3.24	0.839
2	4.54	1.371
3	5.89	1.944
4	7.28	2.543
5	8.70	3.167
6	10.1	3.811
7	11.6	4.472
8	13.1	5.144
9	14.6	5.832
10	16.1	6.528

Table 10.3: Values of χ^2 and A_c for $AOQL$ rectifying inspection plans.

Example 10.25 What sample size is required for a $c = 1$ rectifying inspection single-sampling plan to obtain 1% $AOQL$ if the lot size is $N = 300$? For what value of incoming fraction defective will AOQ be a maximum?
Solution: From Equation 10.76 with $A_1 = 0.839$ the required sample size is

$$n = \frac{1}{\frac{0.01}{0.839} + \frac{1}{300}} = 66.$$

AOQ will be at its maximum value, $AOQL$, when the incoming fraction defective is

$$p_c = \frac{\chi_1^2}{2n} = \frac{3.24}{2 \times 66} = 0.0245.$$

The average total inspection (ATI) is given by Equation 10.67 where P_A depends on the specified value of the fraction defective p. The preferred sampling plan is the one that minimizes the ATI for the historical defective rate. ATI is transcendental in c, so the easiest way to find the optimal sampling plan is to calculate ATI for each c using a spreadsheet and then to identify the sampling plan with the smallest ATI.

Example 10.26 Determine the sampling plan that minimizes the ATI for lots of size $N = 1000$ with $AOQL = 0.02$ when the historical defective rate is $p = 0.01$.
Solution: A spreadsheet was used to solve for n and ATI as a function of c using Equations 10.76 and 10.67. The results from the spreadsheet, shown in Table 10.4, indicate that the sampling plan that minimizes ATI is given by $n = 65$ and $c = 2$. By comparison, this is exactly the same plan indicated in the Dodge-Romig tables for these conditions.

10.4. Acceptance Sampling

c	A_c	n	$P_A\vert_{p=0.01}$	ATI
0	0.368	19	0.827	192.0
1	0.839	41	0.936	105.3
2	**1.371**	**65**	**0.972**	**93.3**
3	1.944	89	0.987	102.0
4	2.543	113	0.994	119.1
5	3.167	137	0.997	139.9

Table 10.4: Solution for the $AOQL$ sampling plan that minimizes ATI.

10.4.3 Variables Sampling Plans for Defectives

The purpose of a variables sampling plan for defectives is to control the defective rate in lots using variables data to make lot accept/reject decisions. A plan must have a high probability $(1 - \alpha)$ of accepting lots with a low fraction defective p_0 (or AQL) and a low probability (β) of accepting lots that have a high fraction defective p_1 (or RQL). These conditions define two points on an OC curve that has a unique sample size and acceptance condition. The decision to accept or reject H_0 is made by comparing the mean of the sample to the critical accept/reject value $\bar{x}_{A/R}$.

The analysis that follows assumes that the population is normally distributed with unknown population mean and that there is one relevant specification limit. The normality of the population being sampled is especially important because even a minor distortion in the distribution's tail nearest to the specification limit will significantly affect the defective rate. If the distribution is not normal, it must be mathematically transformed to at least approximate normality before this method can be considered.

10.4.3.1 Single Sampling Plans for Variables - σ Known

For variables sampling to control the defective rate p, the hypotheses to be tested are $H_0 : p = p_0$ versus $H_A : p > p_0$. With specified values of p_0, p_1, σ, and upper specification limit (USL), there are two corresponding population means μ_0 and μ_1 related by

$$USL = \mu_0 + z_{p_0}\sigma_x = \mu_1 + z_{p_1}\sigma_x. \tag{10.77}$$

This relationship is shown in Figures 10.7a and 10.7b. For these values of μ_0 and μ_1 and the corresponding specified values of α and β, the critical accept/reject value for sample means is

$$\bar{x}_{A/R} = \mu_0 + z_\alpha \sigma_{\bar{x}} = \mu_1 - z_\beta \sigma_{\bar{x}} \tag{10.78}$$

where $\sigma_{\bar{x}} = \sigma_x/\sqrt{n}$. This relationship is shown in Figures 10.7c and 10.7d. When Equations 10.77 and 10.78 are combined and solved for n, the required sample

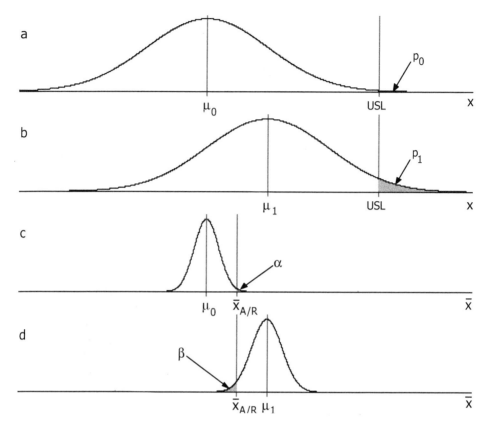

Figure 10.7: Distributions of x and $\bar{x}$ under H_0 and H_A.

size for the sampling plan is

$$n = \left(\frac{z_\alpha + z_\beta}{z_{p_0} - z_{p_1}} \right)^2. \tag{10.79}$$

The same sample size equation is obtained for a one-sided lower specification limit (LSL). Once the sample size has been determined, $\bar{x}_{A/R}$ can be determined from Equation 10.78.

Example 10.27 Find the single sampling plan for variables that will accept 95% of the lots with 1% defectives and reject 90% of the lots with 4% defectives when $\sigma = 30$ and the specification is one-sided with $USL = 700$.
Solution: The two specified points on the OC curve are $(p_0, 1-\alpha) = (0.01, 0.95)$ and $(p_1, \beta) = (0.04, 0.10)$. From Equation 10.79 the required sample

10.4. Acceptance Sampling

size is

$$n = \left(\frac{z_{0.05} + z_{0.10}}{z_{0.01} - z_{0.04}}\right)^2$$

$$= \left(\frac{1.645 + 1.282}{2.33 - 1.75}\right)^2$$

$$= 26.$$

The critical value of $\bar{x}_{A/R}$ is

$$\bar{x}_{A/R} = \mu_0 + z_\alpha \sigma_{\bar{x}}$$

$$= (USL - z_{p_0}\sigma_x) + z_\alpha \frac{\sigma_x}{\sqrt{n}}$$

$$= (700 - 2.33 \times 30) + 1.645 \frac{30}{\sqrt{26}}$$

$$= 640.$$

Equations 10.79 and 4.22 provide an opportunity to compare the sample sizes required for variables and attributes sampling plans for defectives. In the former case, the accept/reject decision is based on the observed value of $\bar{x}$. In the latter case, the accept/reject decision is based on the defective count.

For two points on an OC curve, $(p_0, 1 - \alpha)$ and (p_1, β), the ratio of the attributes- to variables-based inspection sample sizes is given by

$$\frac{n_{attributes}}{n_{variables}} = \frac{\left(\frac{z_\alpha \sqrt{p_0(1-p_0)} + z_\beta \sqrt{p_1(1-p_1)}}{p_1 - p_0}\right)^2}{\left(\frac{z_\alpha + z_\beta}{z_{p_0} - z_{p_1}}\right)^2}. \qquad (10.80)$$

For the special case of $\alpha = \beta$ and when p_0 and p_1 are both small, say, less than about 10%, this ratio simplifies and approximates to

$$\frac{n_{attributes}}{n_{variables}} \simeq \frac{1}{4}\left(\frac{z_{p_0} - z_{p_1}}{\sqrt{p_1} - \sqrt{p_0}}\right)^2. \qquad (10.81)$$

Example 10.28 Determine the sample size ratio for attributes and variables inspection plans that will accept 95% of the lots with 0.1% defectives and reject 95% of the lots with 0.4% defectives.
Solution: The two points on the OC curve are $(p_0 = 0.001, 1 - \alpha = 0.95)$ and $(p_1 = 0.004, \beta = 0.05)$. Because $\alpha = \beta = 0.05$ and both p_0 and p_1 are relatively small, from Equation 10.81 the ratio of the attributes- to variables-based sample

sizes is approximately

$$\frac{n_{attributes}}{n_{variables}} \simeq \frac{1}{4}\left(\frac{z_{0.001} - z_{0.004}}{\sqrt{0.004} - \sqrt{0.001}}\right)^2$$

$$\simeq \frac{1}{4}\left(\frac{3.090 - 2.652}{\sqrt{0.004} - \sqrt{0.001}}\right)^2$$

$$\simeq 48.$$

So, the attributes plan sample size will have to be about 48 times larger than the variables plan sample size to obtain the same performance for acceptable and rejectable quality levels!

10.4.3.2 Single Sampling Plans for Variables - σ Unknown

The decision to reject $H_0 : p \leq p_0$ in favor of $H_A : p > p_0$ is based on how far the sample mean falls from the specification limit given by

$$t = \frac{\bar{x} - LSL}{s} \tag{10.82}$$

or

$$t = \frac{USL - \bar{x}}{s} \tag{10.83}$$

for lower and upper specification limits, respectively. In both cases, H_0 is rejected if t is less than a critical value k that occurs when $\bar{x}$ is too close to the specification limit. The operating characteristic curve (p, P_A) for the sampling plan is characterized by the noncentral t distribution

$$t_{P_A, df, \phi} = -k\sqrt{n} \tag{10.84}$$

where P_A is the probability of accepting H_0, $df = n - 1$ is the degrees of freedom, and

$$\phi = -z_p\sqrt{n} \tag{10.85}$$

is the t distribution noncentrality parameter. For two specified points on the OC curve, $(p, P_A) = (p_0, 1 - \alpha)$ and $(p, P_A) = (p_1, \beta)$, the same values of k and n apply, so the unique sample size must satisfy the nightmarish condition

$$t_{1-\alpha, n-1, -z_{p_0}\sqrt{n}} = t_{\beta, n-1, -z_{p_1}\sqrt{n}}, \tag{10.86}$$

which must be solved iteratively. The noncentral t distribution is asymmetric about $t = 0$, so be careful to honor the signs in Equation 10.86. After the sample size has been determined, the required value of k can be calculated.

10.4. Acceptance Sampling

Example 10.29 Find the single sampling plan for variables that will accept 95% of the lots with 1% defectives and reject 90% of the lots with 4% defectives. The specification is one-sided and σ is unknown.

Solution: The two points on the OC curve are $(p_0, 1 - \alpha) = (0.01, 0.95)$ and $(p_1, \beta) = (0.04, 0.10)$. From Equation 10.86, the condition that determines the sample size is

$$t_{0.95,n-1,-z_{0.01}\sqrt{n}} = t_{0.10,n-1,-z_{0.04}\sqrt{n}},$$

which is satisfied by $n = 78$ because

$$t_{0.95,77,-20.58} = t_{0.10,77,-15.46} = -17.75.$$

The accept/reject value of k for the test is

$$k = \frac{17.75}{\sqrt{78}} = 2.01.$$

10.4.3.3 ANSI/ASQ Z1.9

The purpose of this section is to describe briefly the ANSI/ASQ Z1.9 acceptance sampling standard [4]. ANSI/ASQ Z1.9 is organized the same way as ANSI/ASQ Z1.4. See Section 10.4.1.3 for a brief review of ANSI/ASQ Z1.4. See the ANSI/ASQ Z1.9 standard or one of the many excellent references if you require more information about its use.

The principles used to design MIL-STD-414, which was the predecessor to ANSI/ASQ Z1.9, are presented in *Mathematical and Statistical Principles Underlying Military Standard 414* [66]. MIL-STD-414 was modified in 1974 to align its sample size code letters and operating characteristic curves with MIL-STD-105, which was the predecessor to ANSI/ASQ Z1.4. Appendix E of ANSI/ASQ Z1.9 compares the match of ANSI/ASQ Z1.4 to ANSI/ASQ Z1.9.

ANSI/ASQ Z1.9 is a sampling system for controlling the defective rate of lots using normally distributed variables data. The standard contains methods for single and double specification limits with variability known and unknown. When the variability is unknown, the standard allows the variability to be estimated by the sample standard deviation or the range, although the range method is rarely used today. Accept/reject decisions for lots are made using either Form 1 (also known as the k method), which is based on estimates of $(USL - \mu)/\sigma$ or $(\mu - LSL)/\sigma$, or Form 2 (also known as the M method), which is based on the lot fraction defective. Form 2 can be used with single or double specification limits. Use of Form 1 is limited to single specification limits. The standard gives exact OC curves for the standard deviation method using Form 1; these are calculated from Equations 10.84 and 10.85. These same curves approximate the performance of the other plans in the standard.

ANSI/ASQ Z1.9 contains single sampling plans linked with switching rules (i.e., sampling schemes) indexed by lot size, inspection level, and acceptable quality level (AQL). The discrete $AQLs$ range from 0.1% to 10% with a multiplicative factor of about $10^{1/5} = 1.585$ between adjacent $AQLs$. The probability of accepting lots of AQL quality increases monotonically from 89% for code letter B to 98% for code letter P. Tables of sample size and acceptance condition are provided for normal, tightened, and reduced inspection for each sampling scheme. The sample size for tightened inspection is equal to the sample size for normal inspection within a sampling scheme. Reduced sampling uses a smaller sample size.

Figure 10.8 shows a plot of $\log(n)$ versus $\log(\overline{N})$ for general inspection level II, Form 1 and Form 2, normal and tightened inspection, where $\overline{N}$ is the mean lot size for the indicated lot size code letter. There is a similar relationship between sample size and lot size for the other plans in the standard.

Example 10.30 Plot the OC curves for the normal, tightened, and reduced sampling plans under ANSI/ASQ Z1.9 using a one-sided specification with Form 1, code letter F, and $AQL = 1\%$.

Solution: The sampling plans determined from the standard were normal ($n = 10, k = 1.72$), tightened ($n = 10, k = 1.84$), and reduced ($n = 4, k = 1.34$). The operating characteristic curves were calculated using Equations 10.84 and 10.85. For example, the OC curve for normal inspection is given by

$$t_{P_A, df, -z_p\sqrt{n}} = -k\sqrt{n}$$
$$t_{P_A, 9, -z_p\sqrt{10}} = -1.72\sqrt{10}$$
$$t_{P_A, 9, -z_p\sqrt{10}} = -5.439.$$

The OC curves are shown in Figure 10.9 and are in excellent agreement with the OC curves in the standard.

10.5 Gage R&R Studies

There are two sources of variation in measurements: real variation in the units being measured and variation associated with noise in the measurement system. If the noise in the measurement system is sufficiently small compared to the tolerance of the units being measured or some other appropriate basis for comparison, then the measurement system is acceptable. If the noise in the measurement system is too large, then the measurement system must be improved by taking appropriate corrective action such as by repairing or replacing the measurement instrument and/or by retraining the operators.

Noise associated with the measurement instrument is referred to as *repeatability* and noise associated with biases between operators is referred to as *reproducibility*. An experiment to quantify repeatability and reproducibility for the

10.5. Gage R&R Studies

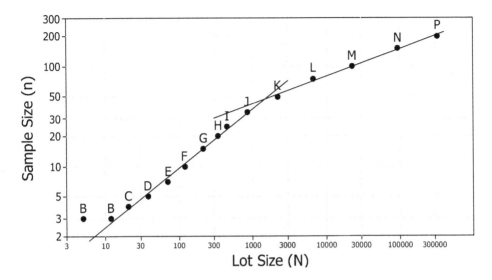

Figure 10.8: Sample size to lot size relationship for ANSI/ASQ Z1.9 sampling plans, level II, normal and tightened inspection.

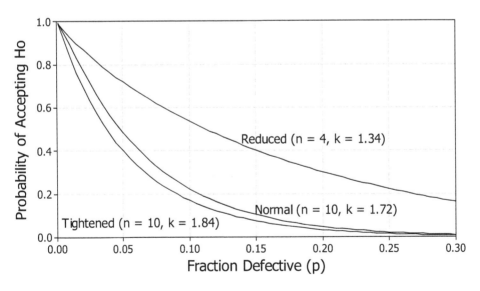

Figure 10.9: OC curves for normal, tightened, and reduced inspection under ANSI/ASQ Z1.4 using the k method with code letter F and $AQL = 1\%$.

Source	df	MS	E(MS)	F
Operator (O)	$o-1$	MS_O	$\sigma_\epsilon^2 + np\sigma_O^2$	$\frac{MS_O}{MS_\epsilon}$
Part (P)	$p-1$	MS_P	$\sigma_\epsilon^2 + no\sigma_P^2$	$\frac{MS_P}{MS_\epsilon}$
Error (ϵ)	$opn - o - p + 1$	MS_ϵ	σ_ϵ^2	
Total	$opn - 1$			

Table 10.5: ANOVA table for the gage RR crossed design.

purpose of validating a measurement system is called a *gage error, gage repeatability and reproducibility,* or *gage R&R study*. In the most common form of gage error study, called the *crossed* design, two or more operators measure the same parts two or more times each in a balanced two-way factorial design. The data are analyzed by random effects ANOVA; then, the standard deviations associated with biases between operators, parts, and the operator by part interaction are estimated using variance components analysis. Measurement repeatability or equipment variation (EV) is estimated from the standard error of the ANOVA:

$$\widehat{EV} = 6\hat{\sigma}_\epsilon. \tag{10.87}$$

Measurement reproducibility or appraiser variation (AV) is estimated from the standard deviations associated with operators and the operator by part interaction:

$$\widehat{AV} = 6\sqrt{\hat{\sigma}_O^2 + \hat{\sigma}_{OP}^2}. \tag{10.88}$$

When the interaction term is not statistically significant, it is usually dropped from the model leaving

$$\widehat{AV} = 6\hat{\sigma}_O. \tag{10.89}$$

The most common gage R&R study acceptance criterion is that, if both $\widehat{EV}$ and $\widehat{AV}$ are less than 10% of the tolerance, then the measurement system is acceptable.

EV and AV are so important for the validation of measurement systems that the precision of their estimates deserves consideration. The purpose of the analysis that follows is to calculate approximate confidence intervals for EV and AV that may be used to develop guidelines for designing gage error studies. The approximate methods used here are not accurate enough for general use, but they are sufficient for the limited purpose of developing gage R&R study design guidelines. See Burdick et al. for a complete analysis of gage error study parameter estimation [10].

The main effects-only ANOVA table for the gage R&R study crossed design with o operators, p parts, and n trials is shown in Table 10.5. The assumption that the operator by part interaction is negligible is not always true, but the assumption simplifies this analysis and renders the same results with respect to developing guidelines for designing gage R&R studies.

10.5. Gage R&R Studies

The point estimates for the operator, part, and error variances are determined by variance components analysis, that is, by solving the simultaneous system of equations obtained by equating the experimental mean squares (MS_i) with their expected mean squares ($E(MS_i)$):

$$MS_i = E(MS_i) \text{ for } i = O, P, \epsilon. \quad (10.90)$$

The resulting point estimates for the variances are

$$\hat{\sigma}_O^2 = \frac{MS_O - MS_\epsilon}{np} \quad (10.91)$$

$$\hat{\sigma}_P^2 = \frac{MS_P - MS_\epsilon}{no} \quad (10.92)$$

$$\hat{\sigma}_\epsilon^2 = MS_\epsilon. \quad (10.93)$$

The one-sided upper $(1 - \alpha)$ 100% confidence interval for σ_ϵ^2 is given by

$$P\left(0 < \sigma_\epsilon^2 < \frac{df_\epsilon MS_\epsilon}{\chi^2_{\alpha, df_\epsilon}}\right) = 1 - \alpha \quad (10.94)$$

where the χ^2 distribution has $df_\epsilon = opn - o - p + 1$ degrees of freedom (see Section 3.1.1.1). Then, from Equations 10.87 and 10.93, the one-sided upper $(1 - \alpha)$ 100% confidence interval for equipment variation is

$$P\left(0 < EV < \sqrt{\frac{df_\epsilon}{\chi^2_{\alpha, df_\epsilon}}} \widehat{EV}\right) = 1 - \alpha \quad (10.95)$$

where $\widehat{EV}$ is the experimental point estimate for the parameter EV.

An exact confidence interval for σ_O^2 does not exist, but an approximate confidence interval can be developed that is sufficient for the purpose of this analysis. The exact one-sided upper $(1 - \alpha)$ 100% confidence interval for $\sigma_\epsilon^2 + np\sigma_O^2$ is given by

$$P\left(0 < \sigma_\epsilon^2 + np\sigma_O^2 < \frac{df_O MS_O}{\chi^2_{\alpha, df_O}}\right) = 1 - \alpha \quad (10.96)$$

where $df_O = o - 1$. If $\sigma_\epsilon^2 \ll np\sigma_O^2$, then an approximate conservative one-sided upper $(1 - \alpha)$ 100% confidence interval for σ_O^2 is

$$P\left(0 < \sigma_O^2 < \frac{df_O MS_O}{np\chi^2_{\alpha, df_O}}\right) = 1 - \alpha. \quad (10.97)$$

With $MS_O \simeq np\hat{\sigma}_O^2$ and $\widehat{AV} = 6\hat{\sigma}_O$, the approximate one-sided upper $(1 - \alpha)$ 100% confidence interval for appraiser variation is then

$$P\left(0 < AV < \sqrt{\frac{df_O}{\chi^2_{\alpha, df_O}}} \widehat{AV}\right) = 1 - \alpha. \quad (10.98)$$

df	$\sqrt{df/\chi^2_{0.05}}$	df	$\sqrt{df/\chi^2_{0.05}}$
1	15.95	15	1.437
2	4.415	20	1.358
3	2.920	25	1.308
4	2.372	30	1.274
5	2.089	40	1.228
6	1.915	50	1.199
7	1.797	60	1.179
8	1.711	80	1.150
9	1.645	100	1.100
10	1.593	300	1.050

Table 10.6: Factors for one-sided 95% upper confidence limits for EV and AV.

The multipliers for calculating approximate 95% one-sided upper confidence limits for EV and AV using Equations 10.95 and 10.98 are shown in Table 10.6.[7] For a gage error study with three operators, ten parts, and two trials, the ANOVA's error degrees of freedom is $df_\epsilon = 49$, so the upper confidence limit for EV will be about 20% larger than its point estimate. For the same study, however, the degrees of freedom for operators will be only $df_O = 2$, so the upper confidence limit for AV will be a factor of 4.415 times its point estimate!

Inspection of the equations for the EV and AV confidence intervals, their associated degrees of freedom, and Table 10.6 lead to the following guidelines for designing gage error studies that will deliver reasonable precision for the EV and AV estimates:

1. The number of parts affects EV but not AV. Use enough parts to challenge the operators.

2. Use as many operators as possible — certainly more than two or three. With seven operators ($df_O = 6$), the upper confidence limit on AV will be about twice the point estimate.

3. The number of trials affects EV but not AV. Two trials are usually sufficient. Three may be a waste of time.

Burdick et al. recommend the use of ten to twenty parts, at least six operators, and two or three trials [10].

---

[7] Factors for two-sided 95% confidence limits are given in Table 3.1.

Chapter 11

Resampling Methods

The analytical methods from the preceding chapters — methods for calculating confidence intervals, performing hypothesis tests, and calculating sample size and power — require knowledge of the distribution of the test statistic being studied. Computer simulation or *resampling* methods relax that requirement, so they can be used to confirm the results of the analytical methods or provide alternative methods of analysis when the analytical methods are invalid. Resampling methods are especially valuable when

- one or more assumptions of the analytical method is violated.

- the distribution of the statistic being studied is not known or is not well behaved.

- the statistical analysis is too complicated to derive an analytical form.

- the experiment must satisfy simultaneous power requirements for two or more responses.

The purpose of this chapter is to demonstrate how *Monte Carlo* and *bootstrap* resampling methods are used to perform sample size and power calculations. In the Monte Carlo method, many new random samples, called *resamples*, are drawn from an assumed parametric distribution. The statistic of interest is calculated from each resample, and the collection of all of those statistics from the many resamples, called the *resampling distribution*, is used to estimate the sampling distribution of the statistic. In the bootstrap method, the resamples are drawn with replacement from an original sample.

There are too many different resampling methods available to cover more than one or two of them here in any detail. See Efron and Tibshirani [20] for more information about these methods and their application to sample size and power calculations.

11.1 Software Requirements

A spreadsheet or statistical software package must have the following capabilities to be useful for power and sample size calculations by resampling methods:

- Create random samples from specified probability distributions.
- Perform random sampling with replacement from a sample data set.
- Calculate and provide access to the required test statistics.
- Repeat a series of operations many times.
- Summarize data in statistical, tabular, or graphical form.

Most statistical software packages support resampling methods.

11.2 Monte Carlo

11.2.1 Sample Size for Confidence Intervals

The following procedure describes how to estimate the sample size to obtain a confidence interval for a population parameter with specified confidence interval half-width using the Monte Carlo method:

1. Identify the parameter to be estimated and the required precision of the estimate.
2. Identify the quantity that will be observed and used to estimate the parameter.
3. Identify the distribution that characterizes the observed quantity.
4. Draw many random samples with the same sample size from the distribution. Calculate the statistic that estimates the parameter for each sample.
5. Use the empirical distribution of the statistic obtained in Step 4 to estimate the sampling distribution of the statistic.
6. Use the $(\alpha/2)\,100\%$ and $(1-\alpha/2)\,100\%$ percentiles to estimate the confidence interval half-width.
7. If the confidence interval half-width does not meet the precision requirements of the experiment, adjust the sample size and return to Step 4.

11.2. Monte Carlo

Example 11.1 How many samples should be drawn from a Poisson population to estimate the mean with 20% precision and 95% confidence? The mean is expected to be $\lambda = 5$.

Solution: The sample size must be sufficient to deliver the following confidence interval for λ:

$$P\left(0.8\widehat{\lambda} < \lambda < 1.2\widehat{\lambda}\right) = 0.95$$
$$P(4 < \lambda < 6) = 0.95$$

where $\widehat{\lambda} = x/n$ and x is the number of counts observed in n sampling units. A MINITAB macro was used to draw 10000 random samples, all of size n, from a Poisson distribution with $\lambda = 5$, and calculate $\widehat{\lambda}$ for each sample. The 2.5^{th} and 97.5^{th} $\widehat{\lambda}$ percentiles were used to estimate the 95% confidence limits. Figure 11.1 shows the Monte Carlo confidence limits as a function of sample size. The sample size that delivers the desired confidence interval width is $n = 19$, which is in good agreement with the sample size given by Equation 5.11:

$$\begin{aligned} n &= \frac{1}{\lambda}\left(\frac{z_{\alpha/2}}{\delta}\right)^2 \\ &= \frac{1}{5}\left(\frac{1.96}{0.2}\right)^2 \\ &= 20. \end{aligned}$$

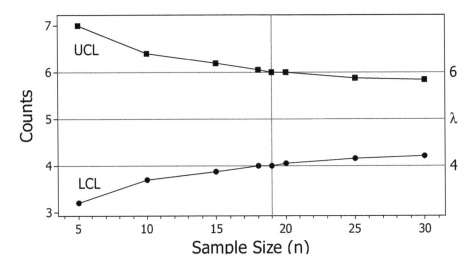

Figure 11.1: Confidence limits for the Poisson mean versus sample size.

Example 11.2 Use the Monte Carlo method to confirm the answer to Example 9.25, that samples of size $n = 31$ are required to estimate, with 95% confidence, the exponential–exponential interference failure rate with 50% precision. Consider wide ranges of exponential distribution means and ratios of the means.

Solution: A MINITAB macro was written that draws random samples of size $n = 31$ for exponential load and strength distributions, where the load distribution has mean μ_L, where μ_L comes from a uniform distribution with $1 \leq \mu_L \leq 10$, and the strength distribution has mean $\mu_S = k\mu_L$, where k comes from a uniform distribution with $3 \leq k \leq 10$. The macro loops 1000 times, collecting the ratio of the empirical interference failure rate $\hat{f}$ to the known failure rate f from each pass. Figure 11.2 shows the histogram of $\hat{f}/f$. About 97% of the observed $\hat{f}$ values fall within ±50% of their known f values, which is consistent with the intended 95% confidence level. The agreement between the two methods improves for larger values of the lower limit of the ratio of the exponential means.

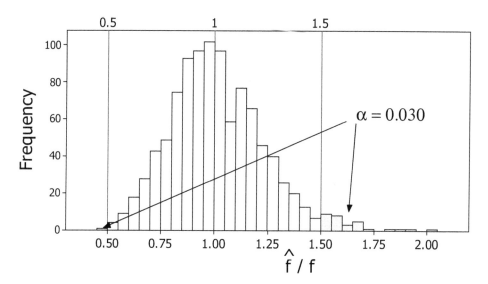

Figure 11.2: Monte Carlo distribution of the ratio of observed to known interference failure rates.

11.2.2 Power and Sample Size for Hypothesis Tests

The following procedure can be used to perform power and sample size calculations for a hypothesis test by the Monte Carlo method:

1. Formulate the hypotheses for the parameter being studied.

2. Identify the variable to be observed and used to estimate the parameter.

3. Use a random number generator to draw many (≥ 1000) resamples from the H_0 distribution using the same sample size for each resample.

4. Calculate the test statistic for each resample. The collection of all of these statistics from all of the resamples is the resampling distribution.

5. From the resampling distribution under H_0, determine the critical values of the statistic that distinguish the accept and reject regions such that the desired level of protection against type I errors is provided.

6. Use a random number generator to draw many (≥ 1000) resamples from the H_A distribution.

7. The power of the test is the fraction of the statistics calculated from the resamples that fall in the H_0 rejection region determined in Step 5

8. If the power obtained in Step 7 is not satisfactory, adjust the sample size and repeat the process starting from Step 3.

Example 11.3 An experiment is proposed to study the defective rate (p) of a process using a sample of size $n = 300$ units. The process is considered to be acceptable when $p \leq 0.01$, but $p \geq 0.03$ is unacceptable. Determine the acceptance number and power for the sampling plan.
Solution: The hypotheses to be tested are $H_0 : p \leq 0.01$ versus $H_A : p > 0.01$. The decision to reject H_0 or not is based on the observed number of defectives x in a sample of size $n = 300$. The critical value of x that distinguishes the accept and reject regions should be chosen such that $\alpha \leq 0.05$ for $p = 0.01$. To determine this critical value, MINITAB was used to select 10000 random samples from a binomial population with $n = 300$ and $p = 0.01$. The frequencies and cumulative percentages by x are shown in the H_0 columns of Table 11.1. From the $CumPct$ column, the critical value of x must be set to $x_{A/R} = 6.5$ to provide good protection against type I errors ($\alpha = 1 - 0.9692 = 0.0308$). The frequency statistics under H_A in the Figure were created by selecting 10000 random samples from a binomial population with $n = 300$ and $p = 0.03$. From the $CumPct$ column, the type II error probability relative to $x_{A/R} = 6.5$ is $\beta = 0.206$, so the power to reject H_0 when $p = 0.03$ is $\pi = 1 - \beta = 0.794$. This is in excellent agreement with the power calculated for the one-sample proportion test (see Section 4.1.2), which is $\pi = 0.797$.

	H0		HA		
x	Count	CumPct	Count	CumPct	
0	452	4.52	2	0.02	|
1	1493	19.45	8	0.10	|
2	2253	41.98	49	0.59	|
3	2274	64.72	158	2.17	Accept H0
4	1741	82.13	322	5.39	|
5	1011	92.24	605	11.44	|
6	468	96.92	916	20.60	|
7	198	98.90	1161	32.21	|
8	69	99.59	1353	45.74	|
9	30	99.89	1337	59.11	|
10	8	99.97	1221	71.32	|
11	2	99.99	984	81.16	|
12	1	100.00	726	88.42	|
13			469	93.11	Reject H0
14			302	96.13	|
15			186	97.99	|
16			117	99.16	|
17			55	99.71	|
18			17	99.88	|
19			6	99.94	|
20			1	99.95	|
21			5	100.00	|
N	10000		10000		

Table 11.1: Monte Carlo power calculation for test for one sample proportion.

Example 11.4 Tukey's quick test is a nonparametric two-sample test for location that is easy to perform using dotplots. The samples must be comparable in size and they must be slipped, that is, one sample must have the largest observation and the other sample must have the smallest observation. The test statistic, T, is the number of slipped or nonoverlapping points. The Tukey test rejects $H_0 : \mu_1 = \mu_2$ when $T \geq 7$. For example, in Figure 11.3 $T = 7 + 10 = 17$, so there is sufficient evidence to reject H_0.

Use the Monte Carlo method to determine the power of Tukey's quick test as a function of effect size and sample size assuming that the two populations are normal and homoscedastic.

Solution: A MINITAB macro was written to draw 1000 random samples of equal sample size from two independent homoscedastic normal populations and count the number of times that H_0 was rejected by the Tukey quick test. Figure 11.4 shows that the power of the test is low for all sample sizes until the difference between the means is greater than 1.5 to 2.5 standard deviations. When the sample size is $n \geq 20$, Tukey's quick test has power $\pi \geq 0.90$ for differences between the means of 1.5 standard deviations or greater.

11.2. Monte Carlo

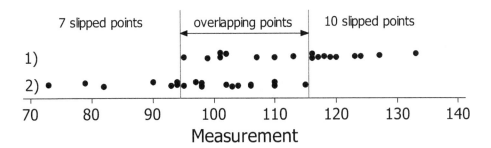

Figure 11.3: Dot plots for use with Tukey's quick test for location.

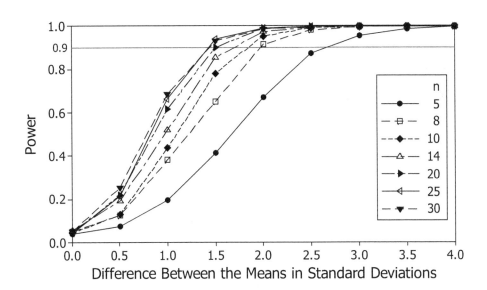

Figure 11.4: Power versus effect size for Tukey's quick test.

11.3 Bootstrap

The bootstrap method is a variation of the Monte Carlo method. The *nonparametric* bootstrap resamples from the original sample data. The *parametric* bootstrap resamples from a parametric distribution fitted to the original sample data. The parametric bootstrap is more stable than the nonparametric bootstrap for some statistics, such as percentiles, especially when the sample size is small; however, the parametric bootstrap requires more assumptions than the nonparametric bootstrap. See Efron and Tibshirani [20] for details about bootstrap methods.

11.3.1 Bootstrap Confidence Intervals

There are simple methods and there are complicated methods for calculating a confidence interval for a population parameter using the bootstrap. The complicated methods are necessary when the bootstrap distribution is skewed or when it is biased, that is, when the mean of the bootstrap distribution is different from the mean of the sample. Two methods for calculating a confidence interval by the bootstrap method are presented: the first assumes that the bootstrap distribution is unbiased and symmetric; the second corrects for bias and skew.

When the bootstrap distribution is symmetric and the bias is small, the confidence interval for the parameter θ is given by the *percentile method*:

$$P\left(\widehat{\theta}^*_{\alpha/2} < \theta < \widehat{\theta}^*_{1-\alpha/2}\right) = 1 - \alpha \tag{11.1}$$

where $\widehat{\theta}^*_{\alpha/2}$ and $\widehat{\theta}^*_{1-\alpha/2}$ are the $(\alpha/2)\,100^{th}$ and $(1-\alpha/2)\,100^{th}$ percentiles of the bootstrap distribution.

When the bootstrap distribution is biased and/or skewed, the confidence interval for θ by the *reflected percentile* or *basic* method is given by

$$P\left(2\widehat{\theta} - \widehat{\theta}^*_{1-\alpha/2} < \theta < 2\widehat{\theta} - \widehat{\theta}^*_{\alpha/2}\right) = 1 - \alpha \tag{11.2}$$

where $\widehat{\theta}$ is the value of the statistic determined from the original sample.

11.3.2 Sample Size for Bootstrap Confidence Intervals

A bootstrap confidence interval is determined by resampling from the original sample using samples of the same size as the original sample. Sample size calculations for bootstrap confidence intervals relax the constraint on sample size, so that confidence interval width can be studied as a function of sample size. The correct sample size for an experiment is the one that delivers the desired confidence interval width, which must be determined by iteration. The confidence interval half-widths for the percentile and reflected percentile methods are both

$$\delta = \frac{\widehat{\theta}^*_{1-\alpha/2} - \widehat{\theta}^*_{\alpha/2}}{2}, \tag{11.3}$$

so the two methods give the same sample size.

11.3. Bootstrap

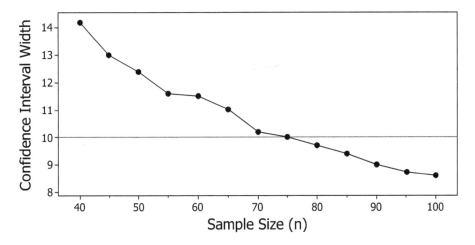

Figure 11.5: Bootstrap confidence interval width versus sample size.

Example 11.5 The following yield strength values (in thousands of psi) for a material were obtained in a pilot study: {56, 23, 25, 68, 35, 31, 13, 15, 48, 37, 57, 69, 50, 76, 50, 19, 88, 33, 10, 21}. What sample size is required to estimate, with 95% confidence, the mean yield strength to within ±5000 psi?

Solution: The bootstrap percentile confidence interval width was studied as a function of sample size using 1000 resamples for sample sizes from $n = 30$ to 100. Figure 11.5 shows the confidence interval width, given by

$$2\delta = \widehat{\theta}^*_{0.975} - \widehat{\theta}^*_{0.025},$$

versus sample size. The sample size required to obtain the desired precision of the estimate is $n = 75$.

11.3.3 Power and Sample Size for Bootstrap Hypothesis Tests

The procedure for calculating the power for a bootstrap hypothesis test is analogous to the procedure used in the analytical methods, except that the bootstrap method uses the bootstrap distribution of the statistic instead of the parametric sampling distribution.

The procedure for calculating the power using the bootstrap method is to

1. obtain a representative sample from the population.

2. create the bootstrap distribution of the statistic $\widehat{\theta}^*$ under the null hypothesis $H_0 : \theta = \theta_0$.

3. determine the critical accept/reject values for H_0 from the $(\alpha/2)\,100^{th}$ and $(1-\alpha/2)\,100^{th}$ percentiles of the H_0 bootstrap distribution.

4. create the bootstrap distribution of the statistic $\widehat{\theta}^*$ under the alternative hypothesis $H_A : \theta \neq \theta_0$ or another appropriate alternative.

5. determine the power to reject H_0 for the specified H_A is given by the fraction of the bootstrap distribution under H_A that falls in the rejection region for H_0.

The sample size for a bootstrap hypothesis test is determined by varying the sample size until the desired power value is obtained.

Example 11.6 The following data were obtained from a pilot study: {56, 48, 44, 62, 50, 47, 49, 57, 48, 55, 96, 47, 46, 47, 49, 72, 46, 61}. Determine the sample size required to obtain 90% power to reject $H_0 : \mu = 50$ when $\mu = 52$. The population is not normal, so assume that the experimental data will be analyzed using the bootstrap method.

Solution: If the population distribution is normal, the one-sample Student's t test would be an appropriate method of analysis. Because the normality assumption is not satisfied, the preferred method of analysis using the bootstrap method uses the analogous bootstrap-t distribution given by

$$t^* = \frac{\bar{x}^* - \mu_0}{s^*/\sqrt{n}} \tag{11.4}$$

where $\bar{x}^*$ and s^* are the mean and standard deviation of bootstrap samples. Figure 11.6 shows the bootstrap distributions of t^* with samples of size $n = 18$ under $H_0 : \mu = 50$ and $H_A : \mu = 56$, where the sample data were shifted to the appropriate population means before bootstrapping using transformations of the form

$$x_i' = x_i - \bar{x} + \mu.$$

From the bootstrap-t distribution under H_0, the acceptance interval for H_0 is $-4.29 \leq t^* \leq 1.62$. The acceptance interval is skewed because the original sample is skewed. From the bootstrap distribution under H_A, the power is $\pi = 0.802$, which does not meet the 90% power requirement.

Figure 11.7 shows the bootstrap-t test power versus sample size for samples from size $n = 16$ to 30. (Figure 11.6 was constructed from 1000 bootstrap samples. Each point in Figure 11.7 was constructed from 10000 bootstrap samples to reduce the noise in the power versus sample size plot.) The sample size required to obtain 90% power to reject H_0 when $\mu = 56$ is $n = 23$.

11.3. Bootstrap

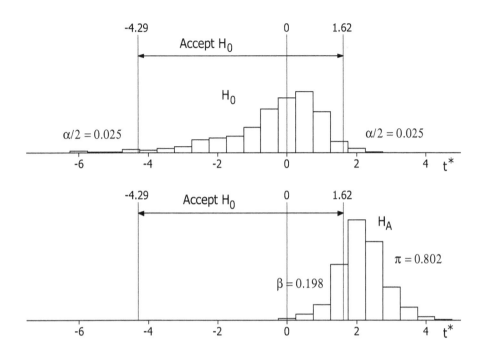

Figure 11.6: Bootstrap-t distributions under $H_0 : \mu = 50$ and $H_A : \mu \neq 50$.

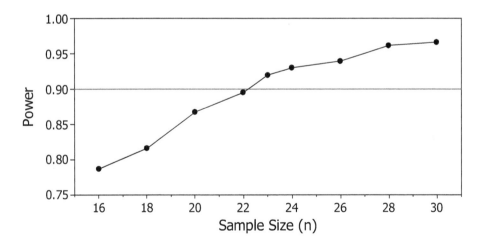

Figure 11.7: Power versus sample size for one-sample test for location using the bootstrap-t method.

Appendix A

Notation

The following notations are used throughout the book, but in certain circumstances it is necessary to deviate from these conventions.

- The symbol x is used to indicate an independent variable, and y is used to indicate a dependent variable.
- The population mean and standard deviation are μ and σ, respectively.
- When there is a chance of ambiguity or confusion, a subscript on a parameter indicates the distribution associated with the parameter. For example, μ_x is the mean of the x distribution and $\mu_{\bar{x}}$ is the mean of the $\bar{x}$ distribution.
- Statistics that estimate parameters may be indicated using the hat notation (e.g., $\hat{\mu}$ and $\hat{\sigma}$) or with common alternatives (e.g., $\bar{x}$ and s). In general, the symbol for the sample standard deviation, s, is used in the context of calculating a test statistic from sample data and the symbol $\hat{\sigma}$ is used to indicate an estimate for σ for a sample size calculation before data are available.
- Sample size is indicated with the symbol n.
- In a designed experiment, the number of replicates of the experiment design is indicated with the symbol r.
- In two-sample problems, the sample size ratio n_1/n_2 may not require specifying the individual n_1 and n_2 values.
- The population proportion is represented by the symbol p or θ.
- The odds are indicated by $O = p/(1-p)$.
- In comparisons of two independent proportions, the risk ratio or relative risk is indicated by $RR = p_1/p_2$ and the odds ratio is indicated by $OR = (p_1/(1-p_1))/(p_2/(1-p_2))$.

- α and β are generally the type I and II errors, respectively, unless indicated otherwise.

- When the value of α is not specified in a problem statement, the value may be assumed to be $\alpha = 0.05$.

- The symbol π indicates the statistical power, i.e., the probability of rejecting H_0 when it is false. The power is always the complement of the type II error rate, i.e., $\pi = 1 - \beta$, and the power is always associated with an effect size, usually indicated with the symbol δ or Δ followed by the parameter being tested, e.g., $\Delta\mu$.

- δ is the confidence interval half-width or the effect size associated with a power value π in a hypothesis test.

- $P()$ is a probability function where the argument is usually expressed in the form of a compound less-than inequality. The argument may also contain the values of relevant parameters, e.g.: $P(a < x < b; \mu, \sigma)$. When x is a continuous variable, the $<$ signs may be replaced with $\leq$ signs without affecting the meaning of the inequality.

- $\Phi()$ is the cumulative normal probability distribution function where the argument is expressed in the form of a compound less-than inequality or only a z value, e.g.: $\Phi(-1.96 < z < 1.96) = 0.95$ and $\Phi(z = -1.96) = 0.025$.

- The amplitude of the normal probability distribution, that is, its probability density function, is indicated with $\varphi(x; \mu, \sigma)$.

- Subscripts on the symbols representing symmetric probability distributions (e.g., the normal and Student's t distributions) indicate a one-tail area. These values are taken as positive quantities (e.g., $z_{0.025} = 1.96$) and the necessary sign is determined in context.

- Subscripts on the symbols representing asymmetric probability distributions (e.g., χ^2_p and F_p) indicate left tail areas. Degrees of freedom are not generally included in the subscript.

- The F distribution's numerator and denominator degrees of freedom are called out explicitly or indicated with the symbols df_1 and df_2, respectively.

- In the context of acceptance sampling, the commonly used symbol P_A is used to indicate the probability of accepting H_0. When H_0 is false, $P_A = \beta = 1 - \pi$.

- In a hypothesis test for one population, the value of the parameter being tested is indicated with a 0 subscript under H_0 and with a 1 subscript under H_A.

- Degrees of freedom values, such as for Student's t, χ^2, and F distributions, are indicated with the symbols ν or df.

- In power calculations for two-tailed tests, only the contribution of the power in the direction of the shift is considered. The contribution from the other tail is assumed to be negligible, which will not be the case for small shifts.

- $E\,()$ is used to indicate the expectation value function.

- The symbol ϕ is used to indicate a probability distribution's noncentrality parameter.

- The symbol $\simeq$ is used to indicate *approximately equal to*.

- A dotted subscript indicates summation over the dotted indexed variable, e.g.:
$$x_{\bullet j} = \sum_{i=1}^{n} x_{ij}$$

- A horizontal bar over a random variable indicates the sample mean. If the random variable is singly indexed, the subscript may be dropped or retained for clarity, e.g.:
$$\bar{x} = \bar{x}_{\bullet} = \frac{1}{n} \sum_{i=1}^{n} x_i$$

- A prime (') superscript indicates the use of a variable transformation, e.g.: $\mu' = \sqrt{\mu}$.

- For attribute random variables, the critical upper acceptance limit for defectives or defects under H_0 is indicated with the symbol c.

- The critical value of a test statistic that distinguishes the accept and reject regions for the null hypothesis may be denoted by the test statistic with the A/R subscript. For example, in the one-sided test of $H_0 : \mu = \mu_0$ versus $H_A : \mu > \mu_0$ the critical value of the sample mean that defines the accept/reject boundary is $\bar{x}_{A/R} = \mu_0 + z_\alpha \sigma_{\bar{x}}$.

- In some cases the probability density function is indicated by $pdf\,()$, where the function's arguments specify the value of the random variable and the appropriate parameters. The function being invoked is indicated in the problem statement. For example, the amplitude of the χ^2 distribution is indicated as $pdf\,(\chi^2; \nu)$.

- In some cases, the cumulative distribution function is indicated by $cdf\,()$ and the inverse cumulative distribution function is indicated by $invcdf\,()$ where the functions' arguments are the relevant limits of the random variable and the appropriate parameters.

- Paired equations for upper and lower control limits and upper and lower specification limits may be combined into a single expression using the / and ± symbols. For example: $USL/LSL = \mu \pm 4\sigma$.

- UCL/LCL may be used to indicate upper and lower control limits for an SPC control chart or upper and lower confidence limits for a confidence interval.

- A vertical line with a subscript and/or superscript that closes an opening left parenthesis indicates the evaluation of the expression at those values. For example: $(x|_3 = 3$ and $(y+2|_1^3 = 5 - 3 = 2$.

- A numerical value with a positive sign (+) or negative sign (−) superscript indicates a value infinitesimally larger or smaller, respectively, than that numerical value. For example, 0^+ is a positive value infinitesimally larger than 0 and 0^- is a negative value infinitesimally smaller than 0.

- The mean square associated with the error in ANOVA is indicated by MS_ϵ. If a variable A in an experiment is a random variable, then its mean square is MS_A and the error term used to test for an A effect is $MS_{\epsilon(A)}$, which will be different from MS_ϵ.

- In a one-way or multi-way classification experiment design, the biases of levels of a variable from the grand mean are indicated with α_i or some similar symbol. For example, the model for the observations (y_{ij}) in a one-way classification design may be written $y_{ij} = \mu + \alpha_i + \epsilon_{ij}$ where ϵ_{ij} represents the error. The symbol α_i is used whether the levels of the classifying variable are fixed or random. This notation differs from the occasional practice of using greek symbols to indicate biases for fixed variables and arabic symbols to indicate biases for random variables.

- The natural logarithm with base e is indicated by $\ln()$. The common logarithm with base 10 is indicated by $\log()$.

- $\min()$ is a mathematical function that returns the smallest value in the list of inputs. For example, $\min(3, 14, 2, 11) = 2$.

- For resampling methods such as the bootstrap, a population parameter is indicated with a greek letter, such as θ, the sample statistic that estimates it is $\hat{\theta}$, and the value of the statistic determined by resampling the original data set is $\hat{\theta}^*$.

Appendix B

Glossary

The purpose of this appendix is to present a short glossary of terms that might be helpful to readers who are not fully knowledgeable in the language of statistical methods and power and sample size calculations. See the references or online resources for more detailed explanations of these terms.

- Acceptable quality level (AQL) – In acceptance sampling, a low value of the long term average defective rate that should have a high probability of being accepted.

- Appraiser variation (AV) – In gage error studies, a measure of the variation of biases between operators.

- Binary – Having two states

- Central probability distribution – The distribution of a statistic under the null hypothesis

- Confidence interval – A statement about the range of probable values of a population parameter.

- Continuity correction – A correction applied to a discrete random variable when its probability is approximated with a continuous probability distribution.

- Covariate – A quantitative input variable in an experiment that is included in the predictive model for the response.

- Cumulative distribution function – A function that gives the probability that a value is less than or equal to a specified value.

- Degrees of freedom – A discrete measure of the amount of information available to estimate a parameter.

- Dichotomous – Having two states, as in a binary or pass/fail variable.
- Effect size – The difference between the null and true parameter values in a power or sample size calculation.
- Equipment variation (EV) – In gage error studies, a measure of the noise associated with the measurement device.
- Expected value – The probability-weighted sum (discrete random variables) or probability density-weighted integral (continuous random variables) of all possible values of the random variable.
- Experiment – Any activity that involves the collection, analysis, and interpretation of data for the purpose of learning about or managing a process.
- Fixed effect – A discrete independent variable that has all of its possible levels included in an experiment.
- Heteroscedastic – Two distributions that have different standard deviations
- Homoscedastic – Two distributions that have the same standard deviation
- Hypothesis test – A method of using sample data to determine which of two complementary hypotheses about the value of a population parameter is likely to be true.
- Inverse cumulative distribution function – A function that gives the value for which the probability is equal to a specified value.
- Limit of practical equivalence – The largest allowable difference between two values such that the values are still considered to be practically equivalent.
- Lot tolerance percent defective ($LTPD$) – In acceptance sampling, a high value of defective rate in individual lots that should have a low probability of being accepted.
- Monte Carlo – A computer simulation method used to approximate the solution to a complicated problem that can't be solved analytically.
- Noncentral probability distribution – The distribution of a statistic under the alternative hypothesis.
- Noncentrality parameter – A parameter that measures the degree of noncentrality in a noncentral probability distribution. A central distribution has noncentrality parameter equal to 0.
- Omnibus test – A single statistical test that replaces many simultaneous tests (e.g.: the ANOVA F test is an omnibus test for differences between treatments, serving in place of many two-sample tests)

- One-tailed interval – A confidence interval with only one bound, either upper or lower, on the value of the population parameter.

- One-tailed test – In hypothesis testing, an alternative hypothesis that considers a shift in the parameter value in one direction, e.g.: $H_0 : \mu \leq \mu_0$ versus $H_A : \mu > \mu_0$ and $H_0 : \mu \geq \mu_0$ versus $H_A : \mu < \mu_0$.

- Orthogonal – An indicator that the variables in an experiment are independent of each other.

- Parameter – A measure of the location or dispersion for a population.

- Population – The complete set of all possible values of a variable.

- Power – The probability of rejecting the null hypothesis when it is really false.

- Practical significance – A difference too large to ignore.

- Precision – 1) The half-width of a confidence interval, and 2) synonym for the repeatability or error in a measurement.

- Probability density function – The amplitude of the probability distribution of a continuous random variable. For example, the probability density function of the normal distribution is the well known bell-shaped curve.

- Random effect – A discrete independent variable that has a sample of its many possible levels included in an experiment.

- Rejectable quality level (RQL) – In acceptance sampling, a high value of the long term average defective rate that should have a low probability of being accepted.

- Repeatability – Synonym for *measurement precision*; usually associated with inherent instrument errors.

- Reproducibility – Measurement errors associated with biases between operators, instruments, methods, etc.

- Resampling – A method to estimate the sampling distribution of a statistic by drawing many pseudosamples of the same size with replacement from the original sample.

- Sample – A subset of units drawn from a population, usually in a random or planned manner so that the sample is representative of the population from which it was drawn.

- Sampling distribution – The theoretical distribution of a statistic obtained by considering all possible values of the statistic.

- Sigmoidal – A term describing an S-shaped cumulative distribution function.

- Skewness – A measure of asymmetry in a probability distribution function. The normal distribution is not skewed, but the χ^2 and F distributions are skewed.

- Statistic – A measure of location, dispersion, or another summary measure determined from sample data.

- Statistics – 1) The plural of *statistic* or 2) the study of statistics.

- Statistically significant – A difference too large to be due to random chance.

- Symmetric – A probability distribution with mirror image shape about the mean.

- Transcendental – A term describing an equation that cannot be solved explicitly for a variable.

- Two-tailed interval – A confidence interval with both upper and lower bounds on the value of the population parameter.

- Two-tailed test – In hypothesis testing, an alternative hypothesis that allows the null hypothesis to be rejected for parameter shifts to higher and lower values (e.g.: $H_0 : \mu = \mu_0$ versus $H_A : \mu \neq \mu_0$).

- Type I error – The probability of rejecting the null hypothesis when it is really true.

- Type II error – The probability of accepting the null hypothesis when it is really false.

- Universe – A synonym for *population*, the word *universe* has been almost completely replaced with *population*.

Appendix C

Greek Alphabet

Name	Lower Case	Upper Case	Latin
alpha	α	A	a
beta	β	B	b
gamma	γ	Γ	g
delta	δ	Δ	d
epsilon	ϵ or ε	E	e
zeta	ζ	Z	z
eta	η	H	h
theta	θ	Θ	y
iota	ι	I	i
kappa	κ	K	k
lambda	λ	Λ	l
mu	μ	M	m
nu	ν	N	n
xi	ξ	Ξ	x
pi	π	Π	p
rho	ρ	P	r
sigma	σ	Σ	s
tau	τ	T	t
upsilon	υ	Υ	u
phi	ϕ or φ	Φ	f
chi	χ	X	q
psi	ψ	Ψ	c
omega	ω	Ω	w

Appendix D

Probability Distributions

This appendix provides a short introduction to the fundamental probability distributions and distribution notations used in this book. Infrequently used distributions are described in the sections where they are introduced.

D.1 Noncentral Distributions

In hypothesis testing, the familiar t, χ^2, and F distributions under H_0 are called the *central* distributions. However, under H_A these distributions are biased with respect to the central distributions, so they are referred to as *noncentral* distributions. The noncentral distributions are characterized by a *noncentrality parameter*, indicated with the symbol ϕ, which is a measure of the discrepancy between H_0 and H_A. The central distributions are special cases of their respective noncentral distributions with $\phi = 0$.

When a noncentral probability distribution is required, the symbol for or value of the noncentrality parameter is included as a subscript following the usual area-under-the-curve subscript on the distribution symbol. When the noncentrality parameter is omitted, the central distribution is used. For example, in the power calculation for one-way fixed-effects ANOVA, the condition that determines the power is

$$F_{1-\alpha} = F_{1-\pi,\phi} \tag{D.1}$$

where $F_{1-\alpha}$ is the F value of the central F distribution that defines the accept/reject boundary for H_0 and $F_{1-\pi,\phi}$ is the F value of the noncentral F distribution with noncentrality parameter ϕ. In this example, α is the type I error rate, π is the power to reject H_0 when H_A is true, and the central and noncentral F distributions have the same numerator and denominator degrees of freedom that are determined in the context of the problem.

D.2 Hypergeometric Distribution

The probability of finding x successes in n trials drawn randomly and without replacement from a population of size N that contains S successes and $N - S$ failures is given by the hypergeometric probability distribution

$$h(x; S, N, n) = \frac{\binom{S}{x}\binom{N-S}{n}}{\binom{N}{n}} \qquad (D.2)$$

where the $\binom{Q}{q}$ notation indicates the combination operation

$$\binom{Q}{q} = \frac{Q!}{q!(Q-q)!} \qquad (D.3)$$

and $q!$ indicates the factorial of q

$$q! = q(q-1)(q-2)\cdots(3)(2)(1). \qquad (D.4)$$

An equivalent formula for the hypergeometric distribution is given by

$$\begin{aligned} h(x; S, N, n) &= h(x; n, N, S) \\ &= \frac{\binom{n}{x}\binom{N-n}{S-x}}{\binom{N}{S}}. \end{aligned} \qquad (D.5)$$

The mean of the hypergeometric distribution is

$$\mu = nS/N \qquad (D.6)$$

and its standard deviation is

$$\sigma = \sqrt{n\left(\frac{S}{N}\right)\left(1 - \frac{S}{N}\right)\left(1 - \frac{n-1}{N-1}\right)}. \qquad (D.7)$$

There are two conditions under which the hypergeometric distribution may be approximated with the binomial distribution:

- When the sample size is small compared to the lot size, say $n/N < 0.1$, then

$$h(x; S, N, n) \simeq b\left(x; n, p = \frac{S}{N}\right). \qquad (D.8)$$

- When the number of successes in the population is small compared to the size of the population, say $S/N < 0.1$, then

$$h(x; S, N, n) \simeq b\left(x; S, p = \frac{n}{N}\right). \qquad (D.9)$$

D.3 Binomial Distribution

The probability of obtaining x successes in n trials when the probability of a success on any trial is a constant value p is given by the binomial probability distribution

$$b(x; n, p) = \binom{n}{x} p^x (1-p)^{n-x}. \tag{D.10}$$

The mean of the binomial distribution is

$$\mu = np \tag{D.11}$$

and its standard deviation is

$$\sigma = \sqrt{np(1-p)}. \tag{D.12}$$

The binomial distribution is skewed to the right when $p < 0.5$, is symmetric when $p = 0.5$, and is skewed to the left when $p > 0.5$. The distribution of sample binomial proportions $\hat{p} = x/n$ may be transformed to approximate normality using the arcsine transformation

$$p' = 2 \arcsin(\sqrt{p}), \tag{D.13}$$

which has approximate variance

$$\sigma_{p'} \simeq \frac{1}{\sqrt{n}}. \tag{D.14}$$

There are many approximations for the binomial distribution, but two of the common ones are these:

- When $n > 100$ AND $np < 10$,

$$b(x; n, p) \simeq Poisson(x; \lambda = np). \tag{D.15}$$

- When $np > 5$ AND $n(1-p) > 5$,

$$b(x = a; n, p) \simeq \Phi\left(a - \frac{1}{2} < x < a + \frac{1}{2}; \mu = np, \sigma = \sqrt{np(1-p)}\right). \tag{D.16}$$

Approximate values of the cumulative binomial distribution may also be estimated from Larson's nomogram (see Appendix E.1).

D.4 Poisson Distribution

The probability of observing x events in an area of opportunity called the *sampling unit*, when the mean number of events per sampling unit is λ, is given by the Poisson probability distribution:

$$Poisson\,(x;\lambda) = \frac{\lambda^x e^{-\lambda}}{x!} \quad \text{for } x = 0, 1, \ldots. \tag{D.17}$$

The Poisson population mean is

$$\mu_x = \lambda \tag{D.18}$$

and its standard deviation is

$$\sigma_x = \sqrt{\lambda}. \tag{D.19}$$

The Poisson distribution is always skewed to the right, that is, its left tail is always suppressed and its right tail is always exaggerated.

When the Poisson mean is sufficiently large, say $\lambda > 20$, then the Poisson distribution may be approximated with the continuity-corrected normal approximation

$$Poisson\,(x = a; \lambda) \simeq \Phi\left(a - \frac{1}{2} < x < a + \frac{1}{2}; \mu = \lambda, \sigma = \sqrt{\lambda}\right). \tag{D.20}$$

When the Poisson mean is moderate to large, say $\lambda > 10$, then the distribution of square root-transformed Poisson counts $x' = \sqrt{x}$ is approximately normal with standard deviation $\sigma' \simeq \frac{1}{2}$, so

$$Poisson\,(x = a; \lambda) \simeq \Phi\left(\sqrt{a - \frac{1}{2}} < x < \sqrt{a + \frac{1}{2}}; \mu = \sqrt{\lambda}, \sigma = \frac{1}{2}\right). \tag{D.21}$$

The Poisson distribution has the property that the sum of two independent Poisson random variables is also Poisson with mean equal to the sum of the original two distribution means. That is, if x_1 is Poisson with mean λ_1 and x_2 is Poisson with mean λ_2, then $x_1 + x_2$ is Poisson with mean $\lambda_1 + \lambda_2$.

The cumulative Poisson distribution may be expressed in terms of the χ^2 distribution:

$$\chi^2_{1-p,2(X+1)} = 2\lambda \text{ where } p = \sum_{x=0}^{X} Poisson\,(x;\lambda). \tag{D.22}$$

Appendix E.2 shows a graph of cumulative Poisson probabilities given by

$$Poisson\,(c;\lambda) = \sum_{x=0}^{c} \frac{\lambda^x e^{-\lambda}}{x!} \tag{D.23}$$

for $1 \leq \lambda \leq 20$ and $0 \leq c \leq 20$.

D.5 Normal Distribution

The probability density function of the normal or bell-shaped curve for a random variable x with mean μ and standard deviation σ is given by

$$\varphi(x; \mu, \sigma) = \frac{1}{\sqrt{2\pi}\sigma} e^{-\frac{1}{2}\left(\frac{x-\mu}{\sigma}\right)^2}. \tag{D.24}$$

Then the cumulative normal probability that x falls between a and b is

$$\Phi(a < x < b; \mu, \sigma) = \int_a^b \varphi(x; \mu, \sigma)\, dx. \tag{D.25}$$

Practical normal distribution problems are usually solved by transforming from x units into z units using the z transform

$$z = \frac{x - \mu}{\sigma}, \tag{D.26}$$

so

$$\Phi(a < x < b; \mu, \sigma) = \Phi(z_a < z < z_b) \tag{D.27}$$

where $z_a = (a - \mu)/\sigma$ and $z_b = (b - \mu)/\sigma$. Values of $\Phi(-\infty < z' < z)$ as a function of z are presented in Appendix E.3.

D.6 Student's t Distribution

Student's t distribution is used to characterize the distribution of sample means $\bar{x}$ when the population standard deviation σ must be estimated with the sample standard deviation s. For samples of size n drawn from a normal population with mean μ_0 and standard deviation σ, the sampling distribution of

$$t = \frac{\bar{x} - \mu_0}{s/\sqrt{n}} \tag{D.28}$$

follows Student's t distribution with $\nu = n - 1$ degrees of freedom. As the sample size becomes large, say $n > 30$, the t distribution approaches the normal distribution. The t distribution is robust to deviations from normality of the population being sampled, especially as the sample size becomes large. Critical values of t_p for several values of p are shown in Appendix E.4 as a function of the degrees of freedom.

In a hypothesis test of $H_0 : \mu = \mu_0$ versus any alternative, the distribution of t given by Equation D.28 follows the familiar Student's t or *central t distribution* when the population being sampled has mean μ_0. If the population mean is μ_1, where $\mu_1 \neq \mu_0$, then the distribution of t given by Equation D.28 follows the noncentral t distribution with noncentrality parameter:

$$\phi = \frac{\Delta\mu}{\sigma_\epsilon/\sqrt{n}} \tag{D.29}$$

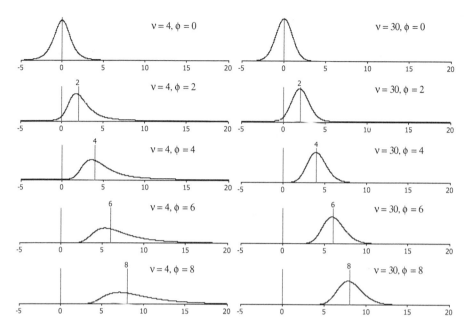

Figure D.1: Probability density functions for Student's t distribution with $\nu = 4, 30$ (columns) and $\phi = 0, 2, 4, 6, 8$ (rows).

where $\Delta\mu = \mu_1 - \mu_0$. The noncentral t distribution is skewed but becomes more symmetric as the sample size increases. Figure D.1 shows the Student's t distribution probability density function for $\nu = 4$ and $\nu = 30$, in the left and right columns, respectively, and $\phi = 0$ to 8 in rows.

See Section 2.2.2.2 for details about power and sample size calculations for the one-sample t test using the noncentral t distribution.

D.7 Chi-square (χ^2) Distribution

The chi-square (χ^2) distribution is used to characterize the distribution of sample variances s^2 for samples drawn from a normal population. The sampling distribution of

$$\chi^2 = \frac{(n-1)s^2}{\sigma^2} \quad \text{(D.30)}$$

where n is the sample size and σ is the population standard deviation, follows the χ^2 distribution with $\nu = n - 1$ degrees of freedom. Values of χ_p^2 as a function of ν are given in Appendix E.5 for selected values of the left-tail probability p. The accuracy of χ^2 distribution-based methods may be severely compromised when the population being sampled is not normal.

The distribution of $\sqrt{2\chi^2}$ is asymptotically normal as $\nu \to \infty$ with mean $\mu' = \sqrt{2\nu - 1}$ and standard deviation $\sigma' = 1$, so

$$\chi_p^2 \simeq \frac{1}{2}\left(z_p + \sqrt{2\nu - 1}\right)^2. \tag{D.31}$$

The χ^2 distribution is also used to analyze one- and two-way tables of count data for goodness of fit and independence, respectively. The χ^2 statistic for a table with k cells is calculated from

$$\chi^2 = \sum_{i=1}^{k} \frac{\left(f_i - \widehat{f}_i\right)^2}{\widehat{f}_i} \tag{D.32}$$

where the f_i are expected cell frequencies and the $\widehat{f}_i$ are observed cell frequencies. The distribution of the χ^2 statistic follows the central χ^2 distribution when $E\left(\widehat{f}_i\right) = f_i$, and the noncentral χ^2 distribution with noncentrality parameter ϕ when $E\left(\widehat{f}_i\right) \neq f_i$. See Section 4.5 for details about sample size and power calculations for one- and two-way tables using the noncentral χ^2 distribution.

D.8 F Distribution

When two samples of size n_1 and n_2 are drawn from normal populations with variances σ_1^2 and σ_2^2, respectively, the distribution of

$$F = \left(\frac{s_1/\sigma_1}{s_2/\sigma_2}\right)^2 = \left(\frac{\sigma_2 \, s_1}{\sigma_1 \, s_2}\right)^2 \tag{D.33}$$

where s_1^2 and s_2^2 are the χ^2-distributed sample variances, follows the central F distribution with $df_1 = n_1 - 1$ numerator and $df_2 = n_2 - 1$ denominator degrees of freedom. The F distribution has a left bound at 0 and is skewed right with mode $(n_2(n_1 - 2))/(n_1(n_2 + 2))$ and mean $\mu = n_2/(n_2 - 2)$. The mode is always less than 1 and the mean is always greater than 1. Values of $F_{0.95}$ are given as a function of df_1 and df_2 in Appendix E.6.

In the F test for homogeneity of two population variances, the hypotheses are $H_0 : \sigma_1^2 = \sigma_2^2$ versus $H_A : \sigma_1^2 \neq \sigma_2^2$ and the F test statistic is

$$F = \left(\frac{s_1}{s_2}\right)^2. \tag{D.34}$$

See Section 3.2.2.1 for details about sample size and power calculations for the two-sample F test for homogeneity of population variances.

In one-way fixed-effects ANOVA to test for differences between k treatment means, the hypotheses are $H_0 : \mu_i = \mu_j$ for all possible pairs of treatments versus

$H_A : \mu_i \neq \mu_j$ for at least one pair of treatments. The ANOVA F test statistic is given by

$$F = \frac{ns_{\bar{x}}^2}{s_\epsilon^2} \tag{D.35}$$

where n is the common sample size for each treatment, $s_{\bar{x}}$ is the standard deviation of the k treatment means, and s_ϵ is the pooled standard error. Under H_0, the F statistic follows the central F distribution with $df_{treatments} = k - 1$ and $df_\epsilon = k(n-1)$ degrees of freedom. Under H_A, the F statistic follows the noncentral F distribution with noncentrality parameter

$$\phi = \frac{n \sum_{i=1}^{k} \tau_i^2}{\sigma_\epsilon^2} \tag{D.36}$$

where the τ_i are the treatment biases relative to the grand mean. See Section 8.1.1 for details about power and sample size calculations for one-way fixed-effects ANOVA using the central and noncentral F distributions.

Appendix E

Probability Tables

- E.1. Larson's Nomogram for the Cumulative Binomial Distribution
- E.2. Cumulative Poisson Probability
- E.3. Standard Normal Probabilities
- E.4. Quantiles of Student's t Distribution
- E.5. Quantiles of the Chi-square (χ^2) Distribution
- E.6. Quantiles of the F Distribution
- E.7. One-sided and Two-sided Tolerance Factors for Normal Distributions
- E.8. Software

Appendix E. Probability Tables

E.1 Larson's Nomogram for the Cumulative Binomial Distribution

Nomogram of the cumulative binomial probability: $P_A = \sum_{x=0}^{c} b(x; n, p)$.

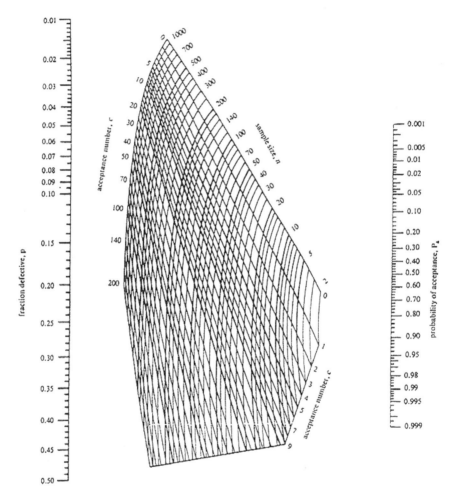

From Larson [34]. Reproduced by permission from the American Society for Quality.

E.2 Cumulative Poisson Probability

Cumulative Poisson probabilities: $Poisson(c; \lambda) = \sum_{x=0}^{c} \frac{\lambda^x e^{-\lambda}}{x!}$.

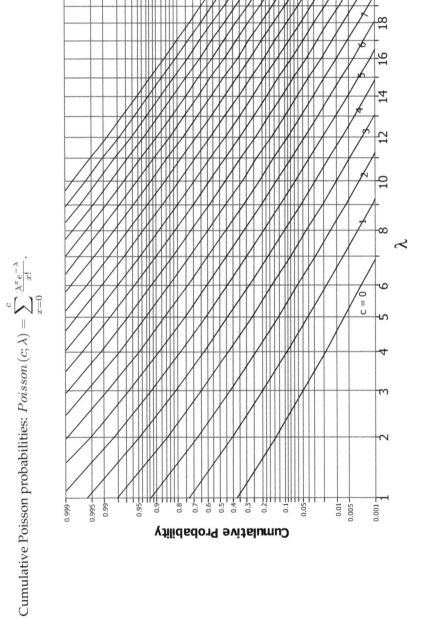

E.3 Standard Normal Probabilities

Standard normal probabilities $p = \Phi(-\infty < z < z_p)$.

z_p	0.00	0.01	0.02	0.03	0.04	0.05	0.06	0.07	0.08	0.09
-3.0	0.0013	0.0013	0.0013	0.0012	0.0012	0.0011	0.0011	0.0011	0.0010	0.0010
-2.9	0.0019	0.0018	0.0018	0.0017	0.0016	0.0016	0.0015	0.0015	0.0014	0.0014
-2.8	0.0026	0.0025	0.0024	0.0023	0.0023	0.0022	0.0021	0.0021	0.0020	0.0019
-2.7	0.0035	0.0034	0.0033	0.0032	0.0031	0.0030	0.0029	0.0028	0.0027	0.0026
-2.6	0.0047	0.0045	0.0044	0.0043	0.0041	0.0040	0.0039	0.0038	0.0037	0.0036
-2.5	0.0062	0.0060	0.0059	0.0057	0.0055	0.0054	0.0052	0.0051	0.0049	0.0048
-2.4	0.0082	0.0080	0.0078	0.0075	0.0073	0.0071	0.0069	0.0068	0.0066	0.0064
-2.3	0.0107	0.0104	0.0102	0.0099	0.0096	0.0094	0.0091	0.0089	0.0087	0.0084
-2.2	0.0139	0.0136	0.0132	0.0129	0.0125	0.0122	0.0119	0.0116	0.0113	0.0110
-2.1	0.0179	0.0174	0.0170	0.0166	0.0162	0.0158	0.0154	0.0150	0.0146	0.0143
-2.0	0.0228	0.0222	0.0217	0.0212	0.0207	0.0202	0.0197	0.0192	0.0188	0.0183
-1.9	0.0287	0.0281	0.0274	0.0268	0.0262	0.0256	0.0250	0.0244	0.0239	0.0233
-1.8	0.0359	0.0351	0.0344	0.0336	0.0329	0.0322	0.0314	0.0307	0.0301	0.0294
-1.7	0.0446	0.0436	0.0427	0.0418	0.0409	0.0401	0.0392	0.0384	0.0375	0.0367
-1.6	0.0548	0.0537	0.0526	0.0516	0.0505	0.0495	0.0485	0.0475	0.0465	0.0455
-1.5	0.0668	0.0655	0.0643	0.0630	0.0618	0.0606	0.0594	0.0582	0.0571	0.0559
-1.4	0.0808	0.0793	0.0778	0.0764	0.0749	0.0735	0.0721	0.0708	0.0694	0.0681
-1.3	0.0968	0.0951	0.0934	0.0918	0.0901	0.0885	0.0869	0.0853	0.0838	0.0823
-1.2	0.1151	0.1131	0.1112	0.1093	0.1075	0.1056	0.1038	0.1020	0.1003	0.0985
-1.1	0.1357	0.1335	0.1314	0.1292	0.1271	0.1251	0.1230	0.1210	0.1190	0.1170
-1.0	0.1587	0.1562	0.1539	0.1515	0.1492	0.1469	0.1446	0.1423	0.1401	0.1379
-0.9	0.1841	0.1814	0.1788	0.1762	0.1736	0.1711	0.1685	0.1660	0.1635	0.1611
-0.8	0.2119	0.2090	0.2061	0.2033	0.2005	0.1977	0.1949	0.1922	0.1894	0.1867
-0.7	0.2420	0.2389	0.2358	0.2327	0.2296	0.2266	0.2236	0.2206	0.2177	0.2148
-0.6	0.2743	0.2709	0.2676	0.2643	0.2611	0.2578	0.2546	0.2514	0.2483	0.2451
-0.5	0.3085	0.3050	0.3015	0.2981	0.2946	0.2912	0.2877	0.2843	0.2810	0.2776
-0.4	0.3446	0.3409	0.3372	0.3336	0.3300	0.3264	0.3228	0.3192	0.3156	0.3121
-0.3	0.3821	0.3783	0.3745	0.3707	0.3669	0.3632	0.3594	0.3557	0.3520	0.3483
-0.2	0.4207	0.4168	0.4129	0.4090	0.4052	0.4013	0.3974	0.3936	0.3897	0.3859
-0.1	0.4602	0.4562	0.4522	0.4483	0.4443	0.4404	0.4364	0.4325	0.4286	0.4247
-0.0	0.5000	0.4960	0.4920	0.4880	0.4840	0.4801	0.4761	0.4721	0.4681	0.4641

E.4 Quantiles of Student's t Distribution

Critical values $(t_{p,\nu})$ of Student's t distribution where $p = P(-\infty < t < -t_{p,\nu})$.

ν	\multicolumn{7}{c}{p}						
	0.001	0.0025	0.005	0.01	0.025	0.05	0.10
1	318.289	127.321	63.656	31.821	12.706	6.314	3.078
2	22.327	14.089	9.925	6.965	4.303	2.920	1.886
3	10.215	7.453	5.841	4.541	3.182	2.353	1.638
4	7.173	5.598	4.604	3.747	2.776	2.132	1.533
5	5.893	4.773	4.032	3.365	2.571	2.015	1.476
6	5.208	4.317	3.707	3.143	2.447	1.943	1.440
7	4.785	4.029	3.499	2.998	2.365	1.895	1.415
8	4.501	3.833	3.355	2.896	2.306	1.860	1.397
9	4.297	3.690	3.250	2.821	2.262	1.833	1.383
10	4.144	3.581	3.169	2.764	2.228	1.812	1.372
11	4.025	3.497	3.106	2.718	2.201	1.796	1.363
12	3.930	3.428	3.055	2.681	2.179	1.782	1.356
13	3.852	3.372	3.012	2.650	2.160	1.771	1.350
14	3.787	3.326	2.977	2.624	2.145	1.761	1.345
15	3.733	3.286	2.947	2.602	2.131	1.753	1.341
16	3.686	3.252	2.921	2.583	2.120	1.746	1.337
17	3.646	3.222	2.898	2.567	2.110	1.740	1.333
18	3.610	3.197	2.878	2.552	2.101	1.734	1.330
19	3.579	3.174	2.861	2.539	2.093	1.729	1.328
20	3.552	3.153	2.845	2.528	2.086	1.725	1.325
21	3.527	3.135	2.831	2.518	2.080	1.721	1.323
22	3.505	3.119	2.819	2.508	2.074	1.717	1.321
23	3.485	3.104	2.807	2.500	2.069	1.714	1.319
24	3.467	3.091	2.797	2.492	2.064	1.711	1.318
25	3.450	3.078	2.787	2.485	2.060	1.708	1.316
26	3.435	3.067	2.779	2.479	2.056	1.706	1.315
27	3.421	3.057	2.771	2.473	2.052	1.703	1.314
28	3.408	3.047	2.763	2.467	2.048	1.701	1.313
29	3.396	3.038	2.756	2.462	2.045	1.699	1.311
30	3.385	3.030	2.750	2.457	2.042	1.697	1.310
40	3.307	2.971	2.704	2.423	2.021	1.684	1.303
50	3.261	2.937	2.678	2.403	2.009	1.676	1.299
60	3.232	2.915	2.660	2.390	2.000	1.671	1.296
80	3.195	2.887	2.639	2.374	1.990	1.664	1.292
100	3.174	2.871	2.626	2.364	1.984	1.660	1.290
∞	3.090	2.807	2.576	2.326	1.960	1.645	1.282

E.5 Quantiles of the Chi-square (χ^2) Distribution

Critical values $(\chi^2_{p,\nu})$ of the χ^2 distribution where $p = P\left(0 < \chi^2 < \chi^2_{p,\nu}\right)$.

					p					
ν	0.005	0.01	0.025	0.05	0.1	0.9	0.95	0.975	0.99	0.995
1	0.00	0.00	0.00	0.00	0.02	2.71	3.84	5.02	6.63	7.88
2	0.01	0.02	0.05	0.10	0.21	4.61	5.99	7.38	9.21	10.6
3	0.07	0.11	0.22	0.35	0.58	6.25	7.81	9.35	11.3	12.8
4	0.21	0.30	0.48	0.71	1.06	7.78	9.49	11.1	13.3	14.9
5	0.41	0.55	0.83	1.15	1.61	9.24	11.1	12.8	15.1	16.7
6	0.68	0.87	1.24	1.64	2.20	10.6	12.6	14.5	16.8	18.5
7	0.99	1.24	1.69	2.17	2.83	12.0	14.1	16.0	18.5	20.3
8	1.34	1.65	2.18	2.73	3.49	13.4	15.5	17.5	20.1	22.0
9	1.73	2.09	2.70	3.33	4.17	14.7	16.9	19.0	21.7	23.6
10	2.16	2.56	3.25	3.94	4.87	16.0	18.3	20.5	23.2	25.2
11	2.60	3.05	3.82	4.57	5.58	17.3	19.7	21.9	24.7	26.8
12	3.07	3.57	4.40	5.23	6.30	18.5	21.0	23.3	26.2	28.3
13	3.57	4.11	5.01	5.89	7.04	19.8	22.4	24.7	27.7	29.8
14	4.07	4.66	5.63	6.57	7.79	21.1	23.7	26.1	29.1	31.3
15	4.60	5.23	6.26	7.26	8.55	22.3	25.0	27.5	30.6	32.8
16	5.14	5.81	6.91	7.96	9.31	23.5	26.3	28.9	32.0	34.3
17	5.70	6.41	7.56	8.67	10.1	24.8	27.6	30.2	33.4	35.7
18	6.26	7.01	8.23	9.39	10.9	26.0	28.9	31.5	34.8	37.2
19	6.84	7.63	8.91	10.12	11.7	27.2	30.1	32.85	36.2	38.6
20	7.43	8.26	9.59	10.85	12.4	28.4	31.4	34.2	37.6	40.0
22	8.64	9.54	11.0	12.3	14.0	30.8	33.9	36.8	40.3	42.8
24	9.89	10.9	12.4	13.9	15.7	33.2	36.4	39.4	43.0	45.6
26	11.2	12.2	13.8	15.4	17.3	35.6	38.9	41.9	45.6	48.3
28	12.5	13.6	15.3	16.9	18.9	37.9	41.3	44.5	48.3	51.0
30	13.8	14.9	16.8	18.5	20.6	40.3	43.8	47.0	50.9	53.7
35	17.2	18.5	20.6	22.5	24.8	46.1	49.8	53.2	57.3	60.3
40	20.7	22.2	24.4	26.5	29.1	51.8	55.8	59.3	63.7	66.8
45	24.3	25.9	28.4	30.6	33.4	57.5	61.7	65.4	70.0	73.2
50	28.0	29.7	32.4	34.8	37.7	63.2	67.5	71.4	76.2	79.5
55	31.7	33.6	36.4	39.0	42.1	68.8	73.3	77.4	82.3	85.8
60	35.5	37.5	40.5	43.2	46.5	74.4	79.1	83.3	88.4	92.0
70	43.3	45.4	48.8	51.7	55.3	85.5	90.5	95.0	100.4	104.2
80	51.2	53.5	57.1	60.4	64.3	96.6	101.9	106.6	112.3	116.3
90	59.2	61.8	65.7	69.1	73.3	107.6	113.2	118.1	124.1	128.3
100	67.3	70.1	74.2	77.9	82.4	118.5	124.3	129.6	135.8	140.2

E.6 Quantiles of the F Distribution

Critical values ($F_{0.95,\nu_1,\nu_2}$) of the F distribution where $P(0 < F < F_{0.95,\nu_1,\nu_2}) = 0.95$.

ν_2 \ ν_1	1	2	3	4	5	6	7	8	9	10	15	20	30	60	120
1	161.4	199.5	215.7	224.6	230.2	234.0	236.8	238.9	240.5	241.9	245.9	248.0	250.1	252.2	253.3
2	18.5	19.0	19.2	19.2	19.3	19.3	19.4	19.4	19.4	19.4	19.4	19.4	19.5	19.5	19.5
3	10.1	9.55	9.28	9.12	9.01	8.94	8.89	8.85	8.81	8.79	8.70	8.66	8.62	8.57	8.55
4	7.71	6.94	6.59	6.39	6.26	6.16	6.09	6.04	6.00	5.96	5.86	5.80	5.75	5.69	5.66
5	6.61	5.79	5.41	5.19	5.05	4.95	4.88	4.82	4.77	4.74	4.62	4.56	4.50	4.43	4.40
6	5.99	5.14	4.76	4.53	4.39	4.28	4.21	4.15	4.10	4.06	3.94	3.87	3.81	3.74	3.70
7	5.59	4.74	4.35	4.12	3.97	3.87	3.79	3.73	3.68	3.64	3.51	3.44	3.38	3.30	3.27
8	5.32	4.46	4.07	3.84	3.69	3.58	3.50	3.44	3.39	3.35	3.22	3.15	3.08	3.01	2.97
9	5.12	4.26	3.86	3.63	3.48	3.37	3.29	3.23	3.18	3.14	3.01	2.94	2.86	2.79	2.75
10	4.96	4.10	3.71	3.48	3.33	3.22	3.14	3.07	3.02	2.98	2.85	2.77	2.70	2.62	2.58
15	4.54	3.68	3.29	3.06	2.90	2.79	2.71	2.64	2.59	2.54	2.40	2.33	2.25	2.16	2.11
20	4.35	3.49	3.10	2.87	2.71	2.60	2.51	2.45	2.39	2.35	2.20	2.12	2.04	1.95	1.90
30	4.17	3.32	2.92	2.69	2.53	2.42	2.33	2.27	2.21	2.16	2.01	1.93	1.84	1.74	1.68
60	4.00	3.15	2.76	2.53	2.37	2.25	2.17	2.10	2.04	1.99	1.84	1.75	1.65	1.53	1.47
120	3.92	3.07	2.68	2.45	2.29	2.18	2.09	2.02	1.96	1.91	1.75	1.66	1.55	1.43	1.35

E.7 One- and Two-sided Tolerance Factors for Normal Distributions

One-sided[1] (k_1) and two-sided[2] (k_2) 95% confidence level tolerance factors for normal distributions.

	k_1		k_2	
	Yield		Yield	
n	0.99	0.999	0.99	0.999
3	10.55	13.86	12.86	16.21
4	7.04	9.21	8.30	10.50
5	5.74	7.50	6.63	8.41
6	5.06	6.61	5.78	7.34
7	4.64	6.06	5.25	6.68
8	4.35	5.69	4.89	6.23
9	4.14	5.41	4.63	5.90
10	3.98	5.20	4.43	5.65
11	3.85	5.04	4.28	5.45
12	3.75	4.90	4.15	5.29
13	3.66	4.79	4.04	5.16
14	3.59	4.69	3.96	5.04
15	3.52	4.61	3.88	4.95
16	3.46	4.53	3.81	4.87
17	3.42	4.47	3.75	4.79
18	3.37	4.42	3.70	4.72
19	3.33	4.36	3.66	4.67
20	3.29	4.32	3.62	4.61
25	3.16	4.14	3.46	4.41
30	3.06	4.02	3.35	4.28
40	2.94	3.87	3.21	4.10
50	2.86	3.77	3.13	3.99

[1] Adapted from Lieberman [38]. Reproduced by permission of American Society for Quality.
[2] Adapted from Eisenhart et al [21]. Public domain.

E.8 Software

Published tables of critical values for probability distributions are convenient, but they are always incomplete. Complete functions for most distributions are available in many software packages. The purpose of this section is to summarize the capabilities of Microsoft Excel, MINITAB, PASS, R, and Piface.

E.8.1 Microsoft Excel

Excel has functions to calculate the cumulative distribution function and its inverse for the hypergeometric, binomial, Poisson, normal, Student's t, χ^2 and F distributions. Excel does not provide any functions for the noncentral t, F, or χ^2 distributions. Be careful using Excel because some of its probability functions are indexed by left tail area, some by right tail area, and some by two tails. Use the **Insert Function> Statistical** menu to access Excel's statistical functions.

E.8.2 MINITAB

MINITAB has a comprehensive list of cumulative probability, inverse cumulative probability, and probability density functions for both central and noncentral distributions. Access MINITAB's probability functions from its graphical user interface using the **Calc> Probability Distributions** menu or from the command prompt using the *cdf*, *invcdf*, and *pdf* functions with appropriate subcommands.

E.8.3 PASS

PASS has a comprehensive list of cumulative probability, inverse cumulative probability, and probability density functions for both central and noncentral distributions. Access the NCSS probability functions from the graphical user interface using the **Analysis> Other> Probability Calculator** menu.

E.8.4 R

R has a comprehensive list of cumulative probability, inverse cumulative probability, and probability density functions for both central and noncentral distributions. R has no graphical user interface, so access its probability functions from the command line. R's probability functions use the prefixes d, q, and p to indicate the probability density function, inverse cumulative distribution function, and cumulative distribution function, respectively. Examples of some of R's probability functions are

- normal: *dnorm*, *qnorm*, and *pnorm*.
- Student's t: *dt*, *qt*, and *pt*.

- χ^2: *dchisq, qchisq,* and *pchisq.*
- F: *df, qf,* and *pf.*
- binomial: *dbinom, qbinom,* and *pbinom.*

See the online help documentation for more information about R's probability functions.

E.8.5   Piface

Piface has a graphical user interface to calculate cumulative probability and inverse cumulative probability functions for both central and noncentral distributions for the most important distributions. Access Piface's probability functions from its **Online tables of common distributions** menu.

Appendix F

Identities and Approximations

The following identities and approximations are used throughout the text and might help explain some of the cryptic mathematical operations and unexpected appearances of some probability distributions.

F.1 Identities

z_p and t_p values are always positive, for example: $z_{0.05} = z_{0.95} = 1.645$. χ^2 and F values are indexed by their left tail areas.

$$\sum_{n=0}^{\infty} x^n = \frac{1}{1-x} \tag{F.1}$$

$$\sum_{n=0}^{\infty} (-1)^n x^n = \frac{1}{1+x} \tag{F.2}$$

$$e^x = \sum_{n=0}^{\infty} \frac{x^n}{n!} \tag{F.3}$$

$$z_p = \lim_{\nu \to \infty} (t_{p,\nu}) \tag{F.4}$$

$$z_{p/2} = \sqrt{\chi^2_{1-p,1}} \tag{F.5}$$

$$z_{p/2} = \sqrt{F_{1-p,1,\infty}} \tag{F.6}$$

$$F_{p,\nu_1,\nu_2} = \frac{1}{F_{1-p,\nu_2,\nu_1}} \tag{F.7}$$

$$t_{p/2,\nu} = \sqrt{F_{1-p,1,\nu}} \tag{F.8}$$

$$F_{p,\nu_1,\nu_2} = \left(\chi^2_{p,\nu_1}/\nu_1\right) / \left(\chi^2_{p,\nu_2}/\nu_2\right) \tag{F.9}$$

$$\chi^2_{p,\nu}/\nu = F_{p,\nu,\infty} \tag{F.10}$$

$$\chi^2_{1-p,2(X+1)} = 2\lambda \text{ where } p = \sum_{x=0}^{X} Poisson(x;\lambda) \tag{F.11}$$

$$\theta = \frac{X}{X + (n - X + 1) F_{1-p,2(n-X+1),2X}} \text{ where } p = \sum_{x=0}^{X} b(x;n,\theta) \tag{F.12}$$

$$\frac{d}{d\lambda}\left(\sum_{x=0}^{c} Poisson(x;\lambda)\right) = -Poisson(c;\lambda) \tag{F.13}$$

$$\frac{d}{dp}\left(\sum_{x=0}^{c} Poisson(x;\lambda = np)\right) = -nPoisson(c;\lambda) \tag{F.14}$$

$$h(x;S,N,n) = h(x;n,N,S) \tag{F.15}$$

F.2 Approximations

The more rigorously the indicated conditions are satisfied, the more accurate is the approximation. *AND* and *OR* are Boolean operators.

$$\ln(n!) \simeq n \ln(n) - n + \frac{1}{2}\ln(2\pi n) \text{ if } n > 10 \tag{F.16}$$

$$(1 \pm x)^n \simeq 1 \pm nx \text{ if } |x| \ll 1 \tag{F.17}$$

$$e^x \simeq 1 + x \text{ if } |x| \ll 1 \tag{F.18}$$

$$\prod_{i=1}^{k}(1-\alpha_i) \simeq 1 - \sum_{i=1}^{k}\alpha_i \text{ if } |\alpha_i| \ll 1 \tag{F.19}$$

$$h(x;S,N,n) \simeq b\left(x;n,p=\frac{S}{N}\right) \text{ if } n/N < 0.1 \tag{F.20}$$

$$h(x;S,N,n) \simeq b\left(x;S,p=\frac{n}{N}\right) \text{ if } S/N \ll 0.1 \tag{F.21}$$

$$b(x;n,p) \simeq Poisson(x;\lambda = np) \text{ if } (n > 20 \text{ AND } p \leq 0.05) \tag{F.22}$$

F.2. Approximations

$$b(x_L \leq x \leq x_U; n, p) \simeq$$
$$\Phi\left(x_L - \frac{1}{2} < x < x_U + \frac{1}{2}; \mu = np, \sigma = \sqrt{np(1-p)}\right)$$
$$\text{if } (np > 5 \text{ AND } n(1-p) > 5) \quad \text{(F.23)}$$

$$Poisson(x_L \leq x \leq x_U; \lambda) \simeq$$
$$\Phi\left(x_L - \frac{1}{2} < x < x_U + \frac{1}{2}; \mu = \lambda, \sigma = \sqrt{\lambda}\right) \text{ if } \lambda > 20 \quad \text{(F.24)}$$

$$-\frac{z\varphi(z)}{1+z^2} < \Phi(z) < -\frac{\varphi(z)}{z} \text{ for } z < 0 \quad \text{(F.25)}$$

$$\chi^2_{1-\alpha,\nu} \simeq \nu + z_\alpha \sqrt{2\nu} \text{ if } \nu > 30 \quad \text{(F.26)}$$

$$\chi^2_{1-\alpha,\nu} \simeq \nu + 1 + z_\alpha \sqrt{2\nu} \text{ if } \nu > 10 \quad \text{(F.27)}$$

$$\chi^2_{1-\alpha,\nu} \simeq \nu \left(1 + \frac{z_\alpha}{\sqrt{2\nu}}\right)^2 \text{ if } \nu > 20 \quad \text{(F.28)}$$

Appendix G

The Delta Method

This section will not be of interest to most readers; it is presented for completeness and to satisfy those who need to understand the derivations of the approximate sample size and power calculations.

Probability distributions often have undesirable properties such as skewness or heteroscedasticity. For many of these distributions, a mathematical transformation of the original random variable results in a distribution that is better behaved. For example, the Poisson distribution is always skewed to the right and its standard deviation changes with the mean. However, the square root of Poisson-distributed data is more closely normal with nearly constant standard deviation with respect to the mean. Transforming a badly behaved distribution into a better behaved one allows simpler and better known analysis methods to be used. The delta method is used to estimate the standard deviation of a transformed distribution from the original distribution.

G.1 One Unknown Parameter

Suppose that the distribution of a random variable x has one unknown parameter θ and that we wish to estimate the distribution of $\widehat{\theta}$ for the purpose of making inferences about θ. Furthermore, suppose that the distribution of $\widehat{\theta}$ is not well behaved, but that a transformation exists for which the distribution of $g\left(\widehat{\theta}\right)$ is asymptotically normal for large samples. Then, by the delta or propagation of error method, the mean of the $g\left(\widehat{\theta}\right)$ distribution is

$$\mu_{g(\widehat{\theta})} = g(\theta) \tag{G.1}$$

and the approximate standard deviation of the $g\left(\widehat{\theta}\right)$ distribution is

$$\widehat{\sigma}_{g(\widehat{\theta})} = \sigma_{\widehat{\theta}} \left(\frac{dg}{d\widehat{\theta}} \bigg|_{\widehat{\theta}=\theta} \right). \tag{G.2}$$

When θ is not known, it may be estimated from the sample data to obtain $\widehat{\mu}_{g(\widehat{\theta})}$ and $\widehat{\sigma}_{g(\widehat{\theta})}$.

An obvious application of these results is the formulation of an approximate large-sample confidence interval for $g(\theta)$:

$$\Phi\left(\widehat{\mu}_{g(\widehat{\theta})} - z_{\alpha/2}\widehat{\sigma}_{g(\widehat{\theta})} < g(\theta) < \widehat{\mu}_{g(\widehat{\theta})} + z_{\alpha/2}\widehat{\sigma}_{g(\widehat{\theta})}\right) = 1 - \alpha \quad (G.3)$$

Once this confidence interval has been obtained, the upper and lower confidence limits may be inverse-transformed to give the desired approximate confidence interval for θ. The delta method may also be used to solve hypothesis testing problems.

G.2 Two or More Unknown Parameters

When the distribution of x has two or more unknown parameters, $\widehat{\theta}_1, \widehat{\theta}_2, \cdots$, the mean of the $g\left(\widehat{\theta}_1, \widehat{\theta}_2, \cdots\right)$ distribution is

$$\mu_{g(\widehat{\theta}_1, \widehat{\theta}_2, \cdots)} = g(\theta_1, \theta_2, \cdots) \quad (G.4)$$

and the approximate variance is

$$\widehat{\sigma}^2_{g(\widehat{\theta}_1, \widehat{\theta}_2, \cdots)} = \left(\sigma_{\widehat{\theta}_1}\left(\frac{\partial g}{\partial \widehat{\theta}_1}\bigg|_{\widehat{\theta}=\theta_1}\right)\right)^2 + \left(\sigma_{\widehat{\theta}_2}\left(\frac{\partial g}{\partial \widehat{\theta}_2}\bigg|_{\widehat{\theta}=\theta_2}\right)\right)^2$$
$$+ 2\sigma_{\widehat{\theta}_1,\widehat{\theta}_2}\left(\frac{\partial g}{\partial \widehat{\theta}_1}\bigg|_{\widehat{\theta}=\theta_1}\right)\left(\frac{\partial g}{\partial \widehat{\theta}_2}\bigg|_{\widehat{\theta}=\theta_2}\right) + \cdots \quad (G.5)$$

where $\sigma_{\widehat{\theta}_i,\widehat{\theta}_j}$ are the covariances. The covariances are 0 when $\widehat{\theta}_i$ and $\widehat{\theta}_j$ are independent.

G.3 Applications of the Delta Method

This section presents some of the common applications of the delta method that appear throughout this book. When the delta method is required in a one-time-only situation, it is presented as required in that section.

G.3.1 Arcsine Transform for the Binomial Proportion

For large samples where the distribution of the sample proportions $\widehat{p}$ is binomial with mean $\mu_{\widehat{p}} = p$ and standard deviation $\sigma_{\widehat{p}} = \sqrt{p(1-p)/n}$, the distribution of

G.3. Applications of the Delta Method

$$g(\hat{p}) = 2\arcsin\left(\sqrt{\hat{p}}\right) \tag{G.6}$$

where $g(\hat{p})$ is expressed in radians, is approximately normal with mean

$$\mu_{g(\hat{p})} = 2\arcsin\left(\sqrt{p}\right) \tag{G.7}$$

and approximate standard deviation

$$\begin{aligned}
\hat{\sigma}_{g(\hat{p})} &= \sigma_{\hat{p}} \left(\frac{d}{d\hat{p}} \left(2\arcsin\left(\sqrt{\hat{p}}\right) \right) \right) \bigg|_{\hat{p}=p} \\
&= \left(\sqrt{\frac{p(1-p)}{n}} \right) \left(\frac{1}{\sqrt{p(1-p)}} \right) \\
&= \frac{1}{\sqrt{n}}.
\end{aligned} \tag{G.8}$$

G.3.2 Log Transform for the Binomial Proportion

The sample binomial proportion $\hat{p}$ has mean $\mu_p = np$ and standard deviation $\sigma_p = \sqrt{np(1-p)}$. By the delta method, the mean of the log-transformed sample proportions $\ln(\hat{p})$ is approximately normal with mean $\mu_{\ln(\hat{p})} = \ln(p)$ and approximate standard deviation:

$$\begin{aligned}
\hat{\sigma}_{\ln(\hat{p})} &= \sigma_p \left(\frac{d}{dp} \left(\ln(\hat{p}) \right) \right) \bigg|_{\hat{p}=p} \\
&= \sqrt{\frac{p(1-p)}{n}} \frac{1}{p} \\
&= \sqrt{\frac{1-p}{np}}.
\end{aligned} \tag{G.9}$$

G.3.3 Log Odds Transform for the Binomial Proportion

The odds parameter O is given by $O = p/(1-p)$ where the distribution of the sample proportion $\hat{p}$ is binomial with mean $\mu_{\hat{p}} = p$ and standard deviation $\sigma_{\hat{p}} = \sqrt{p(1-p)/n}$. When the sample size n is very large and p is unknown, the distribution of $\ln\left(\hat{O}\right)$, called the *log odds* or *logit*, is approximately normal with mean

$$\mu_{\ln(\hat{O})} = \ln(O) \tag{G.10}$$

and approximate standard deviation

$$\hat{\sigma}_{\ln(\hat{O})} = \sigma_{\hat{p}} \left(\frac{d}{d\hat{p}} \ln \left(\frac{\hat{p}}{1-\hat{p}} \right) \right) \bigg|_{\hat{p}=p}$$

$$= \sqrt{\frac{p(1-p)}{n}} \frac{1}{p(1-p)}$$

$$= \frac{1}{\sqrt{np(1-p)}}. \tag{G.11}$$

G.3.4 Log Transform for the Risk Ratio

The relative risk or risk ratio parameter is given by $RR = p_1/p_2$ where p_1 and p_2 are binomial distribution parameters. The standard deviation of $\ln(\hat{p})$ was derived in Section G.3.2, so by the delta method the large-sample sampling distribution of the log-transformed risk ratio

$$\ln\left(\widehat{RR}\right) = \ln\left(\hat{p}_1\right) - \ln\left(\hat{p}_2\right) \tag{G.12}$$

is approximately normal with mean

$$\mu_{\ln(\widehat{RR})} = \ln(RR) \tag{G.13}$$

and approximate standard deviation

$$\hat{\sigma}_{\ln(RR)} \simeq \sqrt{\sigma_{\ln(\hat{p}_1)}^2 + \sigma_{\ln(\hat{p}_2)}^2}$$

$$\simeq \sqrt{\frac{1-p_1}{n_1 p_1} + \frac{1-p_2}{n_2 p_2}} \tag{G.14}$$

or, in terms of the actual count data,

$$\hat{\sigma}_{\ln(RR)} \simeq \sqrt{\frac{1}{x_1} - \frac{1}{n_1} + \frac{1}{x_2} - \frac{1}{n_2}}. \tag{G.15}$$

G.3.5 Log Transform for the Odds Ratio

The odds ratio parameter is given by $OR = O_1/O_2$ where $O_i = p_i/(1-p_i)$ and p_1 and p_2 are binomial distribution parameters. The standard deviation of $\ln\left(\hat{O}\right)$ was derived in Section G.3.3, so by the delta method the large-sample sampling distribution of the log odds ratio

$$\ln\left(\widehat{OR}\right) = \ln\left(\hat{O}_1/\hat{O}_2\right) = \ln\left(\hat{O}_1\right) - \ln\left(\hat{O}_2\right) \tag{G.16}$$

is approximately normal with mean

$$\mu_{\ln(\widehat{OR})} = \ln(OR) \tag{G.17}$$

G.3. Applications of the Delta Method

and approximate standard deviation

$$\widehat{\sigma}_{\ln(\widehat{OR})} = \sqrt{\frac{1}{n_1 p_1 (1-p_1)} + \frac{1}{n_2 p_2 (1-p_2)}}, \qquad \text{(G.18)}$$

or, in terms of the actual success (x) and failure $(x' = n - x)$ counts,

$$\widehat{\sigma}_{\ln(OR)} = \sqrt{\frac{1}{x_1} + \frac{1}{x_1'} + \frac{1}{x_2} + \frac{1}{x_2'}}. \qquad \text{(G.19)}$$

G.3.6 Square-Root Transform for Poisson Counts

When the distribution of sample counts x is Poisson with mean $\mu_x = \lambda$ and standard deviation $\sigma_x = \sqrt{\lambda}$, the large-sample sampling distribution of $g(x) = \sqrt{x}$ is approximately normal with mean

$$\mu_{\sqrt{x}} = \sqrt{\lambda} \qquad \text{(G.20)}$$

and approximate standard deviation

$$\begin{aligned}
\widehat{\sigma}_{\sqrt{x}} &= \sigma_x \left(\frac{d}{dx} (\sqrt{x}) \right) \bigg|_{x=\lambda} \\
&= \sqrt{\lambda} \left(\frac{1}{2\sqrt{\lambda}} \right) \\
&= \frac{1}{2}.
\end{aligned} \qquad \text{(G.21)}$$

G.3.7 Difference Between Two Independent Poisson Counts

When the x_1 and x_2 counts are large, the distribution of $\Delta\widehat{\lambda} = \widehat{\lambda}_1 - \widehat{\lambda}_2$ is approximately normal with mean

$$\mu_{\Delta\widehat{\lambda}} = \lambda_1 - \lambda_2 \qquad \text{(G.22)}$$

and approximate standard deviation

$$\widehat{\sigma}_{\Delta\widehat{\lambda}} = \sqrt{\left(\sigma_{x_1}\frac{\partial}{\partial x_1}\left(\widehat{\lambda}_1 - \widehat{\lambda}_2\right)\right)^2 + \left(\sigma_{x_2}\frac{\partial}{\partial x_2}\left(\widehat{\lambda}_1 - \widehat{\lambda}_2\right)\right)^2}$$

$$= \sqrt{\left(\sigma_{x_1}\frac{\partial}{\partial x_1}\left(\frac{x_1}{n_1} - \frac{x_2}{n_2}\right)\right)^2 + \left(\sigma_{x_2}\frac{\partial}{\partial x_2}\left(\frac{x_1}{n_1} - \frac{x_2}{n_2}\right)\right)^2}$$

$$= \sqrt{\left(\sqrt{x_1}\frac{1}{n_1}\right)^2 + \left(-\sqrt{x_2}\frac{1}{n_2}\right)^2}$$

$$= \sqrt{\left(\frac{x_1}{n_1^2}\right)\bigg|_{x_1=n\lambda_1} + \left(\frac{x_2}{n_2^2}\right)\bigg|_{x_2=n_2\lambda_2}}$$

$$= \sqrt{\frac{\lambda_1}{n_1} + \frac{\lambda_2}{n_2}}. \qquad (G.23)$$

In terms of the actual event counts, the approximate standard deviation is

$$\widehat{\sigma}_{\ln(\widehat{\lambda}_1/\widehat{\lambda}_2)} = \sqrt{\frac{x_1}{n_1^2} + \frac{x_2}{n_2^2}}. \qquad (G.24)$$

G.3.8 Ratio of Two Independent Poisson Counts

When the x_1 and x_2 Poisson counts are large, the distribution of $\ln\left(\widehat{\lambda}_1/\widehat{\lambda}_2\right)$ is approximately normal with mean

$$\mu_{\ln(\widehat{\lambda}_1/\widehat{\lambda}_2)} = \ln(\lambda_1/\lambda_2) \qquad (G.25)$$

and approximate standard deviation

$$\widehat{\sigma}_{\ln(\widehat{\lambda}_1/\widehat{\lambda}_2)} = \sqrt{\left(\sigma_{x_1}\frac{\partial}{\partial x_1}\ln\left(\frac{\widehat{\lambda}_1}{\widehat{\lambda}_2}\right)\right)^2 + \left(\sigma_{x_2}\frac{\partial}{\partial x_2}\ln\left(\frac{\widehat{\lambda}_1}{\widehat{\lambda}_2}\right)\right)^2}$$

$$= \sqrt{\left(\sigma_{x_1}\frac{\partial}{\partial x_1}(\ln(x_1))\right)^2 + \left(\sigma_{x_2}\frac{\partial}{\partial x_2}(\ln(-x_2))\right)^2}$$

$$= \sqrt{\left(\sqrt{x_1}\frac{1}{x_1}\right)^2 + \left(\sqrt{x_2}\frac{1}{x_2}\right)^2}$$

$$= \sqrt{\left(\frac{1}{x_1}\right)\bigg|_{x_1=n\lambda_1} + \left(\frac{1}{x_2}\right)\bigg|_{x_2=n_2\lambda_2}}$$

$$= \sqrt{\frac{1}{n_1\lambda_1} + \frac{1}{n_2\lambda_2}}. \qquad (G.26)$$

G.3. Applications of the Delta Method

In terms of the actual event counts, the approximate standard deviation is

$$\hat{\sigma}_{\ln(\hat{\lambda}_1/\hat{\lambda}_2)} = \sqrt{\frac{1}{x_1} + \frac{1}{x_2}}. \tag{G.27}$$

G.3.9 Standard Normal $\hat{z}$ Statistic

When a sample of size n drawn from a normal population delivers sample mean $\bar{x}$ and sample standard deviation s, the distribution of

$$\hat{z} = \frac{x - \bar{x}}{s} \tag{G.28}$$

is approximately normal with mean

$$\mu_{\hat{z}} = \frac{x - \mu_x}{\sigma_x} \tag{G.29}$$

and approximate standard deviation

$$\begin{aligned}
\hat{\sigma}_{\hat{z}} &= \sqrt{\left(\frac{\partial z}{\partial \mu_x} \sigma_{\bar{x}}\right)^2 + \left(\frac{\partial z}{\partial \sigma_x} \sigma_s\right)^2} \\
&= \sqrt{\left(-\frac{1}{\sigma_x} \frac{\sigma_x}{\sqrt{n}}\right)^2 + \left(-\frac{x - \mu_x}{\sigma_x^2} \frac{\sigma_x}{\sqrt{2n}}\right)^2} \\
&= \sqrt{\frac{1}{n} + \frac{1}{2n}\left(-\frac{x - \mu_x}{\sigma_x}\right)^2} \\
&= \sqrt{\frac{1}{n}\left(1 + \frac{\hat{z}^2}{2}\right)}. \tag{G.30}
\end{aligned}$$

See Chakraborti and Li [11] for a comparison of six normal percentile estimation methods including exact and small sample methods.

G.3.10 Normal Probability at Specified x Value

When x is expected to follow a normal distribution with mean and standard deviation estimated from a sample of size n by $\bar{x}$ and s, respectively, then by the delta method the large-sample sampling distribution of the normal probability $\hat{\Phi}(x; \bar{x}, s)$ (i.e., the left tail area under the normal curve relative to x) is approximately normal with mean $\Phi(z)$ where $z = (x - \mu)/\sigma_x$ and approximate standard deviation

$$\begin{aligned}
\widehat{\sigma}_{\widehat{\Phi}(x)} &= \sqrt{\left(\frac{\partial \Phi(x)}{\partial \mu}\right)^2 \sigma_{\bar{x}}^2 + \left(\frac{\partial \Phi(x)}{\partial \sigma}\right)^2 \sigma_s^2} \\
&= \sqrt{\left(-\frac{\varphi(\widehat{z})}{\sigma_x}\right)^2 \left(\frac{\sigma_x^2}{n}\right) + \left(\frac{\widehat{z}\varphi(\widehat{z})}{\sigma_x}\right)^2 \left(\frac{\sigma_x^2}{2n}\right)} \\
&= \varphi(\widehat{z}) \sqrt{\frac{1}{n}\left(1 + \frac{1}{2}\widehat{z}^2\right)}
\end{aligned} \qquad (G.31)$$

where $\varphi(\widehat{z})$ is the standard normal probability density function evaluated at $\widehat{z} = (x - \bar{x})/s$.

Equations G.30 and G.31 are simply related by

$$\begin{aligned}
\widehat{\sigma}_{\widehat{\Phi}(x)} &= \widehat{\sigma}_{\widehat{z}} \frac{d\Phi(\widehat{z})}{dz} \\
&= \widehat{\sigma}_{\widehat{z}} \varphi(\widehat{z}).
\end{aligned} \qquad (G.32)$$

G.3.11 Sample Standard Deviation

When samples of size n are drawn from a normal population with variance σ_x^2, the distribution of the sample variances s^2 is χ^2 with $n - 1$ degrees of freedom and the corresponding variance of the sample variances is

$$\sigma_{s^2}^2 = \frac{2(n-1)}{n^2}\sigma_x^4. \qquad (G.33)$$

For large samples the sampling distribution of $s = \sqrt{s^2}$ is approximately normal with mean $\mu_s \simeq \sigma_x$ and approximate standard deviation by the delta method

$$\begin{aligned}
\widehat{\sigma}_s &= \left.\frac{d\sqrt{s^2}}{d(s^2)}\right|_{s=\sigma_x} \sigma_{s^2} \\
&= \left.\frac{1}{2s}\right|_{s=\sigma_x} \sqrt{\frac{2(n-1)}{n^2}\sigma_x^4} \\
&= \frac{1}{2\sigma_x}\sigma_x^2\sqrt{\frac{2}{n}} \\
&= \frac{\sigma_x}{\sqrt{2n}}.
\end{aligned} \qquad (G.34)$$

G.3.12 Logarithmic Transform for the Standard Deviation

When samples of size n are drawn from a normal population with variance σ_x^2, the distribution of the sample variances s^2 is χ^2 with $n - 1$ degrees of freedom.

G.3. Applications of the Delta Method

When n is very large, the distribution of $g(s) = \ln(s)$ is approximately normal with mean

$$\mu_{\ln(s)} = \ln(\sigma_x) \qquad (G.35)$$

and approximate standard deviation

$$\begin{aligned}
\widehat{\sigma}_{\ln(s)} &= \sigma_s \left(\frac{d}{ds}(\ln(s))\right)\bigg|_{s=\sigma_x} \\
&= \frac{\sigma_x}{\sqrt{2n}} \left(\frac{1}{\sigma_x}\right) \\
&= \frac{1}{\sqrt{2n}} \qquad (G.36)
\end{aligned}$$

where we used the large-sample normal approximation to the χ^2 distribution to estimate σ_s:

$$\sigma_s \simeq \frac{\sigma_x}{\sqrt{2n}}. \qquad (G.37)$$

The log transformation can also be applied directly to the variance, resulting in

$$\mu_{\ln(s^2)} = \ln(\sigma_x^2) = 2\ln(\sigma_x) \qquad (G.38)$$

and

$$\widehat{\sigma}_{\ln(s^2)} = \sqrt{\frac{2}{n}}. \qquad (G.39)$$

G.3.13 Logarithmic Transform for the Ratio of Two Standard Deviations

When samples of size n_1 and n_2 are drawn from two independent normal populations with $\sigma_1^2 = \sigma_2^2$, then the ratio of their sample variances $\widehat{F} = s_1^2/s_2^2$ follows the F distribution with $\nu_1 = n_1 - 1$ and $\nu_2 = n_2 - 1$ degrees of freedom. When n_1 and n_2 are large, the distribution of $\ln(s_1/s_2)$ is approximately normal with mean

$$\mu_{\ln(s_1/s_2)} = \ln\left(\frac{\sigma_1}{\sigma_2}\right) \qquad (G.40)$$

and approximate standard deviation

$$\begin{aligned}
\widehat{\sigma}_{\ln(s_1/s_2)} &= \sqrt{\sigma_{s_1}^2 \left(\frac{\partial}{\partial \sigma_1}(\ln(\sigma_1/\sigma_2))\right)^2 + \sigma_{s_2}^2 \left(\frac{\partial}{\partial \sigma_2}(\ln(\sigma_1/\sigma_2))\right)^2} \\
&= \sqrt{\frac{\sigma_1^2}{2n_1}\left(\frac{1}{\sigma_1}\right)^2 + \frac{\sigma_2^2}{2n_2}\left(-\frac{1}{\sigma_2}\right)^2} \\
&= \sqrt{\frac{1}{2}\left(\frac{1}{n_1} + \frac{1}{n_2}\right)}. \qquad (G.41)
\end{aligned}$$

The log transformation can also be applied to the variance ratio or $\widehat{F}$ statistic, resulting in

$$\mu_{\ln(\widehat{F})} = \ln\left(\frac{\sigma_1^2}{\sigma_2^2}\right) = 2\ln\left(\frac{\sigma_1}{\sigma_2}\right) \tag{G.42}$$

and

$$\widehat{\sigma}_{\ln(\widehat{F})} = \sqrt{2\left(\frac{1}{n_1} + \frac{1}{n_2}\right)}. \tag{G.43}$$

G.3.14 Coefficient of Variation

For large samples by the delta method, the distribution of the sample coefficient of variation given by

$$\widehat{CV} = \frac{s}{\bar{x}} \tag{G.44}$$

is approximately normal with mean

$$\mu_{CV} = \mu_x/\sigma_x \tag{G.45}$$

and approximate standard deviation

$$\begin{aligned}
\widehat{\sigma}_{CV} &= \sqrt{\left(\left(\frac{\partial CV}{\partial \bar{x}}\bigg|_{\bar{x}=\mu_x}\right)^2 \sigma_{\bar{x}}^2 + \left(\frac{\partial CV}{\partial s}\bigg|_{s=\sigma_x}\right)^2 \sigma_s^2} \\
&= \sqrt{\left(-\frac{\sigma_x}{\mu_x^2}\right)^2 \frac{\sigma_x^2}{n} + \left(-\frac{1}{\mu_x}\right)^2 \frac{\sigma_x^2}{2n}} \\
&= CV\sqrt{\frac{1}{n}\left(CV^2 + \frac{1}{2}\right)}. \tag{G.46}
\end{aligned}$$

A more accurate expression for $\widehat{\sigma}_{CV}$ is obtained by replacing n in Equation G.46 with $n-1$; however, Equation G.46 is easier to use and sufficiently accurate for the purpose of sample size and power calculations.

G.3.15 Process Capability Statistic C_{pk}

For large samples, by the delta method the distribution of the c_{pk} statistic

$$\widehat{c}_{pk} = \frac{|NSL - \widehat{\mu}|}{3\widehat{\sigma}} \tag{G.47}$$

is approximately normal with mean

$$\mu_{\widehat{c}_{pk}} = \frac{|NSL - \mu|}{3\sigma} \tag{G.48}$$

and approximate standard deviation

G.3. Applications of the Delta Method

$$\begin{aligned}
\widehat{\sigma}_{c_{pk}} &= \sqrt{\left(\frac{\partial c_{pk}}{\partial \mu_x}\right)^2 \sigma_{\bar{x}}^2 + \left(\frac{\partial c_{pk}}{\partial \sigma_x}\right)^2 \sigma_s^2} \\
&= \sqrt{\left(-\frac{1}{3\sigma_x}\right)^2 \frac{\sigma_x^2}{n} + \left(-\frac{|NSL - \mu|}{3\sigma_x^2}\right)^2 \frac{\sigma_x^2}{2n}} \\
&= \sqrt{\frac{1}{9n} + \frac{c_{pk}^2}{2n}} \\
&= \widehat{c}_{pk} \sqrt{\frac{1}{n}\left(\frac{1}{9\widehat{c}_{pk}^2} + \frac{1}{2}\right)}.
\end{aligned} \qquad (G.49)$$

Bibliography

[1] Abramowitz, Milton, and Irene Stegun, eds. (1965). *Handbook of Mathematical Functions with Formulas, Graphs, and Mathematical Tables*. New York: Dover.

[2] Agresti, Alan (2002). *Categorical Data Analysis*. New York: Wiley Interscience.

[3] American Society for Quality Control Standards Committee (1993). *ANSI/ASQ Z1.4-1993: Sampling Procedures and Tables for Inspection by Attributes*. Milwaukee: ASQC.

[4] American Society for Quality Control Standards Committee (1993). *ANSI/ASQ Z1.9-1993: Sampling Procedures and Tables for Inspection by Variables for Percent Nonconforming*. Milwaukee: ASQC.

[5] ASTM International (2007). *ASTM E122-07 Standard Practice for Calculating Sample Size to Estimate, with Specified Precision, the Average for a Characteristic of a Lot or Process*. West Conshohocken, PA: ASTM International.

[6] Automotive Industry Action Group (2002). *Measurement Systems Analysis*. Southfield, MI: AIAG.

[7] Bausell, R. Barker, and Yu-Fang Li (2006). *Power Analysis for Experimental Research: A Practical Guide for the Biological, Medical and Social Sciences*. New York: Cambridge University Press.

[8] Beran, R. (1986). "Simulated Power Functions." *Annals of Statistics* 14(1): 151-173.

[9] Brown, Barry, James Lovato, and Kathy Russell (1997). "The Asypow S(plus) Library for Asymptotic Power Calculations." *Journal of Statistical Software* 2(2).

[10] Burdick, Richard, and Franklin Graybill (1992). *Confidence Intervals on Variance Components*. New York: Marcel Dekker.

[11] Chakraborti, S., and J. Li (2007). "Confidence Interval Estimation of a Normal Percentile." *The American Statistician* 61(4): 331-336.

[12] Chow, S., J. Shao, and H. Wang (2003). *Sample Size Calculation in Clinical Research.* New York: Marcel Dekker.

[13] Cohen, Jacob (1988). *Statistical Power Analysis for the Behavioral Sciences.* Hillsdale, N.J.: Lawrence Erlbaum.

[14] Cowan, Glen (1998). *Statistical Data Analysis.* New York: Oxford University Press.

[15] Dattalo, Patrick (2008). *Determining Sample Size: Balancing Power, Precision, and Practicality.* New York: Oxford University Press.

[16] Davison, A. C., and D. V. Hinkley (1997). *Bootstrap Methods and Their Application.* New York: Cambridge University Press.

[17] Dodge, H. F., and H. G. Romig (1998). *Sampling Inspection Tables: Single and Double Sampling.* New York: Wiley.

[18] Donner A., and J. J. Koval (1982). "Design Considerations in the Estimation of Intraclass Correlation." *Annals of Human Genetics* 46: 271–277.

[19] Duncan (1965). *Quality Control and Industrial Statistics*, 3rd Edition, Homewood, IL: Richard D. Irwin)

[20] Efron, Bradley, and R. J. Tibshirani (1993). *An Introduction to the Bootstrap.* New York: Chapman and Hall/CRC.

[21] Eisenhart, C., M. Hastay, and W. Wallis (1947). *Techniques of Statistical Analysis.* New York: McGraw-Hill.

[22] Fleiss, Joseph L., B. Levin, and M. Paik (2003). *Statistical Methods for Rates and Proportions.* Hoboken, N.J.: Wiley.

[23] Grissom, Robert J., and John J. Kim (2005). *Effect Sizes for Research: A Broad Practical Approach.* Mahwah, N.J.: Lawrence Erlbaum.

[24] Hahn, Gerald J., and William Q. Meeker (1991). *Statistical Intervals: A Guide for Practitioners.* N.Y.: Wiley.

[25] Hanley, James (1987). "Standard Error of the Kappa Statistic." *Psychological Bulletin* 102(2): 315-321.

[26] Hanley, James, and Barbara McNeil (1982). "The Meaning and Use of the Area Under a Receiver Operating Characteristic (ROC) Curve." *Radiology* 143: 29-36.

[27] Hoenig, John, and Dennis Heisey (2001). "The Abuse of Power: The Pervasive Fallacy of Power Calculations for Data Analysis." *Journal of the American Statistical Association* 55: 1-6.

[28] Howe, W. G. (1969). "Two-sided Tolerance Limits for Normal Populations – Some Improvements." *Journal of the Americal Statistical Association* 64: 610-620.

[29] Hsieh, F. Y., D. A. Bloch, and M. D. Larsen (1998). "A Simple Method of Sample Size Calculation for Linear and Logistic Regression." *Statistics in Medicine* 17(14): 1623-1634.

[30] Hsu, Jason (1995). *Multiple Comparisons Theory and Methods*. Upper Saddle River, N.J.: Chapman & Hall/CRC.

[31] Kraemer, Helena and Sue Thiemann (1987). *How Many Subjects? Statistical Power Analysis in Research*. Newbury Park, Calif.: Sage.

[32] Krishnamoorthy, K., and Jessica Thomson (2004). "A More Powerful Test for Comparing Two Poisson Means." *Journal of Statistical Planning and Inference* 119: 23-35.

[33] Lachin, J. M. (1981). "Introduction to Sample Size Determination and Power Analysis for Clinical Trials." *Controlled Clinical Trials* 2: 91-113.

[34] Larson, Harry (1966). "A Nomograph of the Cumulative Binomial Distribution." *Industrial Quality Control* 23(6): 270-278.

[35] Lemeshow, Stanley, David W. Hosmer, Janelle Klar, and Stephen K. Lwanga (1990). *Adequacy of Sample Size in Health Studies*. Hoboken, N.J.: Wiley.

[36] Lenth, Russell (2001). "Some Practical Guidelines for Effective Sample Size Determination." *The American Statistician* 55(3): 187-193.

[37] Lenth, Russell (2006-2009). *Java Applets for Power and Sample Size* (computer software). Retrieved 3 August 2009 from http://www.stat.uiowa.edu/~rlenth/Power.

[38] Lieberman, Gerald J. (1958). "Tables for One-sided Tolerance Limits". *Industrial Quality Control*, 14(10).

[39] Lipsey, Mark W. (1990). *Design Sensitivity: Statistical Power for Experimental Research*. Minneapolis: Sage.

[40] Lipson, Charles, and Narendra Sheth (1973). *Statistical Design and Analysis of Engineering Experiments*. New York: McGraw-Hill.

[41] Lloyd, David K., and Myron Lipow (1984). *Reliability: Management, Methods, and Mathematics*. Milwaukee: ASQC.

[42] Mathews, Paul G. (2004). *Design of Experiments with MINITAB*. New York: ASQ Quality Press.

[43] Meeker, W.Q., and L.A. Escobar (1998). *Statistical Methods for Reliability Data*, N.Y.: Wiley.

[44] Millard, Steven P. (2001). *EnvironmentalStats for S-Plus*. Boca Raton, Fla.: CRC.

[45] Montgomery, Douglas (2008). *Design and Analysis of Experiments*. Hoboken, N.J.: Wiley.

[46] Mooney, Christopher Z., and Robert Duval (1993). *Bootstrapping: A Nonparametric Approach to Statistical Inference*. Newbury Park, Calif.: Sage.

[47] Morris, R. L., and E. J. Riddle (2008). "Determination of Sample Size to Detect Quality Improvement in p-charts." *Quality Engineering* 20: 281-286.

[48] Murphy, Kevin R., and Brett Myors (2004). *Statistical Power Analysis: A Simple and General Model for Traditional and Modern Hypothesis Tests*. Mahwah, N.J.: Lawrence Erlbaum.

[49] Natrella, Mary Gibbons (2005). *Experimental Statistics NBS Handbook 91*. Mineola, N.Y.: Dover.

[50] Nelson, Wayne B. (2003). *Applied Life Data Analysis*. N.Y.: Wiley-Interscience.

[51] Neter, John, M. Kutner, C. Nachtscheim, and W. Wasserman (1996). *Applied Linear Statistical Models*. Chicago: Irwin.

[52] Parker, Robert A., and Nancy G. Berman (2003). "Sample Size: More than Calculations." *The American Statistician* 57(3): 166-170.

[53] Rabinovich, Semyon (2000). *Measurement Errors and Uncertainties Theory and Practice*. New York: AIP Press.

[54] Ryan, Thomas P. (2007). *Modern Engineering Statistics*. Hoboken, N.J.: Wiley.

[55] Sagan, Carl, and Ann Druyan (1996). *The Demon-Haunted World: Science as a Candle in the Dark*. New York: Ballantine.

[56] Sahai, Hardeo, and Mario Miquel Ojeda (2004). *Analysis of Variance for Random Models*. Boston: Birkhauser.

[57] Sahai, Hardeo, and Mohammed Ageel (2000). *The Analysis of Variance: Fixed, Random, and Mixed Models*. Boston: Birkhauser.

[58] Scheiner, Samuel M., and Jessica Gurevitch (2001). *Design and Analysis of Ecological Experiments*. New York: Oxford University Press.

[59] Schilling, Edward G., and Dean V. Neubauer (2008). *Acceptance Sampling in Quality Control*. Upper Saddle River, N.J.: Chapman & Hall/CRC.

[60] Sheskin, David (2004). *Handbook of Parametric and Nonparametric Statistical Procedures*. Boca Raton, Fla.: Chapman & Hall/CRC.

[61] Shoukri M. M., M. H. Asyali, and A. Donner (2004). "Sample Size Requirements for the Design of Reliability Studies: Review and New Results." *Statistical Methods in Medical Research* 13: 251-271.

[62] Shrout, P.E., and J. L. Fleiss (1979). "Intraclass Correlations: Uses in Assessing Rater Reliability." *Psychological Bulletin* 86: 420-428.

[63] Snedecor, George and William Cochran (1989). *Statistical Methods*. Ames, Iowa: Iowa State University Press.

[64] Squeglia, Nicholas L. (1994). *Zero Acceptance Number Sampling Plans*. Milwaukee: ASQ Quality Press.

[65] Tosteson, Tor D., Jeffrey S. Buzas, Eugene Demidenko, and Margaret Karagas (2003). "Power and Sample Size Calculations for Generalized Regression Models with Covariate Measurement Error." *Statistics in Medicine* 22(7): 1069-1082.

[66] U. S. Department of Defense. Office of the Assistant Secretary of Defense. Supply and Logistics (1964). *Mathematical and Statistical Principles Underlying Military Standard 414*. Washington, D.C.: U.S. Government Printing Office.

[67] van Belle, Gerald (2002). *Statistical Rules of Thumb*. New York: Wiley.

[68] Walter S. D., M. Eliasziw, and A. Donner (1998). "Sample Size and Optimal Designs for Reliability Studies." *Statistics in Medicine* 17: 101-110.

[69] Wellek, Stefan (2002). *Testing Statistical Hypotheses of Equivalence*. Boca Raton, Fla.: Chapman and Hall/CRC.

[70] Zar, Jerold H. (1996). *Biostatistical Analysis*. Upper Saddle River, N.J.: Prentice-Hall.

Index

acceptable quality level, 237
acceptance sampling, 236
    ANSI/ASQ Z1.4, 243
    ANSI/ASQ Z1.9, 255
    AOQL plan, 249
    attributes plan, 237
    LTPD plan, 247
    rectifying inspection, 246
    type A sampling plan, 241
    type B sampling plan, 238
    variables sampling plan, 251
alternate hypothesis, 6
ANOVA
    fixed effect, 172
    full factorial design, 169
    nested design, 175
    one-way, 163
    random effect, 174
    randomized block design, 167
    two-level factorial design, 176
ANSI/ASQ Z1.4, 243
ANSI/ASQ Z1.9, 255
AOQL sampling plan, 249
approximation, 12, 302
arcsine transformation, 100, 306
area under ROC curve, 158
attribute agreement analysis, 155
attribute sampling plan, 237
average outgoing quality, 249
average outgoing quality limit, 249
average run length, 227
average total inspection, 247, 250

bad practice, 22
binomial distribution, 73, 238, 285, 292
Bland-Altman plot, 161

blocking efficiency, 168
Bonferroni's method, 50
bootstrap method, 268
Box-Behnken design, 187

center cells in factorial design, 184
central composite design, 187
central limit theorem, 7
chi-square distribution, 58, 288, 296
coefficient of variation, 69, 314
Cohen's effect size, 17
Cohen's kappa, 153
completely randomized design, 163
confidence interval, 2
    bootstrap method, 268
    coefficient in two-level factorial design, 180
    coefficient of variation, 70
    correlation, 142
    counts, 115
    cp, 229
    cpk, 230
    exponential mean, 191
    exponential percentile, 192
    exponential reliability, 193
    factors that affect, 13
    half-width, 3, 13
    interference failure rate, 212
    intraclass correlation, 149
    kappa (Cohen's), 154
    mean, 2, 28
    mean, difference between two, 35
    mean, with measurement error, 29
    Monte Carlo method, 262
    normal percentile, 198
    normal probability, 198

INDEX

one-sided, 5
Poisson mean, 115
Poisson means, two, 120
proportion, 74, 86
proportions, two, 92
ROC curve's AUC, 158
slope, in linear regression, 134
standard deviation, 57
standard deviation ratio, 64
two-sided, 5
variance component, 175
Weibull percentile, 196
Weibull reliability, 197
Weibull scale parameter, 194
Weibull shape parameter, 195
confidence limits, 3
contingency table, 111
contrast, 48
control chart
    defectives (np) chart, 224
    defects (c) chart, 225
    run rule, 221
    x-bar chart, 226
control, comparisons with, 54
correlation, 141
    confidence interval, 142
    hypothesis test, 143
counts
    confidence interval for mean, 115
    hypothesis test, 118
cp, 229
cpk, 230, 314

defectives chart, 224
defects chart, 225
delta method, 305
Dunnett's test, 55

effect size, 7, 16
    practically signficant, 8
    relative, 17
equivalence test, 44
    limit of practical equivalence, 45
    means, 45
    proportions, 106
    two one-sided tests (TOST), 45
Excel, 299
exponential distribution, 191
exponential-exponential interference, 216

F distribution, 64, 124, 164, 172, 174, 289, 297
family error rate, 50, 222
finite population correction, 27, 78
Fisher's exact test, 97
Fisher's Z transform, 141
fixed effects model, 172
fractional factorial design, 181
full factorial design, 169

gage R&R study, 258
good practice, 23
goodness of fit, 113, 185
greek alphabet, 281

hypergeometric distribution, 85, 241, 284
    approximations to, 86
hypothesis test, 6
    bootstrap method, 269
    chi-square, 110
    coefficient of variation, 70, 71
    contingency table, 111
    correlation, 143
    correlation, two, 144
    counts, 118
    cp, 231
    cpk, 231
    effect size, 7, 16
    exponential means, two, 207
    factors that affect, 16
    Fisher's exact test, 97
    goodness of fit, 113
    intraclass correlation, 151
    kappa (Cohen's), 155
    log-rank test, 208
    McNemar's test, 104
    mean, 6, 30
    means, difference between two, 39
    Monte Carlo method, 265
    multiple comparisons tests, 50

multiple correlation, 145
Poisson mean, 118
Poisson means, many, 128
Poisson means, two, 123
proportion, 78, 90
reliability, 202
reliability demonstration test, 200
reliability percentile, 205
ROC curve's AUC, 160
slope, in linear regression, 137
standard deviation, 60
standard deviation ratio, 66
type I error, 7
type II error, 7

interference, 211
intraclass correlation, 147

kappa, 153

Lachin's method, 210
lack of fit, 185
Larson's nomogram, 82, 292
limit of practical equivalence, 45
linear regression, 133
log odds ratio, 102, 308
log odds transformation, 307
log-rank test, 208
logistic regression, 139
LTPD sampling plan, 247

McNemar's test, 104
mean
    confidence interval, 2, 28
    equivalence test, 45
    hypothesis test, 6, 30
MINITAB, 25, 299
mixed effects model, 172
modified least squares method, 175
Monte Carlo method, 262
multiple comparisons tests, 50
multiple correlation, 145

nested design, 175
noncentral probability distribution, 32, 110, 164, 283

noninferiority test, 45
normal distribution, 287, 294
normal-normal interference, 212
notation, 273
null hypothesis, 6

odds ratio, 95, 102
one-sided confidence interval, 5
one-tailed test, 9
one-way ANOVA, 163
operating characteristic curve, 18, 83, 237

p value, 9
paired observations
    Bland-Altman plot, 161
    McNemar's test, 104
    paired-sample t test, 33
PASS, 25, 299
Piface, 25, 300
Plackett-Burman design, 183
point estimate, 2
Poisson distribution, 115, 286, 293
Poisson mean
    background count correction, 130
    confidence interval, 115
    confidence interval for two, 120
    hypothesis test, 118
    hypothesis test for many, 128
    hypothesis test for two, 123
power, 7
    graphical presentation, 18
probability distribution
    binomial, 73, 285, 292
    chi-square, 288, 296
    exponential, 191
    F, 289, 297
    hypergeometric, 85, 284
    noncentral, 32, 110, 164, 283
    normal, 287, 294
    Poisson, 115, 286, 293
    Student's t, 287, 295
    Weibull, 194
process capability, 229, 314
proportion

INDEX

confidence interval, 74, 86
confidence interval for two, 92
equivalence test, 106
Fisher's exact test, 97
hypothesis test, 78, 90
hypothesis test for two, 97
McNemar's test, 104
proportional hazards, 208

R, 25, 299
random effects model, 172
randomized block design, 167
rectifying inspection, 246
rejectable quality level, 237
relative effect size, 17
relative risk, 94, 101
reliability
    parameter estimation, 191
    two-sample tests, 206
reliability demonstration test, 199
    location parameter, 200
    percentile, 205
    reliability, 202
repeatability, 256
reproducibility, 256
resampling, 261
response surface design, 187
risk ratio, 94, 101, 308
ROC curve, 158
rule of three, 76, 205, 225
run rule, 221

Sagan, Carl, 6
Satterthwaite's method, 175
Schoenfeld's method, 209
significance level, 9
slope, in linear regression
    confidence interval, 134
    hypothesis test, 137
software, vii, 25, 299
    Microsoft Excel, 299
    MINITAB, 25, 299
    PASS, 25, 299
    Piface, 25, 300
    R, 25, 299

square root transform for counts, 117, 118, 126, 128, 309
standard deviation
    confidence interval, 57, 64
    hypothesis test, 60, 66
statistical process control, 221
Student's t distribution, 31, 287, 295
superiority test, 45

tolerance interval, 233
    nonparametric, 233
    normal, 235, 298
transformation
    arcsine, for proportion, 100, 306
    Fisher's Z, 141
    log odds, 307
    log odds ratio, 308
    square root for counts, 118, 309
Tukey's honest significant difference test, 52
Tukey's quick test, 266
two one-sided tests (TOST), 45
two-level factorial design, 176
two-tailed test, 9
type A sampling plan, 237, 241
type B sampling plan, 237, 238
type I error, 7
type II error, 7

variables plan, 251
variance components analysis, 172, 258

Weibull distribution, 194
Weibull-Weibull interference, 218

x-bar chart, 226

zero successes, 89, 248

CPSIA information can be obtained
at www.ICGtesting.com
Printed in the USA
BVHW041209220422
634852BV00003B/113